The Latin American Studies Book Series

The Latin American Studies Book Series promotes quality scientific research focusing on Latin American countries. The series accepts disciplinary and interdisciplinary titles related to geographical, environmental, cultural, economic, political, urban and health-related research dedicated to Latin America. The series publishes comprehensive monographs, edited volumes and textbooks refereed by a region or country expert specialized in Latin American studies.

The series aims to raise the profile of Latin American studies, showcasing important works developed focusing on the region. It is aimed at researchers, students, and everyone interested in Latin American topics.

Submit a proposal: Proposals for the series will be considered by the Series Advisory Board. A book proposal form can be obtained from the Publisher, Andrea Brody-Barre (andrea.brody-barre@springer.com).

Aníbal Rosales Hernández • Barlin Orlando Olivares Campos • Jesús Viloria Rendón • Laila Iskandar Nahas • Pedro García Montero

Agriculture Under Climate Change

Vulnerabilities and Adaptations in Venezuela

Springer

Aníbal Rosales Hernández
Faculty of Agronomy
Central University of Venezuela
Maracay, Venezuela

Jesús Viloria Rendón
Faculty of Agronomy
Central University of Venezuela
Maracay, Venezuela

Pedro García Montero
Orinoco Group
Environmental Consultant
Maracay, Venezuela

Barlin Orlando Olivares Campos
Research Group on Biodiversity
Management
University of Córdoba
Córdoba, Spain

Laila Iskandar Nahas
Andrés Bello Catholic University
Caracas, Venezuela

ISSN 2366-3421 ISSN 2366-343X (electronic)
The Latin American Studies Book Series
ISBN 978-3-032-06367-0 ISBN 978-3-032-06368-7 (eBook)
https://doi.org/10.1007/978-3-032-06368-7

This Springer imprint is published by the registered company Springer Nature Switzerland AG
The registered company address is: Gewerbestrasse 11, 6330 Cham, Switzerland

Prologue

We often think of the foreword as the gateway to a book, whether it be a literary work, an educational text, or a scientific document. In all cases, writing a foreword is a free exercise, one that becomes even more engaging when the author has closely followed the author's laborious creative process. It becomes especially significant when one is challenged to open the book with a broad perspective, capable of inspiring in its readers not only a full and enjoyable reading experience, but also a preliminary understanding of the subject at hand.

The idea for this book emerged during the preparation of Venezuela's Second Academic Report on Climate Change (DRACC), covering the period 2018–2024. This report was sponsored by the Academy of Physical, Mathematical and Natural Sciences of Venezuela (ACFIMAN), under the coordination of the Academic Secretariat for Climate Change (SACC). As the current coordinator of DRACC-SACC, and as someone who has directly witnessed the meticulous and enthusiastic dedication of the authors of the agriculture chapter, who are also the authors of this book, it is a great personal satisfaction to see how the teamwork experience we gained during the ACFIMAN academic reporting process now takes shape in other outputs of equal quality and relevance. These contributions are valuable not only for the scientific community of Venezuela but also for key national and international stakeholders engaged in climate change.

The upcoming pages of **Agriculture Under Climate Change: Vulnerability and Adaptation in Venezuela** result from a thorough review of scientific literature published both nationally and internationally over the past 6 years. This review is enriched by the extensive and well-regarded professional experience of the authors. The book offers a rigorous, comprehensive, and up-to-date scientific contribution to the intersection of agriculture and climate change in Venezuela. It also provides strategic value to agricultural decision-making in the country, offering a documented guide backed by scientific and technical evidence that helps identify the spatial dimensions of the many challenges climate change imposes, challenges that the national scientific community has analyzed and that, albeit quietly, are being considered by official agencies responsible for the issue.

Within the context of climate change, the book demonstrates, scientifically and convincingly, that agriculture can act both as an intensifier of climate impacts and as a powerful means to mitigate them. The contributions of this book become even more compelling when examined through its findings on the role of adaptation as a measure for preserving or building agricultural resilience. This resilience is framed against the multiple and varied vulnerabilities projected by climate scenarios such as RCP-SSP at both global and national scales. The identification of climate vulnerabilities and resilience within Venezuelan agriculture, alongside the design and implementation of effective adaptation strategies, may have far-reaching positive impacts on the present and future well-being of Venezuelans. These issues, along with other key aspects of agriculture, are thoroughly examined across the book's eight chapters.

The references supporting each chapter attest to the rigorous and systematic review of scientific and technical literature that supports this work. The book navigates the frontiers of scientific knowledge, both locally and globally, exploring how a changing climate challenges our natural and constructed abilities to adapt. It proposes forward-looking options that include scientific, technological, and spatial planning and institutional arrangements. These could shape a climate-resilient road map capable of responsibly and effectively reconciling environmental protection with agricultural production.

In its opening chapters, this book offers a solid and enlightening overview of agriculture in Venezuela in the context of climate change. Chapter 1 presents the current state of the agricultural sector, showing how its drastic deterioration is more a result of structural crisis than of mitigation efforts and highlighting the urgent need to transition toward climate-smart agriculture. Chapter 2 rigorously outlines the methodology employed, grounded in international standards and a systematic, transparent process that underpins each of the book's findings. Chapter 3 addresses climate variability and its impacts on national agriculture, emphasizing regional differences and the vulnerability of agricultural systems to shifts in temperature and precipitation. Together, these three chapters build the conceptual and methodological foundation that guides readers toward a deep and critical understanding of the interactions between agriculture and variable climate in our country.

One of the most innovative aspects of this scientific work lies in its regionalized approach to analyzing climate impacts on agriculture. Chapters 4 (Impact of Climate Change on Venezuelan Agricultural Landscapes) and 5 (Regional Variability of Climate Impacts in Venezuela) explore this approach in greater depth, reflecting the authors' commitment to ensuring that proposed mitigation and adaptation measures are well tailored to the physiographic differences of the country's agricultural landscapes. This allows for the development of region-specific climate solutions adapted to both current and future agricultural systems.

Chapter 4 provides a territorial perspective on the impacts of climate change on agricultural landscapes, focusing on long-term climate impacts in two emblematic agricultural zones: mountainous areas (Andes and Coastal Mountainous) and lowland plains (Western, Central, and Eastern Llanos). The analysis shows how changing regional climate conditions affect soil and crop health, ultimately determining agricultural sustainability.

Chapter 5, titled Regional Variability of Climate Impacts in Venezuela, examines climate projections for temperature and precipitation in Venezuela through the twenty-first century, using models such as MPI-ESM1.2-LR (under SSP3-7.0) and others, including MPI, GISS, and HadGEM (under SSP5-8.5). The Andes emerge as a particularly vulnerable region, given the risks of crop displacement due to altitude changes, habitat fragmentation, and soil erosion, risks that threaten key strategic crops such as coffee, cocoa, potatoes, and plantains. The Amazon and low-lying coastal zones are also identified as areas with high potential risk to agriculture and ecosystem health.

In Chap. 6, Climate Change Mitigation Strategies, the authors present mitigation practices that could reduce greenhouse gas emissions in the agricultural sector while fostering sustainability. Drawing on 52 studies, they highlight two key actions: enhancing agricultural production systems to lower emissions and boosting carbon sequestration through agroecological practices. The chapter also briefly addresses technical, social, legal, and educational aspects of mitigation. It includes the use of geospatial and nuclear technologies and underscores the urgency of halting agricultural frontier expansion, especially into carbon-rich forest areas. Likewise, the chapter promotes practices that enhance both resilience and productivity, as well as the protection of ecosystem services. These measures align with Venezuela's commitments to international climate agendas.

Chapter 7, Agriculture Adaptation Strategies to Climate Change, explores practical adaptation strategies that farmers and other stakeholders could implement. A key insight from the literature review and the authors' own experience is the critical importance of efficient irrigation water use for many crops within Venezuela's agricultural matrix. The chapter presents examples of climate-resilient agricultural practices from tropical regions that could be adapted to the Venezuelan context. It also notes that, although research on agriculture and climate change has grown since 2018, contributions to the adaptation field remain limited, especially in areas such as agroforestry and development of crop varieties resistant to water scarcity and high temperatures. These two impacts are among the most detrimental to national agriculture, as evidenced by the findings reported in this book.

Following this introductory overview, I firmly believe that this book is a highly valuable contribution to the country and beyond. It is the result of a dedicated group of Venezuelan agronomic researchers, Aníbal Rosales, Barlin Olivares, Jesús Viloria, Laila Iskandar, and Pedro García, affiliated with both national and international universities and research centers. These scholars are deeply committed to studying human impacts on the environment and to understanding both global and local consequences of climate change. They are individuals genuinely concerned and actively engaged in providing timely responses to Venezuela's environmental challenges and, more recently, to the growing impacts of climate change on national agriculture.

What we see in this volume is but a small part of their enduring effort, service-oriented vocation, and love for our country, values that have been cultivated over decades of work. This scientific contribution is both rigorous in content and accessible in language, reaching a wide range of readers. I am confident that these two strengths will be recognized and appreciated by future readers, who I foresee becoming key advocates for this book, recommending it to researchers, professionals, university students, and decision-makers involved in agriculture and climate change in Venezuela.

Simón Bolívar University
Caracas, Venezuela
Academy of Physical, Mathematical
and Natural Sciences of Venezuela
Caracas, Venezuela

Alicia Villamizar

Acknowledgments

This book is the result of the efforts of a group of Venezuelan agronomy professionals, selected by the Academic Secretariat on Climate Change (SACC) of the Academy of Physical, Mathematical and Natural Sciences of Venezuela (ACFIMAN). These individuals came together as an *ad hoc* working group to contribute to the "Agriculture and Climate Change" chapter of Venezuela's Second Climate Change Report, scheduled for publication in 2026.

We extend our sincere gratitude to Dr. Alicia Villamizar, Executive Secretary of the SACC, for inviting us to join this initiative and for her unwavering enthusiasm in guiding us through this important challenge. Following the completion of our initial responsibilities, the idea of publishing a book based on the finalized chapter emerged organically within the group. This edition incorporates new content, a refined structure, and an English-language version, driven by our conviction that the knowledge and experiences gained deserve to be shared beyond our borders. We are also deeply appreciative of Dr. Villamizar's generosity in writing the Prologue to this book, an initiative unanimously supported by the authors.

As authors, we are grateful for the collaborative and constructive spirit that defined our work together. Several factors made this experience especially meaningful. First, we shared a collective consciousness of the complexity of the task. This was particularly important given Venezuela's challenges with documentation and Internet connectivity. Second, our commitment was entirely voluntary and sustained over nearly 3 years. We were driven solely by the desire to contribute to the national and global understanding of agriculture and climate change in Venezuela. Third, we were united by a common determination to generate new knowledge. This was achieved through research carried out during the project, especially on the regionalization of climate change impacts across the country. We also worked to prepare articles for publication in internationally recognized journals.

We are especially thankful for the support of Germán Briceño, President of the Institute of Agricultural Policies (IPAF) of the Confederation of Agricultural Producers' Associations of Venezuela (FEDEAGRO), whose careful review and timely comments on the chapter "Venezuelan Agriculture under Climate Change and Economic Contraction: An Introduction" were invaluable.

A fundamental ally in this process has been Springer. We are sincerely grateful to the publishing house for recognizing the value of our work and accepting the manuscript for publication. We extend special thanks to the editorial and production teams responsibles for overseeing the production of this book.

Lastly, we wish to acknowledge the Faculty of Agronomy at the Central University of Venezuela, where all of the authors earned their initial degrees. This institution played a decisive role in our professional formation and opened the paths that have shaped our individual and collective journeys.

About This Book

Venezuelan agriculture faces a context marked by limited scientific production, fragmented knowledge, and severe data gaps regarding the impacts of climate change on its diverse agroecological regions. Despite observable increases in temperature, shifts in precipitation patterns, and rising frequency of extreme climatic events affecting crops, soils, water resources, and livelihoods, there is no comprehensive, systematized body of literature that integrates these dynamics to guide adaptation and mitigation efforts.

Scientific production on agriculture and climate change in Venezuela remains significantly lagging behind other Latin American countries. While nations such as Brazil and Mexico publish over 500 annual papers on this thematic intersection, Venezuela has averaged fewer than 20 per year over the past decade, with a marked concentration in basic sciences and minimal integration with applied and social research. This deficit has created critical information gaps, particularly in the evaluation of regional vulnerabilities, adaptation strategies, and crop-specific impacts under changing climatic conditions. Moreover, until now, no book has systematically addressed the complex interactions between climate change and agriculture in Venezuela.

Ideally, the country should have robust, accessible, and context-specific scientific syntheses that inform policies, practices, and research agendas, enabling agricultural systems to adapt sustainably and ensure food security under climatic risks. The importance of this ideal lies in reducing vulnerability, strengthening resilience and food security, and supporting informed decision-making in a country already facing economic, political, and institutional crises.

However, the current lack of integrated literature is a problem because it perpetuates decision-making based on incomplete evidence, limits the development of climate-smart agricultural strategies, and hinders the formulation of effective policies for adaptation and mitigation. The consequences include continued exposure of farmers and ecosystems to climate threats, inefficient allocation of scarce resources, and missed opportunities to build resilience in the sector. Alternatives to address this problem involve conducting systematic and narrative reviews that synthesize existing knowledge, identify gaps, and propose strategic research and policy directions.

This leads to the guiding research question of this book: How does the existing scientific literature characterize the vulnerabilities, impacts, and adaptation and mitigation strategies of Venezuelan agriculture under climate change?

This book emerged during the preparation of Venezuela's Second Academic Report on Climate Change (DRACC), covering the period 2018–2024, sponsored by the Academy of Physical, Mathematical and Natural Sciences of Venezuela (ACFIMAN), and reflects the collective commitment to advance rigorous, context-specific knowledge to inform national climate action.

The novelty of this book lies in being the first editorial effort to comprehensively address the intersection between agriculture and climate change in Venezuela, integrating for the first time systematic analyses of vulnerabilities, impacts, and adaptation strategies for the country's main agricultural systems. Its originality is evident in the combination of rigorous methodologies, region-specific data, and applicable proposals, as well as in the incorporation of case studies and innovative agroecological approaches that have not been previously explored in the national literature.

The relevance of this work is grounded in its contribution to filling critical information gaps by providing scientific evidence to support the formulation of climate-smart public policies and the implementation of resilient agricultural practices. Moreover, it offers practical tools for producers, researchers, and policymakers seeking to ensure food security, environmental sustainability, and climate justice in Venezuela in the face of escalating global challenges.

General Objective

To analyze in an integrated manner the vulnerabilities, impacts, and adaptation strategies of Venezuelan agricultural systems under climate change, combining scientific, methodological, and applied evidence to strengthen the country's agro-food and socio-ecological resilience

Specific Objectives

1. To examine historical and current trends in Venezuelan agriculture, identifying the economic, institutional, and environmental factors shaping its vulnerability and adaptive capacity
2. To describe the systematic and narrative methodological approach used to synthesize evidence on climate change impacts, vulnerabilities, and adaptation strategies in the Venezuelan agricultural sector
3. To analyze regional climate variability, its manifestations in temperature and precipitation, and its influence on agricultural productivity, water availability, and pest and disease dynamics, as well as to visualize the impact of climate change on the socio-economic aspects of producers and the population

4. To evaluate the projected impacts of climate change on strategic crops, water resources, soils, and ecosystems associated with agriculture across different regions of the country
5. To identify the spatial variability of projected temperature and precipitation changes and their effects on agricultural land suitability, with emphasis on maize cultivation in the western plains
6. To synthesize greenhouse gas mitigation strategies in the agricultural sector, including soil management practices, agroforestry, bioenergy, and reductions in chemical inputs
7. To review and analyze adaptation strategies implemented or proposed in Venezuelan agriculture, such as irrigation management, climate-smart agriculture, and agroecological practices
8. To integrate the book's findings, identifying knowledge gaps, innovation opportunities, and strategic priorities to advance toward a resilient, sustainable, and equitable agricultural system

This book is structured into eight chapters, each addressing a critical dimension of Venezuelan agriculture under climate change. Chapter 1 unveils the stark reality of a sector weakened by economic collapse and environmental degradation, underscoring the urgent need for transformative action. Chapter 2 presents a rigorous methodological framework that sets new standards for climate-agriculture assessments in data-scarce contexts. Chapter 3 exposes the profound impacts of climate variability on agricultural productivity and food security, highlighting the escalating risks from droughts, extreme temperatures, and erratic rainfall. Chapter 4 delivers a compelling analysis of projected climate change impacts on landscapes, crops, water resources, and ecosystems, warning of an imminent reconfiguration of Venezuela's agricultural geography. Chapter 5 reveals the pronounced regional disparities in climate impacts, demonstrating how rising temperatures and shifting rainfall patterns will reshape crop suitability and production potential across the country. Chapter 6 synthesizes innovative mitigation strategies, showcasing agroecological practices, agroforestry systems, and soil management approaches with the power to reduce greenhouse gas emissions while enhancing productivity. Chapter 7 explores adaptation strategies that offer hope for resilience, from climate-smart irrigation to sustainable farming practices tailored to local realities. Finally, Chap. 8 integrates the evidence into a strategic vision, calling for urgent multisectoral action to build a resilient, equitable, and climate-just agricultural future for Venezuela.

This book is tailored to meet the needs of a diverse audience, including farmers and agricultural practitioners. This group is directly affected by the challenges posed by climate variability and change. The book provides them with actionable strategies for adapting to erratic precipitation, increasing temperatures, and extreme weather events, emphasizing sustainable practices to build resilience and enhance productivity.

Policymakers, government officials, and NGOs form another crucial segment of the target audience. These stakeholders are instrumental in shaping policies and implementing programs to address the vulnerabilities of the agricultural sector. By offering localized insights and evidence-based solutions, the book serves as a valuable guide for developing climate-smart agricultural policies, improving food security, and promoting environmental sustainability in Venezuela. Researchers, educators, and students in agricultural science, environmental studies, and climatology will find the book particularly beneficial. Its scientific approach, combined with its focus on Venezuela's unique agricultural landscapes, fills a critical gap in the literature and serves as a foundation for further research.

Finally, this book represents a pioneering contribution toward the development of adaptive and sustainable agricultural policies and practices in Venezuela. It is a call for collective action, grounded in rigorous and context-specific science, to address climate challenges and ensure food security and climate justice for the country.

Contents

About the Authors

Barlin Orlando Olivares Campos completed his Ph.D. Cum Laude in Agricultural, Food, Forestry Engineering, and Sustainable Rural Development (International Mention) from the Agrifood Campus of International Excellence (CeiA3) of the University of Cordoba (UCO), Spain (2022). He was awarded a scholarship from the Iberoamerican Postgraduate University Association (AUIP). His outstanding scientific activity earned him the UCO Extraordinary Ph.D. Award in 2022 and the National Award for Relevant Doctoral Thesis from the Spanish National Research Council (CSIC) in 2024.

He is an agricultural engineer from the Faculty of Agronomy of the Central University of Venezuela (2009), obtaining the Recognition for the Best Final Degree Project and a master's degree in Environmental Technology from the International University of Andalusia, Spain (2015), through a Carolina Foundation scholarship.

He started in the scientific field in 2008, when he obtained a scholarship to stay at the Water Center for Arid and Semi-Arid Zones of Latin America and the Caribbean (CAZALAC) based in Chile, within the framework of the UNESCO International Hydrological Program. Later, he was a young researcher at the National Institute of Agricultural Research (INIA) of Venezuela from September 2009 to January 2014, carrying out scientific activities and knowledge transfer to farmers. At the same time, he was an instructor professor in the Agricultural Climatology Department of the National Experimental Polytechnic University of the Bolivarian Armed Forces. He was also a researcher at the Institute for Advanced Studies Foundation, Venezuela (IDEA), during August 2014 for the scientific-technical dissemination of agricultural biotechnology, and finally a researcher at the National Institute of Meteorology and Hydrology, Venezuela (INAMEH) (January 2016–May 2017), generating scientific production of the agroecology zoning of crops. He has more than 15 years of experience in research on agri-food systems and has significantly advanced the generation of knowledge in agricultural sciences through the use of sophisticated algorithmic models of machine learning.

He is currently a member of the Biodiversity Management Research Group (GESBIO) at the UCO, where he works as a data scientist developing the Federated Data Space for Citrus and as a consultant for the FAO in machine learning. Also, he is an honorary associate professor in food bromatology for the bachelor's degree in food science and technology at the UCO.

He is the author of an outstanding scientific production of journal articles, books, and chapters in books, also. He has participated in the supervision of thesis and conference presentations and obtained various fellowships, accreditations, and recognitions, as well as participated in projects for the promotion and dissemination of science in agricultural spaces. He has carried out nine research stays in research institutes and universities of international prestige in Argentina, Chile, Costa Rica, Colombia, Ecuador, Spain, and Panama.

Aníbal Rosales Hernández is an agronomist, graduated from the Central University of Venezuela (UCV) in 1970. He obtained a Master of Science and a Ph.D. from Cornell University, USA, in 1981. He is currently a retired professor of soils at UCV-Agronomy. He served as Coordinator of the Graduate Program in Soil Science and Director of the Soil Institute. He designed and established the national soil collection of Venezuela at UCV. With permission from UCV, he served as General Director of Information and Research at the Ministry of Environment and Renewable Natural Resources of Venezuela, responsible for national inventories of water, soils, and vegetation. He established the National Center for Plant Genetic Resources and Biodiversity and the Amazon Environmental Research Center. He was a member of the Venezuelan Delegation for the preparation of the United Nations Framework Convention on Climate Change and before the Global Environmental Facility (GEF).

He was a professor of Watershed Management at the Center for Development Studies (CENDES) of UCV. He was a member of the Working Teams of the Hydroelectric Basin Evaluation Projects and the Upper Guárico River Basin, which supplies water to the capital city of Caracas. He has conducted and led environmental studies and soil and land evaluations, some for the energy and oil sector. He has been a consultant on environmental projects for the World Bank, the GEF (member of the PIF formulation team for the Canaima National Park project, Venezuela), and the Inter-American Development Bank. After he retired from UCV, he served as Operations Manager for the international environmental firm Ecology and Environment S.A., based in Caracas, leading environmental studies, especially in the oil and gas sector.

He has participated in the following recent publications: (1) coordinator of the agriculture chapter of the Second Climate Change Report of Venezuela (ANIHVEN and ACFIMAN), ongoing until 2025; (2) team member (co-author) of the project "Strategies for Agro-Export in the Context of Future Development Scenarios of Venezuela," Orinoco Group and IICA, 2023; (3) coordinator and co-author of the proceedings of the Forum-Workshop "Harmonization of Agricultural and Urban Uses for the Sustainable Development of the Lake Valencia Depression," Orinoco Group-SVCS-ANIHVEN, 2021; (4) coordinator and co-author of the integrated watershed management plan: case of the upper Guárico River basin, IICA and Orinoco Group, 2020; (5) co-author of the works "Mapping Projected Variations of Temperature and Precipitation Due to Climate Change in Venezuela" and "Projected

Effect of Climate Change on Land Suitability for Maize Cultivation in the Western Plains of Venezuela"; (6) promoter of the GeoVenezuela publication and co-author of the chapter "Soil Geography of Venezuela," Fundación Empresas Polar, 2007; and (7) co-editor of the book "Water in Venezuela: A Scarce Resource" and co-author of the following chapters: "Watersheds and Their Integrated Management," "Research and Human Resource Training Agenda for the Integrated Use of Water," and "Final Message," Fundación Empresas Polar, 2015, Caracas.

Jesús Viloria Rendón is an agricultural engineer graduated from the Central University of Venezuela, where he also earned a Master of Science degree in Soil Science. His academic journey culminated with a Doctor of Philosophy degree from the University of Oxford in 1989, following his specialization in photointerpretation applied to soil studies at CIAF in Colombia. This solid academic background shaped his comprehensive vision to approach soil studies through science, technology, and territorial planning.

For over 25 years, Dr. Viloria served as a professor and researcher at the Faculty of Agronomy of the Central University of Venezuela, teaching undergraduate, master's, and doctoral courses, while also holding leadership roles as Head of the Department of Soil Science and Director of the Institute of Soil Science. His passion for teaching is reflected in his supervision of more than 35 undergraduate and graduate research projects, mentoring generations of professionals committed to the sustainable management and conservation of soil resources.

His intellectual production is extensive, authoring and co-authoring 39 scientific articles published in peer-reviewed journals and 29 technical publications and book chapters and presenting 61 papers at national and international scientific events. These contributions have strengthened knowledge in areas such as soil science, agricultural land suitability analysis, geographic information systems, remote sensing, and quantitative geospatial data analysis.

Currently, he combines his status as an active retired professor with consultancy work for international organizations such as the Inter-American Development Bank, IICA, and PRO-VITA, leading projects on soil analysis, mapping, and sustainable crop management across Latin America and Africa. His commitment to science and society continues to leave a profound mark on research and development of solutions for rational soil and land management.

Laila Iskandar Nahas is an agricultural engineer who graduated from the Faculty of Agronomy at the Central University of Venezuela (UCV) in 1990, with postgraduate studies in ecology at the Faculty of Sciences of the same institution, where she specialized in landscape ecology and watershed conservation. Throughout her academic and professional career, she has played key roles in training and disseminating knowledge on environmental, ecological, and sustainability topics.

She was a founding professor of the Postgraduate Program in Environmental Engineering and the diploma course "Topics in Ecology, Environment, and Sustainability for University Teaching" at Andrés Bello Catholic University (UCAB), where she also established the "Ecology, Environment, and Sustainability"

Institutional Chair, which she has been teaching since 2015. Additionally, she has been a faculty member in the Geography School at UCV, teaching soil science, and at the School of Environmental Engineering at the Maritime University of the Caribbean, where she taught the course on vegetation cover restoration. Her experience as an independent consultant covers various areas related to ecosystem conservation, watershed management, sustainable agriculture, agroforestry systems, and integrated water management.

She has collaborated with institutions such as the Inter-American Institute for Cooperation on Agriculture (IICA), the Sociological Studies Foundation (FUNDES), Nestlé Venezuela, the Provincial Bank Foundation, HIDROCAPITAL, Empresas Polar Foundation, and the Hatillo Municipality, participating in capacity-building projects, water resource management, soil and water conservation, school gardens, and sustainable agricultural practices since 2004.

Since 2015, she has been part of the Orinoco Sustainable Development Group, contributing to regional sustainable development initiatives. In 2024, she coordinated and co-authored the project "Assessment of Land Suitability in the Protected Area of Greater Caracas for Shade-Grown Coffee Production" for IICA, Venezuela. She also participated as a member of the technical team in the project "Strategies for Agro-Export in the Context of Future Development Scenarios for Venezuela" (2022) and was a member of the coordinating team and co-author of the profile for the integrated watershed management plan for the upper Guárico River basin, developed for IICA, Costa Rica (2019–2020). Her international experience includes serving as a consultant on the "Smart Agricultural Practices" project in Belize for the Tropical Agricultural Research and Higher Education Center (CATIE) of Costa Rica in 2018, in response to the impacts of climate change on the Central American agricultural sector.

Additionally, she was a co-editor of the book "Water in Venezuela: A Scarce Resource," published by the Empresas Polar Foundation in 2015. Engineer Iskandar Nahas has contributed to various publications related to ecology, water management, and conservation and has served as a speaker and workshop facilitator for youth and adults on environmental topics. Her work has been presented at multiple scientific conferences and academic publications, including her contribution to the "Venezuela History Dictionary" and to Interciencia journal, where she addressed topics such as ecological carrying capacity. Her commitment to sustainability, environmental conservation, and applied research positions her as a leading figure in ecology, natural resource management, and sustainable agriculture in Venezuela and the region.

Pedro García Montero was born in Venezuela and graduated as an agronomist from the Central University of Venezuela in 1975. From the beginning, he showed a deep interest in soils and natural resources, which led him to specialize in soil science, earning his Master of Science from the University of Florida in 1984. His career was marked by scientific rigor and a strong commitment to the sustainable management of Venezuela's environment, qualities that would define his professional legacy.

Over five decades of experience, Pedro García Montero became a leading figure in soil studies, natural resource inventory and evaluation, environmental planning, baseline environmental studies, and territorial planning. His role as a senior consultant allowed him to lead and advise strategic projects, earning recognition for his integrated vision in environmental assessment and watershed management, always oriented toward sustainable development.

He held high-responsibility positions in key institutions, such as Coordinator of the Natural Resources Inventory Project for the Guayana Region at CVG Tecmin; Director of Soils, Vegetation, and Fauna; Director of Land Conservation and Evaluation; and Director-General of the Autonomous Service for Environmental Development of the Amazon (SADA-AMAZONAS) within the Ministry of Environment and Renewable Natural Resources. Currently, he continues to contribute through the Orinoco Group, reflecting on pathways for a post-oil Venezuela.

His passion for teaching is reflected in 25 years of undergraduate and postgraduate education at Venezuelan universities, including the Central University of Venezuela, Andrés Bello Catholic University, and the Experimental University of Guayana. He mentored several generations of professionals, who highlight his conceptual clarity, commitment to applied science, and constant motivation to connect knowledge with the country's realities.

He has left a valuable bibliographic legacy with over 30 publications, including chapters in internationally recognized books. His contributions include, among others, studies such as "*Tepui Peatlands: Setting and Features*" (Zinck and García 2011) and "*Properties and Classification of the Tepui Peats*" (García et al. 2011), which delve into the peatlands of the Tepuis and their paleoenvironmental significance; "*Territorial Zoning Based on Geopedologic Information: Case Study in the Caroní River Basin, Venezuela*" (García 2016), where he proposes applied methodologies for territorial planning; "*A Vision on the Territorial Planning of the Amazonas State as a Basis for Its Sustainable Development*" (García 2022), which offers a strategic reflection for sustainable development in the Venezuelan Amazon; and *Psammic Peinobiomes*: *Nutrient-Limited Ecosystems of the Upper Orinoco and Rio Negro Basins* (A. Zinck, O. Huber, P. Garcia, and E. Medina, editors (2023)), focusing on the characterization and genesis of these unique Amazonian ecosystems. His work transcends disciplines and borders, integrating geomorphology, pedology, and ecology to benefit environmental management in Venezuela and Latin America.

Chapter 1
Venezuelan Agriculture under Climate Change, Territorial Transformation and Economic Contraction: An Introduction

1.1 Introduction

Agriculture in Venezuela has historically played a fundamental role in the country's development, yet over the past decades, it has undergone a severe and persistent decline. Once considered a key pillar of the national economy before the oil boom, Venezuelan agriculture has been marginalized, suffering from structural weaknesses, adverse economic policies, and profound institutional deterioration (Rodriguez et al. 2022). This transformation has undermined food security and significantly altered the sector's relationship with environmental sustainability.

The economic collapse of Venezuela, particularly after the early 2000s, intensely affected agricultural activities. The contraction of planted and harvested areas, the reduction of the cattle population, and the breakdown of rural infrastructure have translated into a marked decrease in agricultural productivity. As a result, Venezuela has become heavily dependent on food imports, while rural livelihoods have deteriorated amidst increasing poverty, insecurity, and resource scarcity (Anido and García Lobo 2018; Wirtz et al. 2024). This structural decline is crucial to understanding the environmental dynamics that currently characterize Venezuelan agriculture in the context of climate change.

One unexpected consequence of the collapse of the agricultural sector has been the reduction in greenhouse gas (GHG) emissions associated with farming activities, particularly methane (CH_4) emissions from rice cultivation and cattle production (Kılıç et al. 2022). However, this decrease has not been the result of mitigation or adaptation strategies aimed at combating climate change, but rather a by-product of economic and institutional collapse (Ramirez et al. 2025). The absence of deliberate environmental management strategies reveals the vulnerability of the sector to both socioeconomic crises and climatic variability (Lescher Soto et al. 2024).

The relationship between agriculture and climate change in Venezuela is, therefore, complex. On the one hand, reduced agricultural activity has led to lower GHG

A. Rosales Hernández et al., *Agriculture Under Climate Change*, The Latin American Studies Book Series, https://doi.org/10.1007/978-3-032-06368-7_1

emissions in the short term; on the other hand, the country faces significant risks due to the degradation of natural resources (Segnini et al. 2025). Deforestation, land-use change, and the loss of tree cover, particularly in vulnerable ecosystems north of the Orinoco River, have intensified environmental degradation processes, increasing the long-term exposure of the agricultural sector to climate risks (Paredes-Trejo et al. 2023; Arias et al. 2025).

Addressing the future of agriculture in Venezuela requires acknowledging these interrelated crises. Simply reactivating production without integrating sustainability and climate adaptation measures would likely lead to a resurgence of GHG emissions and further environmental degradation. Therefore, the construction of a resilient agricultural sector must be based on the incorporation of climate-smart practices (Madurga-López et al. 2021), sustainable land management (Sánchez Flores et al. 2024), and the protection of natural ecosystems critical to agricultural productivity (Olivares 2023).

This chapter aims to provide a comprehensive overview of the current state of agriculture in Venezuela in the context of climate change. It examines the historical trends, current challenges, and environmental impacts associated with the sector's decline, as well as the opportunities for promoting a sustainable agricultural transformation. Through this analysis, we seek to contribute to the understanding of how socioeconomic and environmental crises intersect in shaping the future of food systems in Venezuela.

The state of agriculture in Venezuela, as will be seen in this Chapter, is not in its best condition, like most economic activities in the country, including the oil industry. The contribution of GHG emissions generated by agricultural activities has decreased significantly because of the decline in planted and harvested areas and the decrease in the cattle population. It is clear that the reduction in GHG emissions is not attributable to deliberate plans to mitigate and/or adapt to climate change in agriculture, hence the justification for including this chapter in the book.

1.2 Overview of Agriculture and Economic Decline in Venezuela

The agricultural sector in Venezuela, once a crucial component of the national economy, has experienced a profound decline over the past century, particularly intensifying in the last 15 years. During the early twentieth century, agriculture was progressively displaced by the booming oil industry, which became the primary driver of economic growth and government revenue. The emergence of an oil-dependent economy led to a systemic neglect of rural development and agricultural modernization (Cabrera 2025). Food imports, financed by oil revenues, became the norm, reducing the competitiveness of local agricultural production and fostering a dependency that would later prove detrimental when oil revenues declined (Eze 2021).

Agriculture in Venezuela has experienced ups and downs over the past 25 years, with adverse consequences for the sector. Three stages can be identified: first, from the end of the last century (1998–1999) to 2008–2009, with production growth in several crops; then, a sustained decline for 12 consecutive years between 2008–2009 and 2020, with the most negative historical peaks. This severe decline has had serious consequences for the sector, which are still being felt even with the recovery from 2021 to 2024. Key factors such as land expropriations, price controls, financial credit restrictions, insecurity in rural areas, and shortages of agricultural inputs severely weakened the sector's productive capacity during those 12 years (Herrera-Cuenca et al. 2022).

The decline in production of staple crops like corn, rice, sugarcane, and coffee, combined with a shrinking cattle population, mirrored the broader economic collapse and the erosion of national food production.

For general information, since there are no precise statistics on the country's agricultural credit portfolio, between 1995 and 1998, 8–12% of total bank loans (mostly to the private sector) went to large- and medium-sized farmers. Between 2003 and 2008, financial contributions to the agricultural sector, mostly in the form of subsidies and foreign exchange facilities, were abundantly provided by official public organizations. The period from 2013 to 2020 saw the greatest credit crisis and collapse of the entire period. Only 2% of total bank loans went to the agricultural sector, and the deficit in this key service for agriculture persists. According to estimates for 2023, no more than 10% of farmers had access to formal bank loans (Banco Central de Venezuela 2024). The lack of chemical fertilizers is equally critical in the country; they have to be imported, and still are, at high prices. The national fertilizer company practically only sells urea (N fertilizer), and the formulations (N, P, K) are unavailable.

This agricultural and economic deterioration has had far-reaching consequences beyond food production. It has contributed to widespread food insecurity, increased rural poverty, and lost critical agricultural knowledge and infrastructure (Tapia et al. 2022). Furthermore, the collapse of agriculture has diminished the sector's ability to adapt to and mitigate climate change impacts, leaving Venezuela more vulnerable to environmental degradation.

Since the emergence of oil as the engine of income for the Venezuelan economy in the first decades of the twentieth century, agriculture began to decline, and oil rentierism allowed the import of food in a competitively unfavorable way for national agricultural production, which Briceño (2018) explains for the last two decades as the sustained fall in production. According to Gutiérrez (2020), in Fig. 1.1, during the period 1998–2020, the GDPpc presented an unstable behavior, but with a decreasing trend, in the order of −5.3% as the average annual growth rate.

Figure 1.2a illustrates the Gross Domestic Product (GDP) distribution for South American countries in 2022, according to data from Bolt and van Zanden (2023), adjusted for inflation and living cost differences. The map reveals significant economic disparities within the region, with Brazil emerging as the dominant economy, represented by the darkest shading. In contrast, Venezuela is shown among the countries with the lowest GDP levels, highlighting the severe economic contraction

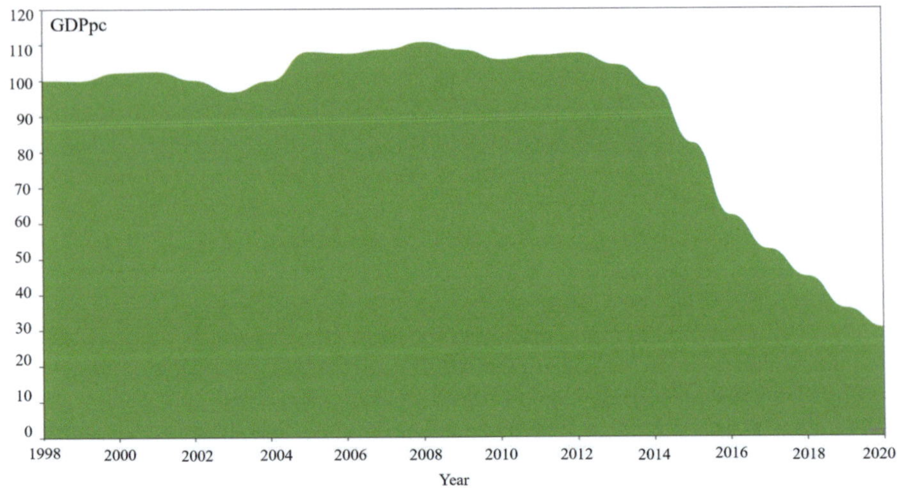

Fig. 1.1 Evolution of the Gross Domestic Agricultural Product per capita (GDPpc) index in Venezuela (1998–2020) according to data from the Central Bank of Venezuela

it has suffered in recent decades. This economic decline has had profound repercussions on sectors such as agriculture, limiting investment capacity, technological innovation, and adaptation to climate change challenges.

Figure 1.2b highlights a long period of modest economic growth until the mid-twentieth century, followed by a pronounced and sustained increase in GDP coinciding with the oil boom. However, starting around 2013, Venezuela experienced a dramatic collapse in its GDP, marking one of the sharpest economic contractions in recent history (Adagunodo 2022). By 2022, the country's GDP had plummeted to levels comparable to those of several decades ago, reflecting the severe economic, political, and social crises that have profoundly affected all sectors, including agriculture (Doocy et al. 2022).

1.3 Agricultural Production, GHG Emissions, and Climate Change

In general, it is important to note that Venezuela ranks as the 35th largest emitter of carbon dioxide equivalent (CO_2-e) gases, with total emissions of 264.40 million tons (Fig. 1.3a), while the country's per capita emissions stand at 8.36 tons per person. By comparison, the United States emits approximately 15 tons per person (World Resources Institute 2025). Energy is the largest emitting sector, and land-use change and forestry, plus agriculture, are the second. The nearly 70% drop in CO_2 equivalent energy emissions between 2014 and 2022 is evident, due to the

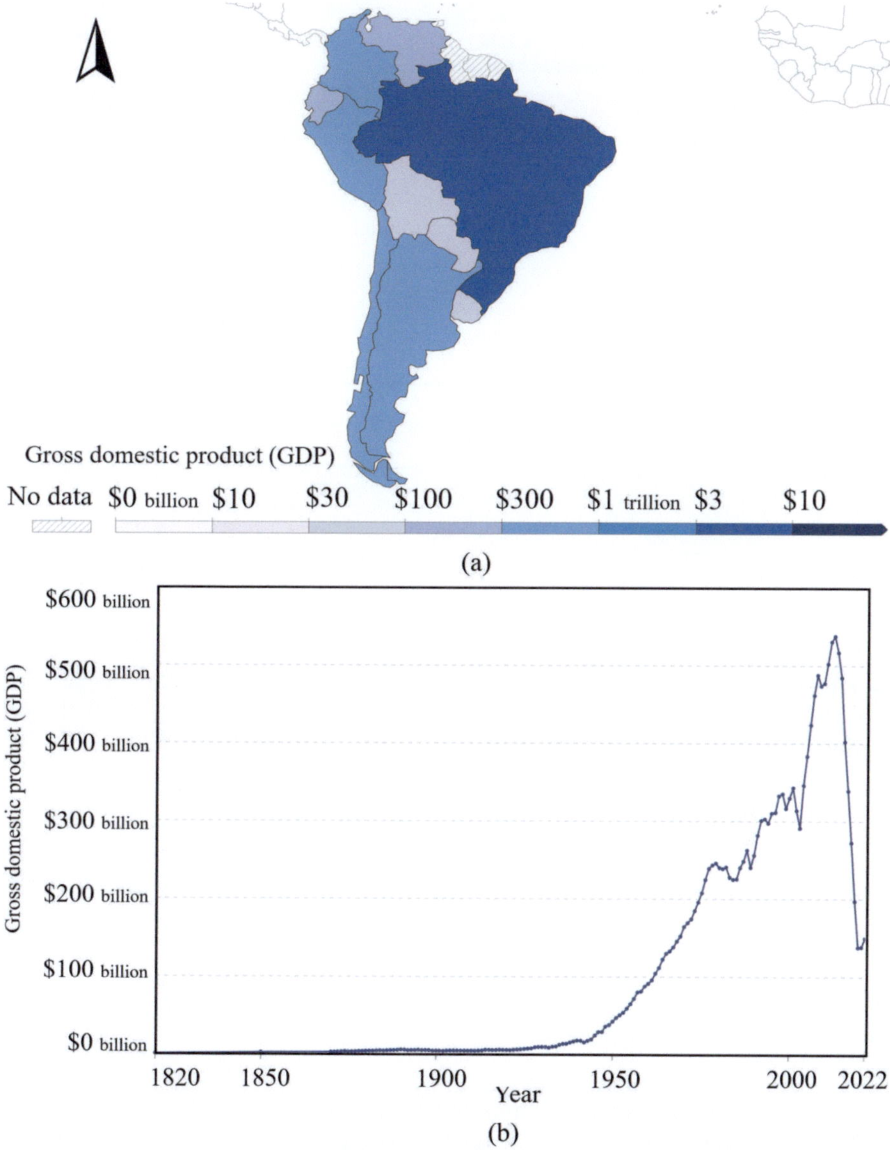

Fig. 1.2 Gross domestic product (GDP) in (**a**) 2022 for South America and (**b**) 1820–2022 for Venezuela. Data is adjusted for inflation and differences in living costs between countries. (Data source: Bolt and van Zanden (2023) available at OurWorldinData.org/economic-growth)

sharp decline of the oil industry, both in crude oil production and refining (Nathan et al. 2024). Agriculture reduced its total equivalent emissions by approximately 3% over the same period, but land-use change, which mostly includes deforestation to increase agricultural land, was substantially reduced between 2014 and 2022.

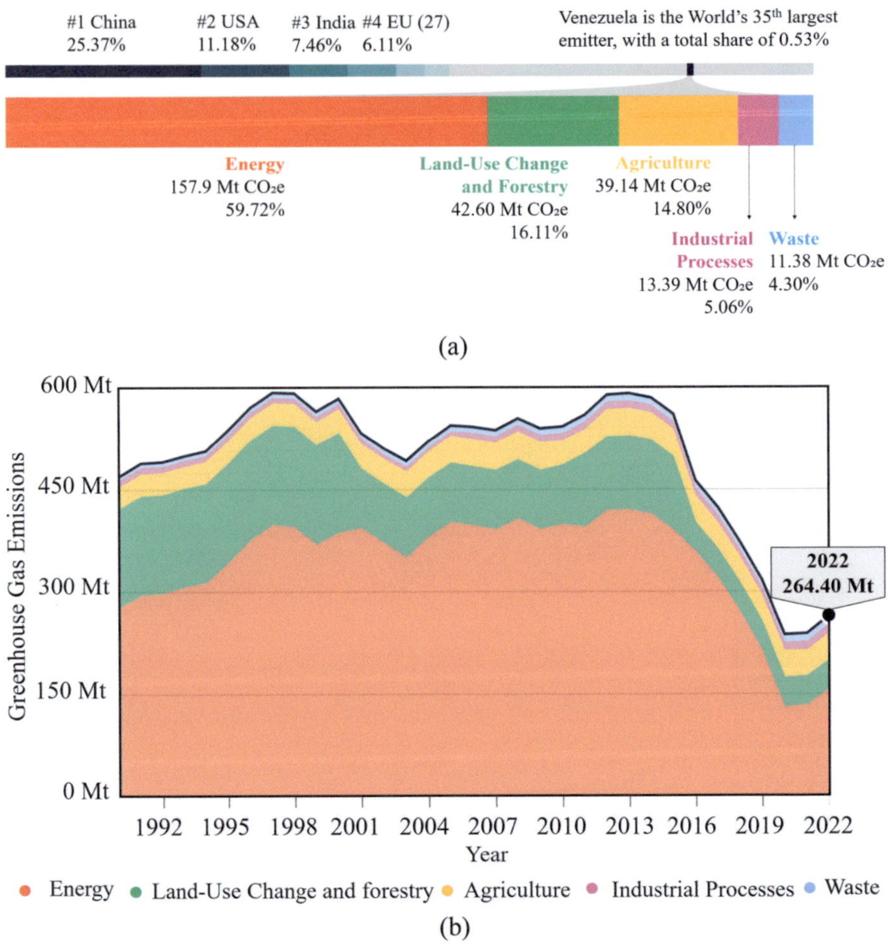

Fig. 1.3 Venezuela's greenhouse gas emissions: (**a**) sectoral breakdown and (**b**) historical trend (1990–2022) according to Climate Watch Historical GHG Emissions (2025) available at https://www.climatewatchdata.org/ghg-emissions

Figure 1.3b illustrates the evolution of CO_2-e emissions in Venezuela from 1990 to 2022. A sharp decline in energy-related emissions between 2014 and 2022 is visible, primarily due to the collapse of the Venezuelan oil industry (Carbonell Betancourt and Scarpellini 2021). Similarly, the reduction in CO_2-e emissions from land-use change, which includes deforestation plus agriculture, is linked to the decline in agricultural activity across the country (López and Olivares 2019).

According to estimates by Fedeagro (2019), shown in Table 1.1, the deplorable state of national agricultural production is evident through its estimates in 12 crops, which together constitute 80% of the value of vegetable agrarian production for the reference year. The country's food production deficit and food insecurity are

Table 1.1 Estimates of crop production for the years 2018–2019 according to Fedeagro (2019)

Crop	National desirable consumption (ton)	Production 2019 (ton)	Production 2018 (ton)	Increase (%) 2019 vs. 2018	% Self-supply
Corn[a,b]	4,000,000	450,000	721,000	−38	11.25
Rice[b]	1,200,000	238,650	418,000	−43	19.89
Sorghum	2,500,000	3253	4822	−33	0.13
Sugarcane[c]	15,000,000	2,300,000	3,382,782	−32	15.33
Coffee[d]	1,800,000	362,600	518,000	−30	20.14
Potato[e]	472,030	26,679	38,930	−31	5.65
Tomato[e]	390,000	52,138	65,172	−20	13.37
Onion[e]	322,050	43,347	55,400	−22	13.46
Bell pepper[e]	168,000	14,523	23,755	−39	8.64
Orange[e]	400,000	31,526	45,000	−30	7.88
Sunflower	720,000	0	3465	−100	0.00
Sesame	N/A	14,000	13,500	4	

[a] Demand for white corn: 1,400,000 tons. For yellow corn: 2,500,000 tons
[b] To calculate the supply of corn and rice, wet production was converted into net conditioned production
[c] Production is from sugarcane; its relationship is 1000 kg of cane = 80 kg of sugar
[d] Figure in quintals (One quintal = 46 kg)
[e] Data in this file correspond to the contribution of domestic production to desirable fresh consumption since these items are not imported in a fresh state

notorious (Herrera-Cuenca et al. 2021), given the scarcity of foreign currency for the import of food due to the breakdown of the national oil industry and the scarce or non-existent economic complexity of the country with non-oil export capacity, which generates foreign exchange (Monaldi 2018).

Nitrous oxide (N_2O) emissions are primarily associated with the application of nitrogen fertilizers and animal manure. Soil bacteria convert added nitrogen into N_2O gas, which is subsequently emitted into the atmosphere as a potent greenhouse gas. Figure 1.4 displays a downward trend in N_2O emissions, which is interpreted as a decline in the use of nitrogen-based fertilizers in the country's agricultural sector. This reduction is likely linked to decreased crop productivity and overall agricultural output.

Considering the current state of agriculture in Venezuela, it is worth highlighting the cases of the pronounced reductions in rice production and the cattle population and their impact on methane emissions (CH_4). GHGs caused by agricultural activities are mainly CH_4 and N_2O; CO_2 contributes to very low emissions. The largest volume of methane emissions from the agricultural sector comes from the cattle population (enteric fermentation) and rice production under flooding (Carmona et al. 2005). The study of Briceño (2021) estimates that in 2021, only 58,500 ha of rice (irrigated and rainfed) were planted, while the average planted area in 1998, 1999, and 2000 was 146,349 ha (MARN-UNDP-GEF 2005), which makes a difference of 87,849 ha in that period.

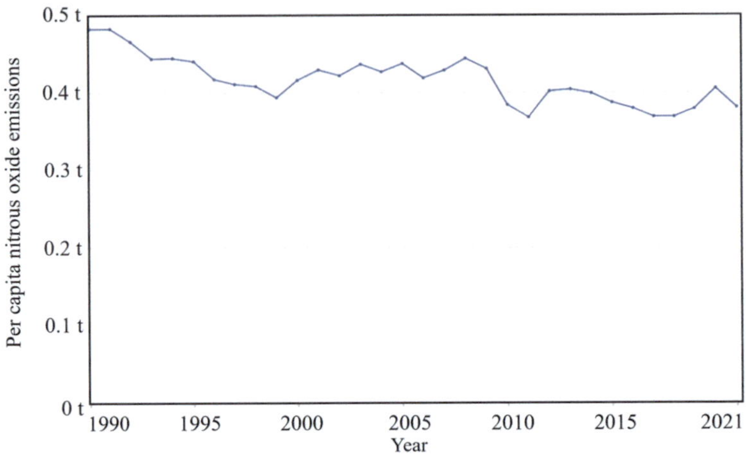

Fig. 1.4 Per capita nitrous oxide (N_2O) emissions from agriculture in Venezuela according to Climate Watch Historical GHG Emissions (2025) available at https://www.climatewatchdata.org/ghg-emissions

The estimate of total CH_4 emissions from rice cultivation in Venezuela (MARN-UNDP-GEF 2005) was 29.3 Gg of CH_4, corresponding to the area of 146,349 ha. For 58,500 ha in 2021, 11.7 Gg would correspond proportionally, assuming the same production conditions of the crop (varieties, agricultural practices, flood levels) for the average planted area in 1998–1999–2000.

Similar to rice, the First National Communication on Climate Change (MARN-UNDP-GEF 2005) reports a cattle population of 13,773,376 head in the 3-year average 1998–2000 and an estimated CH_4 emission from enteric fermentation of this population of 757.2 Gg. It is necessary to point out that it was impossible to identify a source in the scientific literature for the number of heads of cattle today. However, the Venezuelan Federation of Livestock (FEDENAGA) currently reports in press reports an estimated 11 million heads, which in good form coincides with data estimated by the Venezuelan Institute of Milk and Meat (INVELECAR 2021) of 10,842,000 head for the year 2020; the herd fell by 21.3% in the period considered. The emissions of CH_4 from the population of 10,842,000 head would proportionally be 596 Gg, which means a decrease of 161.2 Gg of CH_4 due to population reduction. Similar to rice, the decline is not due to a CH_4 mitigation program in the country's cattle production but to the decline of the sector for different reasons.

Although the figures on land-use changes are not sufficiently clear in Venezuela, such changes represent a significant source of net greenhouse gas emissions and environmental degradation (deforestation, biodiversity loss, soil erosion, acceleration of gas emissions from bare soils), particularly due to the slashing and burning of forest vegetation residues to expand the agricultural frontier (Chacón-Moreno and Llambí 2024).

Figure 1.5a presents the global distribution of total GHG emissions in 2023, expressed in carbon dioxide equivalents (CO_2-e) over a 100-year timescale. The

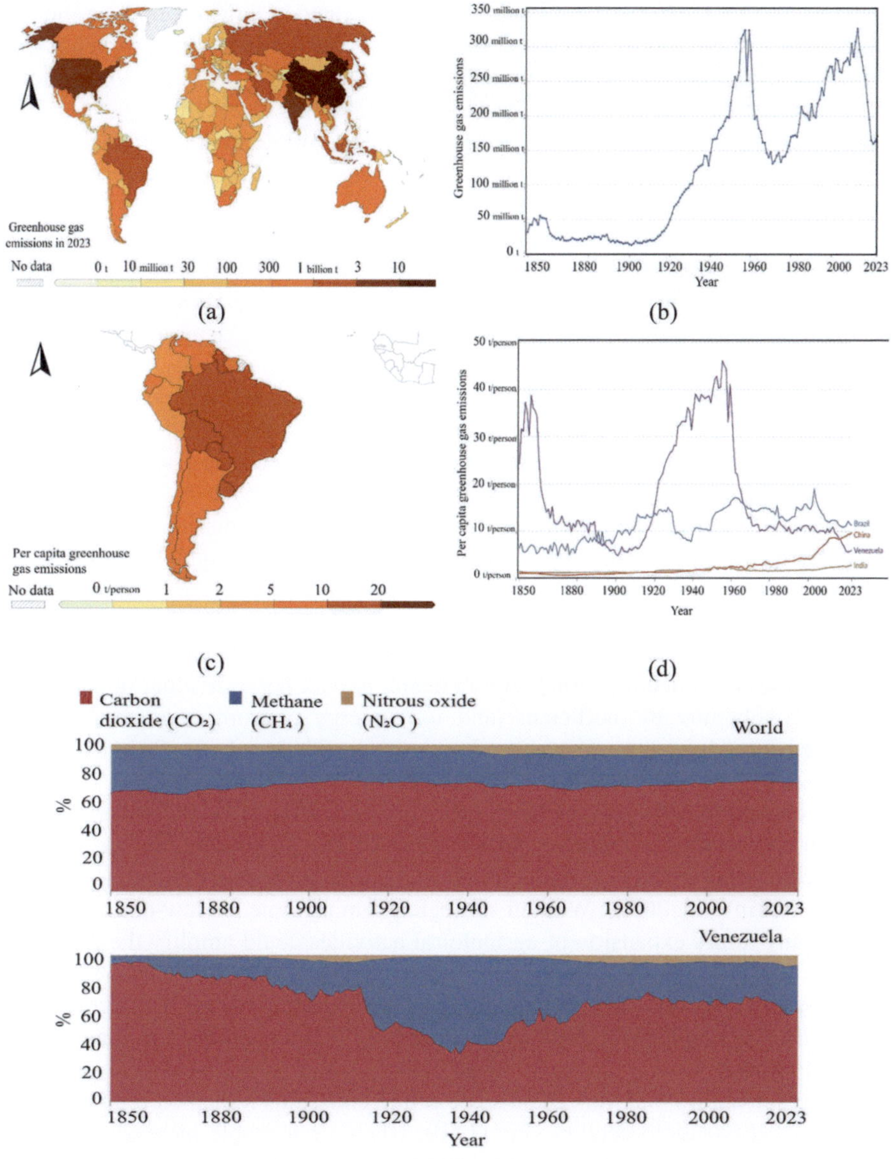

Fig. 1.5 (**a**) Greenhouse gas emissions in 2023 in the world; (**b**) in Venezuela; (**c**) per capita greenhouse gas emissions in 2023 in South America; (**d**) in Venezuela; (**e**) greenhouse gas emissions by gas, 1850–2023 in the World and Venezuela. Note: Greenhouse gas emissions include carbon dioxide, methane, and nitrous oxide from all sources, including land-use change. They are measured in tons of carbon dioxide equivalents over a 100-year timescale. (Data source: Jones et al. (2024) and available at OurWorldinData.org/co2-and-greenhouse-gas-emissions)

map reveals that the highest emissions are concentrated in large economies such as China, the United States, India, and members of the European Union. Venezuela (Fig. 1.5b), in contrast, appears among the countries with moderate to low total emissions compared to global standards. This position is a direct consequence of the overall economic contraction, including the collapse of the agricultural sector, the reduction of industrial activity, and diminished fossil fuel production in recent decades (Ramirez et al. 2025).

Despite the relatively lower total emissions, agriculture remains a critical contributor to Venezuela's GHG profile, particularly through CH_4 emissions from cattle enteric fermentation and flooded rice cultivation, and N_2O emissions from fertilizer use. However, we reaffirm that the observed reductions in agricultural GHG emissions are not the result of strategic mitigation actions but rather a by-product of the decline in agricultural productivity. Without comprehensive adaptation and mitigation strategies, any future recovery of agricultural production could lead to a rapid rebound in GHG emissions, undermining efforts to build climate-resilient food systems and environmental sustainability in the country.

Figure 1.5c shows the distribution of per capita GHG emissions across South America for 2023. Unlike total emissions, this metric highlights the emissions intensity relative to population size. In this case, Venezuela appears with moderate per capita emissions compared to other countries in the region, such as Brazil or Argentina. This indicates that, although Venezuela's overall economic and productive activities have declined, the environmental impact per individual remains significant, likely due to inefficient land use, energy consumption patterns, and continued emissions from the agricultural sector (Nathan et al. 2024; Ramirez et al. 2025).

This context underscores the importance of considering not only total emissions but also the relative environmental footprint of the population. In the case of Venezuelan agriculture, low production levels do not necessarily equate to sustainability (Nahrin et al. 2023). Without strategic planning aimed at low-emission agriculture, any future expansion of agricultural activities could amplify the country's per capita emissions (Madurga-López et al. 2021). Therefore, the adoption of climate-smart practices and technologies is critical to ensuring that agricultural growth does not contribute disproportionately to the intensification of the national carbon footprint.

The graph shows that Venezuela's per capita emissions remained relatively low until the mid-twentieth century (Fig. 1.5d), when a noticeable increase coincides with the peak of the oil economy and industrial expansion. Unlike China and India, where emissions have steadily risen in recent decades, Venezuela's per capita emissions display a decline after the early 2000s, reflecting the economic collapse and reduced industrial and agricultural activity. Despite this decrease, Venezuela's emissions per capita are still higher than those of India and comparable to those of China, signaling inefficiencies in energy and land use even amid economic contraction.

The general context outlined above underlines the complexity of Venezuela's environmental challenges: the reduction in national emissions is not a result of deliberate climate policies or technological innovations but rather an unintended consequence of systemic economic decline. Moreover, as agricultural production recovers, there is a risk that emissions could rebound if sustainability measures are not integrated into the sector's revitalization strategies. Thus, understanding the historical trajectory of emissions is crucial to designing future policies that decouple agricultural growth from increased environmental degradation.

Globally, the profile of GHG emissions has remained relatively stable over time, with CO_2 consistently representing the dominant share, followed by CH_4 and smaller contributions from N_2O (Habeeb et al. 2024). In contrast, Fig. 1.5e shows a very particular pattern of CH_4 emissions in Venezuela between 1920 and 1980, which are not primarily linked to agricultural production. They are mostly associated with crude oil production through the venting of CH_4 gas from the enormous number of new wells that came online in different regions of the country during that period, and new oil refineries.

In Venezuela, CH_4 historically accounted for a significant share of total GHG emissions, particularly between 1920 and 1980, reflecting the importance of oil production and a lower contribution of cattle farming and flooded rice cultivation. This contrasts with global trends, where CO_2 emissions from fossil fuel combustion and industrial processes overwhelmingly dominate (Cabrera 2025). The reduction in the relative weight of methane emissions in Venezuela after the 2000s coincides with the decline of oil as well as agricultural production, and the collapse of the rural economy, rather than with the implementation of active mitigation policies (Nathan et al. 2024).

These patterns emphasize the critical need for targeted strategies in the Venezuelan agricultural sector to manage methane and nitrous oxide emissions effectively. As the sector seeks to recover, there is a risk of an increase in the share of CH_4 and N_2O emissions if traditional practices persist without modernization. Integrating emission-reducing technologies, sustainable livestock management, and improved crop practices will be essential not only for environmental protection but also for aligning agricultural development with global climate goals.

The historical patterns of emissions, particularly the dominance of methane from agricultural activities, highlight the vulnerability of the sector to unsustainable practices. As Venezuela seeks to reactivate its agricultural economy, it faces a critical crossroads: either replicate environmentally harmful models or embrace a transition toward sustainable, low-emission agriculture. The following section explores how environmental degradation, particularly deforestation and land-use change, further compounds these challenges and shapes the prospects for a resilient agricultural future.

1.4 Transformation of Venezuelan Territory: Deforestation and Land-Use Change

Understanding deforestation and land-use change is essential for assessing the environmental transformations that have occurred in Venezuela over recent decades. These dynamics not only reflect the expansion or contraction of productive activities such as agriculture, livestock, and mining, but also reveal the pressure exerted on natural ecosystems, particularly forests and water resources. This section presents a spatial overview of Venezuela's current land cover, allowing the reader to visualize the distribution and extent of key categories, including forest formations, agricultural areas, grasslands and shrublands, non-vegetated zones, and water bodies, based on the most recent national land-use data. By establishing this baseline, the analysis facilitates a deeper understanding of how anthropogenic activity and environmental degradation are reshaping the landscape, with implications for biodiversity conservation, climate regulation, and the resilience of agroecosystems.

Figure 1.6a illustrates the spatial distribution of the main land cover and land-use categories in Venezuela, classified into five major groups: forest formations, grasslands and shrublands, agricultural areas, non-vegetated areas, and water bodies. The dominant land cover is forest formations, shown in green, which are concentrated primarily in the southern region of the country, including the states of Amazonas and Bolívar, as well as in parts of the Andean piedmont and the Coastal Range (Pacheco-Angulo et al. 2021). This spatial pattern reflects the continued presence of largely undisturbed natural ecosystems, consistent with the quantitative data indicating a total of 65.09 million ha, representing 61.63% of the national territory.

Complementarily, Fig. 1.6b quantifies and compares the relative share of each category. Agricultural areas cover approximately 22.07 million ha (20.90%), mostly concentrated in the central, western, and parts of the eastern "Llanos," corresponding to zones used for crop cultivation and pasture (MapBiomas 2025). Grasslands and shrublands, occupying 15.31 million ha (14.50%), are mainly distributed across ecotonal areas or regions with partially degraded natural vegetation (López-Beltrán et al. 2019). Non-vegetated areas (1.20 million ha; 1.14%) and water bodies (1.93 million ha; 1.83%) have a more localized distribution, typically associated with arid environments, urban zones, mostly natural water bodies, and artificial reservoirs.

The historical time series chart illustrates the evolution of Venezuela's major land cover and land-use categories over 39 years (1985–2023) (Fig. 1.6c), measured in million hectares. Forest formations consistently dominate the national territory, although they exhibit a declining trend from approximately 68.6 million ha in 1985 to 65.1 million ha in 2023, indicating a gradual loss of forest cover over time. This decrease is likely associated with deforestation for land conversion for agriculture, and illegal mining in the south of the Orinoco River, meaning both an expansion of anthropogenic frontiers, particularly in regions such as the Orinoco and Amazon basins (Fletcher et al. 2021).

Conversely, agricultural areas (crops and livestock) show a relatively stable yet slightly increasing pattern, rising from around 19 million to over 22 million ha, reflecting the persistence and expansion of agricultural land use. Grasslands and shrublands fluctuate moderately, suggesting sensitivity to land management practices and ecological transitions. Non-vegetated areas and water bodies remain relatively constant in extent, although minor variations may reflect seasonal hydrological dynamics and localized land degradation (Fletcher et al. 2022). Overall, the chart underscores long-term landscape transformation processes in Venezuela, with

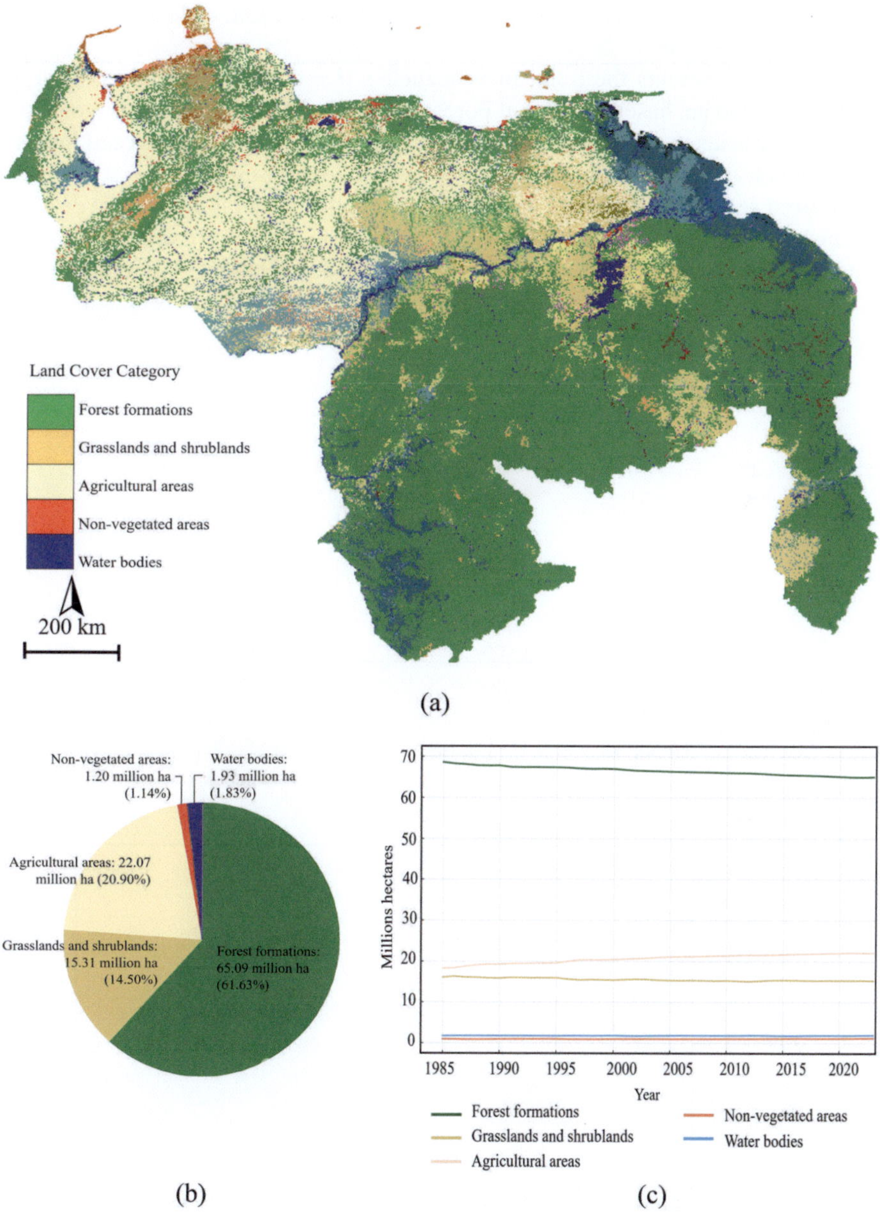

Fig. 1.6 (**a**) Land cover and land use in Venezuela (2023). A spatial representation of dominant land cover types and land-use categories across the national territory; (**b**) surface area (in million hectares) by major land cover and land-use classes; (**c**) temporal evolution (1985–2023) of five major land cover and land-use categories according to MapBiomas (2025) available in https://plataforma.venezuela.mapbiomas.org/

implications for biodiversity conservation, climate regulation, and sustainable land management (Lobo et al. 2022).

Figure 1.7a presents the trend in Venezuela's tree cover loss in 2023, close to 90,000–100,000 ha, based on Global Forest Watch (2024) data. The continuous loss of tree cover leads to a reduction in biodiversity and ecosystem services and intensifies greenhouse gas emissions, especially CO_2 from deforestation processes. For the agricultural sector, this dynamic translates into greater vulnerability to soil erosion, diminished water retention, and local climate alterations, all of which pose additional risks to food security and sustainable rural development. The reduction of

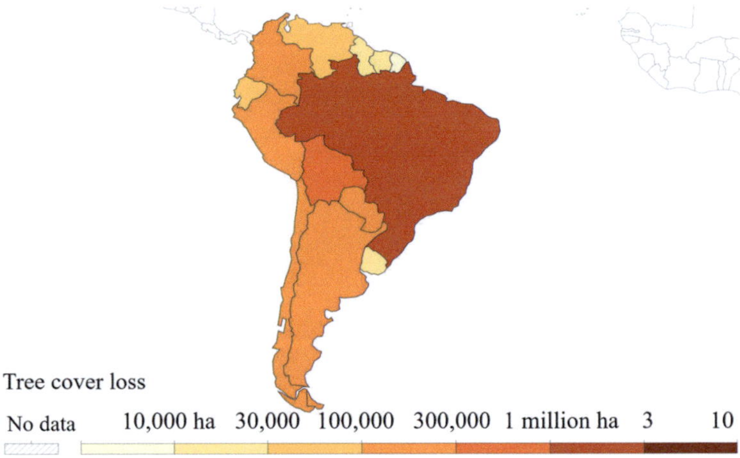

(a)

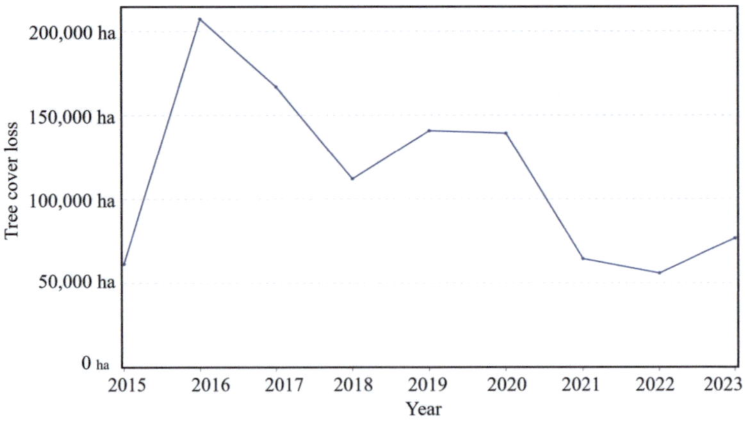

(b)

Fig. 1.7 Tree cover loss in South America 2023 (**a**) and (**b**) 2015–2023 in Venezuela. Tree cover loss is the temporary or permanent loss of vegetation greater than five meters in height. (Data source: Global Forest Watch (2024) and available at OurWorldinData.org/forests-and-deforestation) forested areas contributes directly to increased GHG emissions, particularly CO_2, as stored carbon in vegetation and soils, is released into the atmosphere.

In Venezuela, this environmental degradation aggravates the challenges sector, already weakened by economic decline and climate variability (Omoke et al. 2022). The loss of vegetation cover further exacerbates soil erosion and soil organic matter breakdown, reduces water availability, and intensifies the vulnerability of rural communities, making the urgent implementation of sustainable land management practices a critical priority (Udemba and Yalçıntaş 2023).

The graph highlights a peak of deforestation around 2016, followed by a gradual but irregular decline (Fig. 1.7b). Despite the lower values compared to the peak years, the data for 2023 indicate a loss of 76,614 ha of forest cover, evidencing that the country continues to experience considerable environmental pressure on its forested natural ecosystems. The significant forest loss observed, especially during the peak years, represents an important source of greenhouse gas emissions due to the release of stored carbon. This ongoing environmental degradation exacerbates the challenges faced by the country's agricultural sector, impacting soil health (Lobo et al. 2022; Olivares et al. 2022), water cycles (Viloria et al. 2023), and local climate stability (Paredes-Trejo et al. 2023).

1.5 Temporal Analysis of Venezuela's Water Bodies Under Environmental Pressure

The evolution of surface water bodies serves as a critical indicator of environmental stability and hydrological health, particularly in regions exposed to climatic stress and anthropogenic disturbance. In the case of Venezuela, understanding the temporal dynamics of aquatic systems provides insight into the interplay between climate variability, such as shifts in precipitation and drought regimes, and human-driven landscape transformations, including watershed degradation, hydropower development, and land conversion. This section introduces a long-term assessment (1996–2023) of the country's surface water coverage, highlighting trends in spatial reduction and categorizing the types of water bodies based on their origin and use. By contextualizing these changes across both temporal and geographic dimensions, the analysis offers a foundation for evaluating the impacts of environmental degradation and infrastructure expansion on the nation's hydrological balance and ecological resilience.

The temporal analysis of water bodies in Venezuela reveals a noticeable long-term contraction in their surface extent from 1996 to 2023 (Fig. 1.8c). At the start of the observed period, total water coverage exceeded 2.5 million ha. However, it underwent a consistent reduction, reaching values near 1.9 million ha in recent years. This decline suggests not only the influence of climatic variability, such as prolonged droughts or changes in hydrological regimes, but also structural human interventions, including damming, diversion of watercourses, clogging reservoirs,

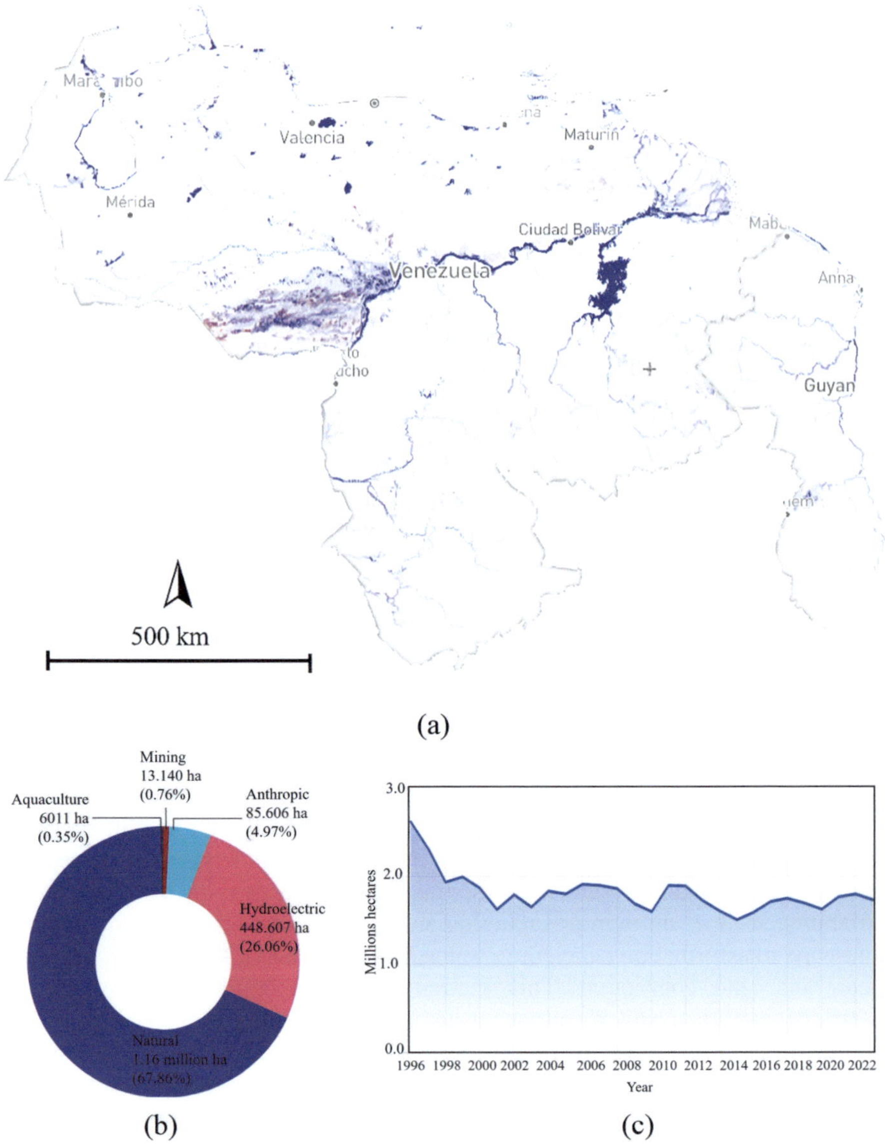

Fig. 1.8 (**a**) Spatial distribution of surface water bodies across Venezuela; (**b**) classification of total water surface area (2023) by origin and use, differentiating natural, hydroelectric, anthropic, aquaculture, and mining-associated bodies; and (**c**) temporal dynamics of surface water bodies in Venezuela (1996–2023) according to MapBiomas (2025) available in https://plataforma.venezuela.mapbiomas.org/

and land-use conversion. The stabilization observed after 2010 may reflect both hydroclimatic adaptation and a slowing pace of anthropogenic transformation.

Complementing the temporal trend, Fig. 1.8b provides a categorical breakdown of current water body types. Natural water bodies dominate, representing approximately 67.86% of the total (1.16 million ha), indicating the country's still significant hydro-ecological endowment. However, 26.06% (448,607 ha) corresponds to water bodies formed or modified by hydroelectric infrastructure, underscoring the substantial transformation of aquatic systems for energy production (MapBiomas 2025). Meanwhile, anthropogenic uses (such as reservoirs, canals, and managed wetlands) make up nearly 5%, and aquaculture, though still minor, represents 0.35%, hinting at the diversification of water use in productive sectors. Mining-related water surfaces, albeit small (0.76%), are environmentally relevant due to potential contamination and hydrological disruption.

Geographically, the distribution of water surfaces is highly heterogeneous across Venezuelan states (Fig. 1.8a). Bolívar stands out with 647,512 ha, largely attributable to the Guri Dam and associated hydrographic networks. States such as Apure and Amazonas also register substantial water bodies (257,209 ha and 160,740 ha, respectively), reflecting their location within major basins such as the Orinoco and their prevalence of floodplains and wetlands. In contrast, states such as Monagas and Falcón have considerably less water coverage, underlining the importance of regional-scale water governance and hydrological planning in addressing spatial disparities in water availability and use (Hidalgo López 2023).

When the anthropogenic factor is considered, as noted by Rosales Hernández and García Montero (2015), watershed systems are usually subject to changes in their components, structure, and functionality. These changes often lead to significant alterations in hydrological flows and the rates of sediment production and transport, both spatially and temporally. In this regard, assessing the sediment production potential of a watershed, whether under natural or human-altered conditions, is essential information for the design and operation of various hydraulic structures and ecosystem management in mountain regions.

The loss of usable water storage capacity in reservoirs designed for irrigation or even for supplying water to populations is a natural phenomenon that occurs over time. However, under certain conditions, such as intense human or natural interventions, as has been the case in some upper watersheds in Venezuela, this process can be so greatly accelerated that total loss of reservoir capacity can occur. In such cases, the water bodies can disappear entirely due to sediment infill (siltation), and this can happen much earlier than expected, rendering the reservoir unable to serve its intended purpose.

Even more critically, suitable dam sites, locations where reservoirs can be built, are rare and unique. Watershed management, therefore, involves, among other activities, reducing soil erosion and sediment production in order to avoid soil destruction as a CO_2 reservoir and GHG emissions and extend the operational lifespan of water reservoirs used for irrigation and domestic water supply. Watershed conservation is essentially a complex process of changing land use from degrading to sustainable use.

Together, these figures reveal that while Venezuela retains a considerable portion of its water bodies in a relatively natural state, there is a significant and increasing footprint of human intervention (Segnini et al. 2025). The prominence of hydroelectric infrastructure, the growing role of aquaculture, and evidence of sustained surface water loss over nearly three decades point to the urgent need for integrated water resource management (Flores et al. 2023). Such a strategy must balance ecological conservation with energy, agricultural, and economic demands, particularly under mounting pressures from climate change and socioenvironmental degradation.

1.6 Agricultural Sector Limitations and Recovery Strategies

Due to the absence of updated agricultural statistics in Venezuela, Fedeagro on its website (2024a) included production estimates in tons of 14 agricultural items made by the production associations affiliated with the Confederation. For example, the estimates of five basic items (corn, rice, sorghum, coffee, and sugarcane) showed a small rebound in production in 2021, 2022, and 2023 (Caraballo et al. 2025). These increases are associated with some changes in policies that strongly restrict production, such as marketing controls by price fixing and exchange rate liberalization. There is a hint of a reduction in tariff-free food imports due to foreign exchange deficits and free imports of agricultural inputs, given the notable insufficiency in domestic markets, such as fertilizers and pesticides.

In the Fedeagro Annual Assembly (2024b), despite the small increases in agricultural production from 2021 to 2023, there was consensus on the need to overcome severe limitations that affect the sector, as shown below.

1.6.1 Decapitalization of Farmers

The Venezuelan agricultural sector has faced a severe process of decapitalization, fundamentally driven by the escalation of production costs. Inputs such as seeds, fertilizers, agrochemicals, and labor have become increasingly expensive, while logistical difficulties have added hidden costs related to transportation and storage. This economic environment has prevented many small and medium farmers from maintaining adequate production levels, forcing them to reduce the cultivated area or abandon agricultural activities altogether.

Additionally, the progressive decline in productivity has further weakened farmers' economic resilience. Factors such as soil degradation, limited access to quality seeds, and a lack of investment in modern technologies and technology transfer programs have led to a drop in yields per hectare. Climate change has exacerbated this problem, introducing greater variability in rainfall patterns, temperature increases, and extreme weather events that undermine the stability of production systems.

The decapitalization has also been fueled by persistently low prices at the producer level. Due to market distortions, import competition, and weak internal demand, farmers are often forced to sell their products at prices that do not cover production costs. This situation generates disincentives to invest in improvements, limits reinvestment in the sector, and perpetuates a vicious cycle of low productivity, reduced profitability, and increased poverty in rural areas.

1.6.2 Credit Restrictions of the Financial System

One of the main structural obstacles facing the Venezuelan agricultural sector is the severe restriction of credit, largely due to the monetary policies imposed by the Central Bank of Venezuela. The high legal reserve requirements limit the ability of private banks to offer financing to productive sectors, especially agriculture. Without access to affordable credit, farmers are unable to finance essential activities such as land preparation, seed acquisition, fertilization, pest control, and harvesting, thus stifling the production cycle from the outset.

The lack of financing particularly affects small- and medium-sized producers, who have fewer resources and are less likely to access informal or alternative financing mechanisms. In addition, the limited availability of long-term credit instruments prevents investment in agricultural infrastructure, such as irrigation systems, warehouses, and cold chains, which are crucial for increasing productivity and reducing post-harvest losses. The absence of formal financial support systems also discourages the adoption of innovations necessary to face climate change.

Moreover, credit restrictions deepen inequalities in the rural economy. Producers with more capital or better access to informal networks can survive, while the most vulnerable are gradually excluded from the market. This dynamic promotes the concentration of land in a few hands and accelerates rural depopulation, weakening social cohesion and compromising food sovereignty. Addressing the financial limitations of the agricultural sector requires urgent policy reforms aimed at democratizing access to credit and promoting inclusive rural development.

1.6.3 Critical and Recurrent Shortage of Fuel

The shortage of fuel in Venezuela has become a chronic problem that severely restricts agricultural operations. Diesel and gasoline are essential for running tractors, irrigation systems, and harvesters and for transporting agricultural inputs and products to markets. In many rural areas, farmers must wait days or even weeks to access limited quantities of fuel, which delays or completely paralyzes critical phases of the production cycle, such as planting and harvesting. This situation not only reduces yields but also increases production costs due to the need to acquire fuel on the black market at inflated prices.

Beyond direct effects on farm operations, the fuel shortage undermines the functioning of supply chains and rural economies. The lack of a reliable fuel supply disrupts the timely delivery of seeds, fertilizers, and machinery parts and prevents the proper distribution of harvested products, leading to significant post-harvest losses. This affects food availability at local and national levels and discourages farmers from expanding or even maintaining their production areas.

The recurrent fuel crisis has a cumulative effect on the agricultural sector's resilience. Producers lose confidence in their ability to plan and execute complete production cycles, which leads to a progressive abandonment of farming activities. Furthermore, the absence of a stable energy infrastructure also limits the potential for mechanization and modernization of agriculture, trapping farmers in low-productivity systems that are highly vulnerable to economic and climate shocks.

1.6.4 Competition from Duty- and Tax-Free Imports and Smuggling

The uncontrolled influx of duty-free and tax-free agricultural imports has been a major blow to Venezuelan farmers. Imported products, often subsidized in their countries of origin, enter the domestic market at prices well below the cost of local production. This unfair competition forces national producers to sell at unprofitable prices or risk losing market share altogether. Over time, this has eroded the viability of traditional farming, weakened local food systems, and deepened the economic decapitalization of rural areas.

Smuggling compounds the problem by introducing agricultural products through informal channels that evade both quality controls and taxation. This practice not only depresses market prices but also undermines food safety standards, exposing consumers to potential health risks. For producers, the effect is devastating: legitimate production cannot compete against illegally imported goods that are sold at a fraction of the cost, creating a distorted market that discourages formal agricultural activities and tax contributions to the state.

The persistence of these dynamics generates a climate of uncertainty that inhibits investment in agriculture. Farmers are reluctant to improve infrastructure, adopt new technologies, or expand cultivated areas when there is no guarantee of fair competition in the marketplace. Overcoming this challenge requires a combination of effective border controls, the reestablishment of tariff protections for strategic products, and the promotion of local consumption policies that prioritize Venezuelan agricultural production.

1.6.5 Contraction of Consumption of Key Foods

The economic crisis in Venezuela has severely contracted the consumption of essential foods that traditionally form the basis of the national diet, such as corn, rice, sugar, coffee, and vegetables. This decline in consumption is closely linked to the reduction in household purchasing power, as hyperinflation and unemployment have diminished the ability of the majority of the population to access a varied and nutritious diet. Consequently, the demand for domestically produced food has shrunk, affecting not only consumers but also the viability of local agricultural production systems.

The decrease in the consumption of key foods generates a vicious cycle for national agriculture. Producers, faced with falling demand, reduce their planting areas or abandon production altogether. This, in turn, leads to greater dependence on imports for food supply and undermines the country's food sovereignty. In many regions, the collapse of the agricultural economy has triggered rural depopulation, as families seek alternative livelihoods in urban centers or abroad.

The impact of this contraction is particularly severe in regions and productive axes that were historically specialized in certain crops. For example, areas dedicated to corn and rice production have seen a dramatic decline in economic activity, affecting not only farmers but also related industries such as transportation, processing, and commercialization. The contraction of food consumption, therefore, has systemic effects, weakening rural economies and social structures.

To reverse this trend, it is essential to implement policies that stimulate the recovery of domestic demand through income improvement, food security programs, and incentives for local production and consumption. Strengthening the internal market is a critical step toward rebuilding a sustainable and resilient agricultural sector capable of supporting Venezuela's food needs.

1.6.6 Lack of Incentives for Agricultural Production, Environmental Protection, Technological Innovation, and Sustainability

The absence of adequate incentives for agricultural production in Venezuela is one of the fundamental obstacles hindering the sector's recovery. Farmers operate in an environment where there are few or no direct subsidies, minimal access to technical assistance, and no guaranteed markets for their produce. Without financial or institutional support, producers are unable to modernize their operations, expand cultivated areas, or adopt sustainable practices that could improve both productivity and environmental outcomes.

Environmental protection in the agricultural sector is similarly neglected. The lack of policies promoting soil conservation, water management, biodiversity protection, and reforestation exacerbates land degradation and increases the sector's vulnerability to climate change. Practices such as slash-and-burn agriculture and uncontrolled land use persist, largely because there are no structured programs to incentivize sustainable land management. Without interventions that link agricultural production with ecological preservation, the country risks deepening the deterioration of its natural resource base.

Technological innovation is also largely absent from the current agricultural landscape. Investment in research, development, and extension services has collapsed, leaving farmers without access to improved seeds, efficient irrigation systems, pest control technologies, or climate-resilient agricultural practices. The technological gap widens each year, making Venezuelan agriculture less competitive regionally and globally. Innovation ecosystems, universities, research centers, and public–private partnerships are either severely weakened or non-existent, limiting the potential for transformation.

Finally, the lack of a coherent sustainability strategy threatens the future of agriculture in Venezuela. Agricultural systems that are not adapted to climate variability will continue to suffer from productivity shocks, soil erosion, and water scarcity. Incentivizing/encouraging sustainable practices, such as agroecology, conservation agriculture, forest plantations, and agroforestry, is not just an environmental imperative but an economic necessity to build a resilient, productive, and competitive agricultural sector. Without strategic incentives, the sector will remain trapped in low-productivity, high-risk cycles.

1.6.7 Depletion of the Fleet of Machinery, Equipment, and Agricultural Implements

The depletion and obsolescence of agricultural machinery and equipment in Venezuela represent a severe structural limitation for the sector. Most of the machinery in use today dates back several decades and suffers from chronic maintenance problems due to the scarcity of spare parts, high repair costs, and lack of technical support. Tractors, harvesters, irrigation pumps, and other essential tools have deteriorated to such an extent that they can no longer guarantee the timely execution of agricultural operations, severely affecting production efficiency and yield.

The shortage of functioning machinery forces many farmers to rely on outdated, manual, or semi-mechanized practices, which significantly increase labor costs and extend production times. In a context of labor shortages and rising input prices, the lack of machinery exacerbates vulnerabilities, especially during critical periods such as planting and harvesting. The mechanization deficit particularly impacts extensive crops like rice, corn, and sugarcane, where productivity heavily depends on the availability of appropriate equipment.

Furthermore, the technological stagnation of the machinery fleet limits the adoption of climate-smart agriculture practices. Modern technologies such as precision agriculture, conservation tillage, and efficient irrigation systems require access to updated machinery and digital tools. Without this technological leap, Venezuelan agriculture will remain tied to practices that are not only less productive but also more environmentally degrading, increasing soil erosion, water wastage, and GHG emissions.

Addressing this limitation requires a comprehensive renewal plan for the agricultural machinery fleet, involving credit mechanisms for equipment acquisition, public–private partnerships to establish maintenance and supply centers, and training programs for farmers on the use and management of modern technologies. A new generation of mechanization, adapted to the realities of climate change, is essential to revitalize production and improve the resilience of Venezuelan agriculture.

1.6.8 Insecurity of Goods and People, Agricultural Roads, Electricity Supply, and Public Services in the Rural Sector

Insecurity has become one of the most critical challenges facing rural areas in Venezuela. Frequent robberies, theft of machinery, livestock, and harvests, and the general lack of protection have severely undermined the safety of agricultural producers. Farmers often face threats not only to their goods but also to their physical integrity, creating an atmosphere of fear and discouragement. This insecurity discourages investment, limits expansion plans, and contributes to the abandonment of productive activities and rural migration.

The precarious state of agricultural roads compounds the challenges for the sector. Poorly maintained and often impassable roads during the rainy season make it extremely difficult to transport inputs to farms and products to markets. As a result, producers face high logistical costs, suffer post-harvest losses, and have limited access to essential services. The deterioration of transportation infrastructure isolates rural communities, weakens the rural economy, and prevents the integration of farmers into national and regional supply chains.

The instability and deficiencies in the electricity supply further aggravate the rural crisis. Frequent blackouts, voltage fluctuations, and the lack of reliable energy services severely impact agricultural activities that depend on electricity, such as irrigation, cold storage, and mechanized processing. Additionally, the poor quality of electrical service limits the potential for introducing new technologies that require stable energy, such as climate monitoring systems, automated irrigation, or value-added agro-industries.

Finally, the general collapse of public services, including water supply, health care, telecommunications, and education, has accelerated the depopulation and marginalization of rural areas. Without access to basic services, retaining skilled labor, attracting young people to agriculture, or fostering rural entrepreneurship becomes increasingly difficult. A sustainable recovery of Venezuelan agriculture must therefore be based on a comprehensive strategy that not only promotes production but also rebuilds the social, economic, and security fabric of rural communities.

1.7 Conclusions

The agricultural sector in Venezuela today reflects the convergence of multiple crises: economic collapse, institutional deterioration, environmental degradation, and climate change vulnerability. Over the past decades, agriculture has shifted from being a strategic pillar of the national economy to a marginalized sector marked by reduced productivity, decapitalization of farmers, and loss of food sovereignty. This decline has not only affected the country's ability to feed its population but has also weakened the resilience of rural communities and ecosystems.

The reports of Briceño (2018), Gutiérrez (2020), and Fedeagro (2024b) agree that the main causal factors of the decline of agricultural activity in the country in the last two and a half decades are: (1) interventions in rural private property (expropriations), (2) the fall in the real prices of agricultural products at the producer level, (3) massive food imports creating unfair competition, (4) insecurity in rural areas, (5) the deficit of agricultural inputs of all kinds (fertilizers, agrochemicals, machinery spare parts, fuels, others), (6) the intervened agricultural marketing and credit restrictions for the agricultural sector.

Despite the reduction in greenhouse gas emissions linked to the contraction of agricultural activities, this trend is neither sustainable nor intentional. Without proactive mitigation and adaptation strategies, any resurgence in production risks reversing these temporary environmental "gains" and exacerbating climate impacts. The persistence of deforestation, methane emissions from livestock and rice cultivation, and inefficient agricultural practices highlight the urgent need for a transformation toward sustainable, climate-smart agriculture.

Reviving Venezuelan agriculture requires more than reactivating production; it demands addressing systemic limitations such as lack of credit, technological obsolescence, insecurity, and weak rural infrastructure. Integrating environmental sustainability, technological innovation, and rural development policies is critical for building a resilient and competitive sector. Only through coordinated efforts can Venezuela achieve a recovery that ensures food security, economic stability, and a genuine contribution to global climate goals.

References

Adagunodo M (2022) The effect of oil receipts and fuel subsidy payments on the current account deficit in Nigeria and Venezuela. Ann Spiru Haret Univ Econ Ser 22(1):137–152. https://oaji.net/articles/2022/5273-1653897465.pdf. Accessed 7 Mar 2025

Anido JD, García Lobo L (2018) Agricultura, sistema alimentario y desarrollo territorial rural en Venezuela entre dos milenios. In: Sosa Cárdenas D (ed) Pilares fundamentales para construir una mejor Venezuela. Universidad Católica Andrés Bello, Caracas, pp 67–109

Arias PA, Fernández-Berrío A, Bedoya-Pineda V, Acevedo-Ortiz MY, Martínez JA (2025) Compound dry and hot extremes and their implications for fire activity over the Orinoco River Basin in northern South America. Clim Dyn 63(4):175. https://doi.org/10.1007/s00382-025-07660-5

Banco Central de Venezuela (2024) Boletines estadísticos 1995–2023. Banco Central de Venezuela, Caracas, Venezuela

Bolt J, van Zanden JL (2023) Maddison Project Database 2023 with major processing by Our World in Data. Gross domestic product (GDP)—Maddison Project Database—in constant international-$. Historical data [dataset]. https://ourworldindata.org/grapher/gdp-maddison-project-database. Accessed 27 Apr 2025

Briceño G (2018) La situación agrícola de Venezuela, una aproximación al problema y líneas de acción para resolverlo en el corto plazo. Red Agroalimentaria de Venezuela. https://es.scribd.com/document/489338261/

Briceño G (2021) Resultado del año 2021. Confederación de Asociaciones de Productores Agropecuarios de Venezuela (Fedeagro). Instituto de Políticas Agrícolas, Caracas. https://fedeagro.org/wp-content/uploads/2022/05/El-2021.pdf

Cabrera K (2025) Venezuela's energy transition: a synergistic approach to blue hydrogen, AI, and LNG development. Int J Educ Pract Eng 2(2):46–56. https://doi.org/10.70504/ijepe.v2i2.13979

Caraballo LJ, Zapata RE, Castillo VS (2025) Comercio agroalimentario entre Colombia y Venezuela: un estudio basado en el modelo de corrección de errores. Rev Agroaliment 31(60):35–64. https://n9.cl/hi9ln0. Accessed 7 Mar 2025

Carbonell Betancourt CM, Scarpellini M (2021) Venezuela, oil and climate change: overcoming nostalgia. In: Handbook of climate change management: research, leadership, transformation. Springer International Publishing, Cham, pp 1–31. https://doi.org/10.1007/978-3-030-22759-3_242-1

Carmona JC, Bolívar DM, Giraldo LA (2005) El gas metano en la producción ganadera y alternativas para medir sus emisiones y aminorar su impacto a nivel ambiental y productivo. Rev Col Cienc Pec 18:1. http://www.scielo.org.co/pdf/rccp/v18n1/v18n1a06.pdf. Accessed 11 Mar 2022

Chacón-Moreno E, Llambí LD (2024) Land-use and climate change impacts in the Venezuelan Andes at multiple spatial–temporal scales. In: Safeguarding mountain social-ecological systems, vol 2, pp 47–52. https://doi.org/10.1016/B978-0-443-32824-4.00022-5

Doocy S, Page KR, Liu C, Hoaglund H, Rodríguez DC (2022) Venezuela: out of the headlines but still in crisis. Bull World Health Organ 100(8). https://doi.org/10.2471/BLT.22.288269

Eze MA (2021) A comparative survey data analysis of declining oil revenue implications on the economic performance of oil exporting countries: Nigeria, Venezuela, and Norway. J Resour Dev Manag 76:22–52. https://doi.org/10.7176/JRDM/76-03

Fedeagro (2019) La Catástrofe Agricola del 2019. Confederación de Asociaciones de Productores Agropecuarios de Venezuela. Fedeagro. Documentos. https://fedeagro.org/resultados-de-la-agricultura-vegetal-del-2018/. Accessed 8 Mar 2022

Fedeagro (2024a) Estimaciones Fedeagro/Producción en Toneladas. Confederación de Asociaciones de Productores Agropecuarios de Venezuela. Página Web. https://fedeagro.org/estimaciones/. Accessed May 2024

Fedeagro (2024b) Declaración de Valencia 2024. Asamblea Anual 2024. Confederación de Asociaciones de Productores Agropecuarios de Venezuela, Valencia, Venezuela. Página Web. https://fedeagro.org/wp-content/uploads/2024/04/Declaracion-de-Valencia-2024-22-de-abril.pdf. Accessed May 2024

Fletcher IK, Hernández-Villena J, Moreno JE, Drakeley C, Jones K, Grillet ME, Lowe R (2021) The effect of environmental degradation and land use change on malaria re-emergence in South Venezuela: a spatiotemporal modelling study. Lancet Planet Health 5:S13. https://doi.org/10.1016/S2542-5196(21)00097-8

Fletcher IK, Grillet ME, Moreno JE, Drakeley C, Hernández-Villena J, Jones KE, Lowe R (2022) Synergies between environmental degradation and climate variation on malaria re-emergence in southern Venezuela: a spatiotemporal modelling study. Lancet Planet Health 6(9):e739–e748. https://doi.org/10.1016/S2542-5196(22)00192-9

Flores S, Forister ML, Sulbaran H, Díaz R, Dyer LA (2023) Extreme drought disrupts plant phenology: insights from 35 years of cloud forest data in Venezuela. Ecology 104(5):e4012. https://doi.org/10.1002/ecy.4012

Global Forest Watch (2024) With major processing by Our World in Data. Tree cover loss [dataset]. Global forest watch, Tree cover loss by dominant driver [original data]. https://ourworldindata.org/grapher/tree-cover-loss. Accessed 27 Apr 2025

Gutiérrez A (2020) Venezuela, la crisis y la seguridad alimentaria: hacia una nueva estrategia. Agroalimentaria 26(51):77–116. https://n9.cl/03crp. Accessed May 2024

Habeeb FA, Habeeb MA, Khaleel YL, Ameen FN (2024) Global analysis and prediction of CO_2 and greenhouse gas emissions across continents. Appl Data Sci Anal 2024:173–188. https://doi.org/10.58496/ADSA/2024/014

Herrera-Cuenca M, Landaeta-Jiménez M, Sifontes Y (2021) Challenges in food security, nutritional, and social public policies for Venezuela: rethinking the future. Front Sustain Food Syst 5:635981. https://doi.org/10.3389/fsufs.2021.635981

Herrera-Cuenca M, Landaeta-Jiménez M, Hernández P, Sifontes Y, Ramírez G, Vásquez M, Maingon T (2022) Exploring food security/insecurity determinants within Venezuela's complex humanitarian emergency. Dialog Health 1:100084. https://doi.org/10.1016/j.dialog.2022.100084

Hidalgo López C (2023) Threat of non-metallic mining on the integrity of the Tocuyo River basin, Lara state, Venezuela. Region Water Conserv 6(1):104–120. https://doi.org/10.32629/rwc.v6i1.1258

INVELECAR (2021) Rebaño ganadero de Venezuela. Instituto Venezolano de la Leche y la Carne. Minuta Agropecuaria. https://n9.cl/3inmx. Accessed May 2024

Jones AD, Collins WD, Edmonds J, Torn MS, Janetos A, Calvin KV, Wise M (2024) With major processing by Our World in Data. Annual greenhouse gas emissions including land use [dataset]. Jones et al., National contributions to climate change 2024.2 [original data]. https://ourworldindata.org/grapher/total-ghg-emissions. Accessed 27 Apr 2025

Kılıç B, Arabacı E, Öz A (2022) Evaluation of total greenhouse gas emissions of countries with tropical climate. Sci J Mehmet Akif Ersoy Univ 5(2):26–32. https://dergipark.org.tr/en/pub/sjmakeu/issue/74842/1186747

Lescher Soto I, Villamizar A, Olivares BO, Gutiérrez ME, Nagy GJ (2024) Navigating the uncertain terrain: Venezuela's future using the Shared Socioeconomic Pathways framework—a systematic review. Climate 12(7):98. https://doi.org/10.3390/cli12070098

Lobo D, Lozano Z, Toro M, Valencia IH, Rey JC (2022) Soil biodiversity under different uses and managements in Venezuela. In: Sustainable soil management as a key to preserving soil biodiversity and stopping its degradation, vol 240. https://n9.cl/lab4u2

López M, Olivares B (2019) Normalized Difference Vegetation Index (NDVI) applied to the agricultural indigenous territory of Kashaama, Venezuela. UNED Res J 11(2):112–121. https://doi.org/10.22458/urj.v11i2.2299

López-Beltrán M, Olivares B, Lobo-Luján D (2019) Changes in land use and vegetation in the agrarian community Kashaama, Anzoátegui, Venezuela: 2001–2013. Rev Geogr Am Cent 2(63):269–291. https://doi.org/10.15359/rgac.63-2.10

Madurga-López I, Dutta Gupta T, Castellanos AS, Martínez-Barón D, Pacillo G (2021) Climate security nexus in Latin America and the Caribbean: Venezuela and Colombia. CGIAR FOCUS Climate Security. https://cgspace.cgiar.org/items/dfc3cb13-3749-4d81-9452-b83002f292fb

MapBiomas (2025) Collection 2 of the annual series of Land Cover and Surface Water Bodies, Maps of Venezuela, Provita. Consulted on May 17, 2025. https://plataforma.venezuela.mapbiomas.org/

MARN-PNUD-GEF (2005) Primera Comunicación Nacional de Cambio Climático. Ministerio del Ambiente y los Recursos Naturales-Programa de la Naciones Unidas para el Desarrollo-Global Enviromental Facility. https://unfccc.int/sites/default/files/resource/vennc01.pdf. Accessed 9 Mar 2022

Monaldi F (2018) The collapse of the Venezuelan oil industry and its global consequences. Atlantic Council, Washington, DC. https://n9.cl/lblri

Nahrin R, Rahman MH, Majumder SC, Esquivias MA (2023) Economic growth and pollution nexus in Mexico, Colombia, and Venezuela countries: the role of renewable energy in carbon dioxide emissions. Energies 16(3):1076. https://doi.org/10.3390/en16031076

Nathan B, Maasakkers JD, Naus S, Gautam R, Omara M, Varon DJ, Aben I (2024) Assessing methane emissions from collapsing Venezuelan oil production using TROPOMI. Atmos Chem Phys 24(11):6845–6863. https://doi.org/10.5194/acp-24-6845-2024

Olivares B, Rey JC, Lobo D, Navas-Cortés JA, Gómez JA, Landa BB (2022) Machine learning and the new sustainable agriculture: applications in banana production systems of Venezuela. Agric Res Updates 42:133–157

Olivares BO (2023) Banana production in Venezuela: novel solutions to productivity and plant health, The Latin American studies book series. Springer Nature. https://doi.org/10.1007/978-3-031-34475-6

Omoke PC, Nwachukwu T, Ibrahim A, Nwachukwu O (2022) Asymmetric impact of financial development, trade openness, and environmental degradation on economic growth in Venezuela. Environ Sci Pollut Res 29(18):27411–27420. https://doi.org/10.1007/s11356-021-18421-2

Pacheco-Angulo C, Plata-Rocha W, Serrano J, Vilanova E, Monjardin-Armenta S, González A, Camargo C (2021) A low-cost and robust Landsat-based approach to study forest degradation and carbon emissions from selective logging in the Venezuelan Amazon. Remote Sens 13(8):1435. https://doi.org/10.3390/rs13081435

Paredes-Trejo F, Olivares BO, Movil-Fuentes Y, Arevalo-Groening J, Gil A (2023) Assessing the spatiotemporal patterns and impacts of droughts in the Orinoco River basin using Earth observations data and surface observations. Hydrology 10(10):195. https://doi.org/10.3390/hydrology10100195

Ramirez CAP, Rodríguez EVV, Cimetta SDS, De León MG (2025) Energy balance and GHG emissions from oil and natural gas production in Venezuela. Dyna 92(235):100–109. https://doi.org/10.15446/dyna.v92n235.116401

Rodriguez F, Llamedo M, Belhaj H, Belhaj A (2022) Challenges associated with the acid gases production and capture in hydrocarbon reservoirs: a critical review of the Venezuelan cases. In: SPE thermal well integrity and production symposium. OnePetro. https://doi.org/10.2118/212146-MS

Rosales Hernández A, Garcia Montero P (2015) Las cuencas hidrográficas y su gestión integral. In: Gabaldon A, Rosales A, Buroz E, Cordova JR, Uzcategui G, Iskandar L (eds) Agua en Venezuela: una riqueza escasa, Capítulo 22, vol 2. Fundación Empresas Polar, pp 866–913. https://bibliofep.fundacionempresaspolar.org/media/1378209/agua_tii_cap22.pdf

Sánchez Flores E, Olivares BO, Vinueza Chérrez R, Serrano Rodríguez R (2024) Are we doing enough to understand our relationship with the environment? A bibliometric analysis of the instruments that evaluate pro-environmental behavior. Sustain Dev 32(6):6555–6578. https://doi.org/10.1002/sd.3023

Segnini S, Cressa C, Rincón JE, Montoya JV, Castillo MM, Barrios M, Scott-Frías J (2025) Biodiversity and ecological processes in running waters of Venezuela. Inland Waters:1–13. https://doi.org/10.1080/20442041.2024.2431404

Tapia MS, San-Blas G, Machado-Allison C, Carmona A, Landaeta-Jiménez M (2022) Food security and food safety challenges in Venezuela. Front Sustain Food Syst 5:812955. https://doi.org/10.3389/fsufs.2021.812955

Udemba EN, Yalçıntaş S (2023) Economic and environmental implications of resource rent: a dual analysis of Venezuela's sustainability. Int Soc Sci J 73(247):215–231. https://doi.org/10.1111/issj.12378

Viloria JA, Olivares BO, García P, Paredes-Trejo F, Rosales A (2023) Mapping projected variations of temperature and precipitation due to climate change in Venezuela. Hydrology 10:96. https://doi.org/10.3390/hydrology10040096

Wirtz AL, Stevenson M, Guillén JR, Ortiz J, Barriga Talero MÁ, Page KR, Spiegel PB (2024) Persistent food insecurity and material hardships: a latent class analysis of experiences among Venezuelan refugees and migrants in urban Colombia. Nutrients 16(7):1060. https://doi.org/10.3390/nu16071060

World Resources Institute (2025) Climate watch historical GHG emissions. World Resources Institute, Washington, DC. https://www.climatewatchdata.org/ghg-emissions. Accessed 7 May 2025

Chapter 2
Methodological Framework for a Climate-Agriculture Review in Venezuela

2.1 Introduction

Understanding the vulnerabilities and adaptation pathways of agriculture under climate change in Venezuela requires a rigorous and transparent methodological approach. As climate impacts become more spatially heterogeneous and temporally dynamic, the synthesis of scientific evidence through systematic methods provides a reliable foundation for informing policy and guiding adaptation strategies. This chapter outlines the methodological framework adopted for conducting a comprehensive review of primary studies, reports, and relevant literature that assess climate change impacts, vulnerabilities, and adaptive responses in Venezuelan agriculture.

Systematic reviews are recognized for their capacity to minimize bias and improve reproducibility by applying structured and replicable procedures to the identification, selection, and appraisal of evidence (Cumpston et al. 2019; Olivares et al. 2023). Unlike narrative reviews, which may be influenced by the subjective interpretation of findings, systematic approaches adhere to predefined criteria that enhance objectivity and transparency. In this work, we draw upon the guidance provided in the Cochrane Handbook for Systematic Reviews of Interventions (Cumpston et al. 2019), adapting it to the context of environmental and agricultural research, while maintaining the fundamental principles of methodological rigor and analytical consistency.

The Preferred Reporting Items for Systematic Reviews and Meta-Analyses (PRISMA) framework served as a cornerstone for structuring this review. Originally developed to improve the clarity and transparency of systematic reviews in health research (Moher et al. 2015), PRISMA has been increasingly applied in interdisciplinary contexts, including environmental sciences and agricultural systems (Liberati et al. 2009; Lescher-Soto et al. 2024; Sánchez-Flores et al. 2024). The

A. Rosales Hernández et al., *Agriculture Under Climate Change*, The Latin American Studies Book Series, https://doi.org/10.1007/978-3-032-06368-7_2

updated PRISMA 2020 statement (Page et al. 2021) guided the formulation of the review protocol, ensuring comprehensive reporting across each stage of the review process, from identification to synthesis.

The scope of this review encompassed literature published in both peer-reviewed and grey literature domains, reflecting the diversity of sources available in the Venezuelan context. The decision to include technical reports, institutional documents, and local case studies aimed to capture underrepresented knowledge from non-academic sources, which is crucial for understanding climate vulnerabilities in data-scarce regions. This approach aligns with recent calls for the integration of diverse knowledge systems in systematic synthesis processes (Page et al. 2021).

The methodological strategy also considered the challenges posed by regional and temporal heterogeneity. Venezuela's varied agroecological zones, altitudinal gradients, and sociopolitical dynamics necessitate a multiscale and interdisciplinary analytical lens. Accordingly, the selection criteria and data extraction procedures were designed to account for spatial, temporal, and sectoral variations, enabling a more nuanced interpretation of climate-agriculture interactions across different contexts.

Ultimately, this chapter serves to document the procedures followed to ensure that the resulting synthesis is not only methodologically sound but also relevant and applicable to stakeholders engaged in climate adaptation planning. The methodological choices made herein respond to the dual need for scientific credibility and contextual relevance in addressing one of the most pressing challenges to sustainable development in Venezuela: the resilience of its agricultural systems in the face of a changing climate.

2.2 Review Objectives and Scope

This systematic review was developed as part of the contributions to the Second Academic Report on Climate Change in Venezuela (DRACC), within the framework of Working Group II: Impacts, Vulnerability, and Adaptation of the Venezuelan academic network. The objective of the review was to synthesize and critically examine the available scientific evidence regarding the impacts of climate change on agriculture in Venezuela, with a particular focus on identifying vulnerabilities and documented adaptation strategies across diverse agroecological and sociopolitical contexts. The aim was to generate an evidence-based foundation to support the national academic contribution to climate policy and planning.

The scope of the review spanned both biophysical and socioeconomic dimensions of agricultural vulnerability to climate change. It included literature addressing the impacts of climatic variability and long-term climate trends on crop systems, livestock, soil degradation, water availability, pests and diseases, fisheries, and rural livelihoods. Furthermore, studies that documented or proposed adaptation strategies, such as technological innovations, agroecological transitions, institutional responses, or traditional knowledge practices, were considered central to the analytical framework.

Geographically, the review encompassed all agroecological regions of Venezuela, including the Andean highlands, Llano's plains, Caribbean coast, Amazon basin, and Orinoco delta, recognizing the high territorial heterogeneity in both climate exposure and agricultural practices. Temporally, the review prioritized studies published between 2018 and 2025, reflecting a period of growing national and international attention to the relationship between climate and agriculture, as well as the emergence of local research under evolving political, economic, and environmental conditions.

Due to the interdisciplinary nature of the topic, the review incorporated findings from a wide range of sources, including peer-reviewed journal articles, technical reports, theses, conference proceedings, and institutional publications. This broad scope aligned with the recommendations of Page et al. (2021) and Cumpston et al. (2019), who emphasized the importance of capturing diverse forms of evidence, especially in regions where scientific production is unevenly distributed and much relevant knowledge exists outside conventional academic publishing.

The review was structured around two core research questions: (1) What were the documented impacts of climate change on agricultural systems in Venezuela? and (2) What forms of vulnerability and adaptation had been identified or proposed across different agroecological regions?

This approach enabled a comparative and thematic synthesis of the available literature, while also allowing for the identification of evidence gaps, emerging trends, and context-specific dynamics relevant to both research and policy. By aligning the objectives of the review with the priorities established by DRACC Working Group II, this chapter contributed to a nationally grounded and methodologically robust assessment of the impacts of climate change on Venezuelan agriculture.

2.3 Search Strategy

The search strategy implemented in this review followed the principles of transparency, breadth, and replicability as recommended in the Cochrane and PRISMA guidelines for systematic reviews (Cumpston et al. 2019; Page et al. 2021). The objective was to identify a comprehensive body of literature addressing the impacts of climate change on agriculture in Venezuela, including documented vulnerabilities and adaptive responses. In line with recommendations from Booth et al. (2016), the strategy integrated both academic databases and grey literature sources to maximize coverage and minimize selection bias.

Electronic searches were conducted in the databases Scopus, Web of Science, Scielo, and Google Scholar, the latter due to its proven utility in capturing grey literature and region-specific academic output (Haddaway et al. 2015). In addition, searches were extended to repositories of international and national organizations such as FAO, IPCC, CLACSO, Redalyc, and local institutions like INE, MINEC, INIA, and Venezuelan Universities. This hybrid strategy enabled the inclusion of valuable scientific, technical, and institutional knowledge often excluded from mainstream indexing platforms.

Boolean logic and keyword clusters were applied across thematic and geographic dimensions. Search terms were organized into three main categories: (1) climate-related terms (e.g., "climate change," "climate variability," "drought," "El Niño phenomenon," "flooding"), (2) agriculture-related terms (e.g., "crops," "yields," "food systems"), and (3) impact and response terms (e.g., "impact," "adaptation," "mitigation," "resilience"). Typical queries included combinations such as "climate variability" AND "crops" AND "Venezuela," "drought" AND "yields" AND "impact," "El Niño phenomenon" AND "agriculture," and "flooding" AND "mitigation" AND "food production." These terms were adjusted as needed for English- and Spanish-language publications to ensure contextual relevance across different agroecological regions of the country.

The review covered publications from 2018 to 2025, as the First Academic Report on Climate Change in Venezuela had already compiled literature up to 2018. However, several studies published before 2018 were also included in this review, particularly when they had not been cited in the previous academic report. This decision aimed to ensure the inclusion of relevant but previously overlooked evidence, especially foundational studies that contribute to understanding long-term trends, regional baselines, or early adaptation experiences. By extending the temporal range selectively, the review enhanced both its historical depth and thematic completeness.

No language restrictions were imposed, though Spanish and English were predominant. All references were managed using Zotero, facilitating deduplication, classification, and traceability. To enhance completeness, backward and forward citation snowballing was employed, as recommended by Wohlin (2014), allowing the identification of additional sources cited by or citing key publications that might have escaped initial keyword searches.

Furthermore, grey literature was systematically incorporated using search strategies adapted from Godin et al. (2015), recognizing its critical role in Latin American environmental research, particularly where peer-reviewed publications are sparse or politically constrained. Theses, policy documents, institutional reports, and working papers were included based on their relevance and methodological transparency. This inclusive approach ensured a broader representation of regional knowledge systems, including case studies and local-level evidence often absent from academic publishing (Haddaway et al. 2015).

2.3.1 Methodological Note on Grey Literature and Snowballing

Given the fragmented nature of agricultural and climate research in Venezuela, special attention was paid to non-indexed but academically relevant sources. Grey literature was appraised for credibility, source transparency, and analytical depth, even when lacking peer review, as such documents often provide rich contextual data and grounded observations from field-based initiatives. In parallel, snowballing techniques, both backward (reference lists) and forward (citation indexes), proved effective for discovering high-impact studies and emerging thematic clusters beyond the scope of initial searches (Wohlin 2014). Together, these methods enhanced the comprehensiveness and contextual sensitivity of the evidence base included in this review.

2.4 Eligibility Criteria

To ensure consistency and transparency in the selection of studies, a set of inclusion and exclusion criteria was defined a priori, by PRISMA and Cochrane guidelines (Cumpston et al. 2019; Page et al. 2021). These criteria served as the basis for screening titles, abstracts, and full texts, allowing for the identification of relevant literature that addressed the impacts of climate change on agriculture in Venezuela, with particular emphasis on vulnerabilities, adaptation strategies, and agroecological heterogeneity. Inclusion criteria (Table 2.1) were as follows:

(a) Geographic focus: Studies had to refer explicitly to Venezuela or border regions with comparable biophysical and socioeconomic conditions (e.g., Andean corridor, Amazon basin).
(b) Thematic relevance: Studies had to address one or more of the following topics: climate variability or change; agricultural production systems; impacts on crops, livestock, soils, or water resources; agroecosystem vulnerabilities; and adaptation or mitigation strategies.
(c) Types of sources: Peer-reviewed journal articles, technical reports, doctoral and master's theses, institutional publications, book chapters, and national climate reports were considered eligible.

Table 2.1 Inclusion and exclusion criteria used in the review

Inclusion criteria	Exclusion criteria
Studies explicitly focused on Venezuela or its agroecological regions	Studies focused exclusively on non-Venezuelan contexts without relevant regional comparability
Addressed topics related to climate change/variability and agriculture	Lacked explicit linkage between climate variables and agriculture
Included peer-reviewed articles, technical reports, theses, and institutional documents	Publications such as editorials, opinion pieces, or news articles without research content
Published between 2018 and 2025, with exceptions for earlier studies not cited in the first Academic Report	Duplicated records or documents are inaccessible after exhaustive retrieval attempts
Written in Spanish or English	Written in languages other than Spanish or English, without available translation

(d) Time frame: Publications from 2018 to 2025 were prioritized. However, earlier works were included when not cited in the First Academic Report on Climate Change in Venezuela, especially if they provided original data, foundational insights, or addressed gaps in the existing evidence base.
(e) Language: Publications in Spanish or English were accepted.

Exclusion criteria (Table 2.1) included the studies focused solely on regions outside Venezuela without clear relevance or comparability to the Venezuelan context; publications lacking empirical or conceptual content (e.g., opinion pieces, editorials, or news articles); duplicated records and inaccessible documents after exhaustive retrieval attempts and studies not addressing climate-agriculture linkages explicitly or focusing only on unrelated environmental variables without discussion of agricultural impacts or responses.

A two-stage screening process was applied. First, titles and abstracts were reviewed for relevance and completeness. In a second stage, full texts were evaluated against the eligibility criteria. Discrepancies or ambiguous cases were resolved through consensus among reviewers. The use of structured screening protocols minimized subjective bias and ensured reproducibility of selection decisions (Booth et al. 2016).

The resulting body of literature reflected the interdisciplinary and context-specific nature of climate-agriculture dynamics in Venezuela, encompassing multiple regions, temporal scales, and disciplinary approaches. This selection process ensured that the synthesis included the most pertinent and methodologically sound contributions available for the period under review.

2.5 Screening and Selection Process

The process of screening and selecting studies was conducted in two sequential stages: an initial review of titles and abstracts, followed by a full-text assessment of potentially eligible documents. This approach, consistent with PRISMA standards (Page et al. 2021), was designed to ensure both methodological transparency and reproducibility in the identification of relevant literature.

In the first stage, all records retrieved through the search strategy were imported into the reference management software Zotero. Duplicate entries were identified and removed automatically and manually. The remaining records were screened by examining titles and abstracts against the predefined inclusion and exclusion criteria (see Table 2.1). Studies clearly unrelated to the climate-agriculture nexus, or with no Venezuelan relevance, were excluded at this phase.

The second stage consisted of a detailed full-text review of all documents deemed potentially eligible during the initial screening. In this phase, studies were critically assessed for their methodological quality, thematic alignment, and completeness of information. Particular attention was given to studies with unclear abstracts or titles to avoid premature exclusion of relevant materials. When necessary, reviewers

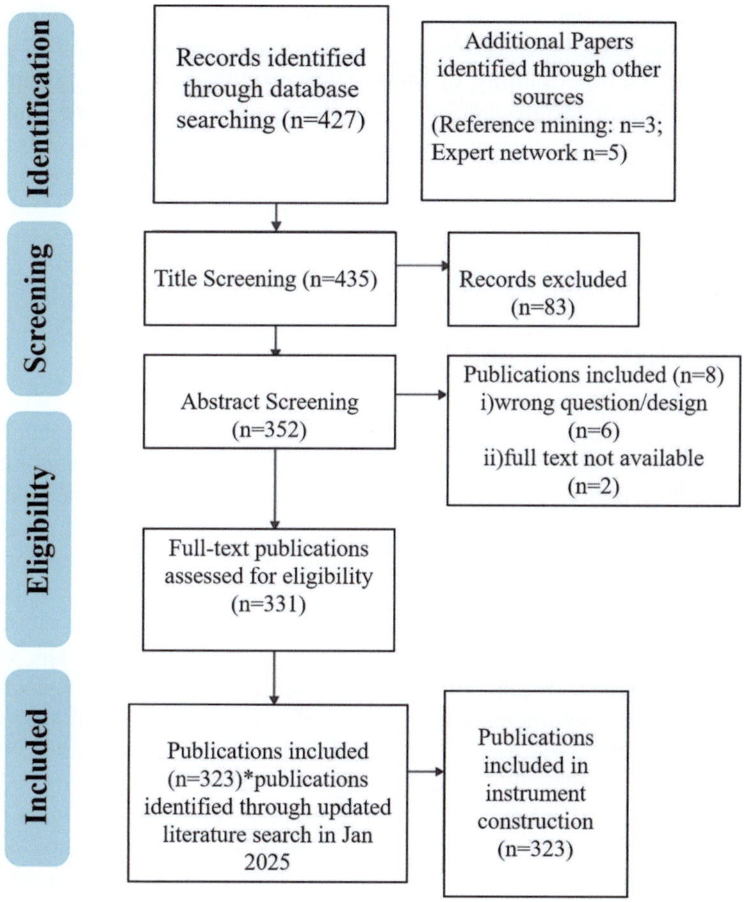

Fig. 2.1 PRISMA flow diagram of study selection for the systematic review on climate change and agriculture in Venezuela (2018–2025)

sought clarification through auxiliary searches (e.g., checking citations or related works).

Discrepancies in study eligibility were resolved through consensus among the review team. If disagreements persisted, a third reviewer was consulted to ensure consistency. This collaborative review process helped mitigate individual bias and strengthened the reliability of selection decisions, as recommended by systematic review methodology literature (Booth et al. 2016).

The screening process was documented by the PRISMA 2020 flow diagram (Fig. 2.1), which captures the number of records identified, screened, excluded, assessed for eligibility, and ultimately included in the final synthesis. Initially, 427 records were identified through database searching. An additional 8 records were included from other sources, such as reference mining ($n = 3$) and expert consultation ($n = 5$), bringing the total to 435. During the title screening phase, 83 records were excluded, leaving 352 records for abstract screening. Following this, 21 records were excluded based on abstract relevance. Subsequently, 331 full-text articles were assessed for eligibility, of which 8 were excluded—6 due to inappropriate research questions or design and 2 because full texts were not accessible. Ultimately, 323 publications met the inclusion criteria and were retained for the review. All 323 studies were also considered for the development of data extraction instruments. This selection process reflects a rigorous and transparent screening strategy aligned with PRISMA guidelines.

2.6 Data Extraction and Management

Following the screening and eligibility assessment, a structured and standardized data extraction process was carried out to ensure consistency, reproducibility, and transparency throughout the review. The extraction strategy was aligned with guidance provided by Cumpston et al. (2019) and the PRISMA 2020 framework (Page et al. 2021), and it was tailored to the interdisciplinary nature of climate-agriculture interactions within the Venezuelan context.

A data extraction template was developed and pilot-tested using a subset of ten publications to ensure clarity, adequacy, and comprehensiveness. The template included bibliographic details (authors, year, journal), study setting (geographical region, ecosystem type), methodological characteristics (study design, type of data, analytical tools), thematic focus (e.g., variability, mitigation, adaptation, impact), type of agricultural system (e.g., rainfed, irrigated, livestock), climate-related stressors addressed (e.g., droughts, floods, temperature rise, El Niño), and main findings or outcomes. Particular attention was given to whether the study addressed smallholder systems, indigenous knowledge, or policy implications, which are critical components of the vulnerability and adaptation discourse.

All extracted data were entered into a centralized spreadsheet hosted on a secure institutional cloud service with restricted access. Double data entry was performed independently by two reviewers to minimize human error and bias. Discrepancies were resolved through discussion and, when necessary, by consulting a third reviewer. This approach ensured the reliability and validity of the dataset used in the synthesis phase.

The extracted data were organized thematically to facilitate subsequent coding, analysis, and visualization. Categories such as "climatic drivers," "agricultural impacts," and "adaptation responses" were established based on recurring topics in the literature and complemented by inductive codes emerging during the review. In line with undefined Egger et al. (1997), potential sources of reporting bias were documented, including selective reporting of outcomes or lack of methodological clarity in older studies.

To ensure transparency and reproducibility, the data extraction file, along with the coding protocol and list of included studies, was archived as supplementary material and will be made publicly available through the publication repository associated with the DRACC initiative. This documentation will allow future researchers to replicate or build upon the present review.

2.7 Synthesis Methods

The synthesis of the extracted data was carried out using a narrative and thematic approach, complemented by comparative tables and spatial-temporal aggregation techniques. Given the heterogeneity of study designs, data sources, and disciplinary orientations among the included publications, a meta-analysis was deemed inappropriate. Instead, a qualitative synthesis was conducted to integrate findings across different contexts and methodological frameworks, following recommendations from Booth et al. (2016) and Page et al. (2021).

The narrative synthesis involved the systematic organization and textual summarization of evidence around key thematic axes, such as climate-related stressors, agricultural impacts, vulnerability profiles, and adaptation measures. Each study was reviewed in detail to extract major conclusions, contextual variables, and sectoral relevance. This process allowed for the identification of converging patterns, contradictions, and knowledge gaps, with special attention to regional differences and temporal trends.

To enhance comparability, summary tables were developed to align studies by variables such as agroecological zone, climate risk (e.g., drought, flooding, and temperature extremes), crop or livestock system, and type of adaptation response. These tables enabled the cross-tabulation of impacts and responses, supporting the identification of recurring dynamics across Venezuelan regions. They also facilitated the comparison between sectors (e.g., crop agriculture, forestry, fisheries, and livestock) and between types of producers (e.g., subsistence farmers vs. commercial operations).

A thematic coding process was applied to the qualitative data using a combination of deductive codes, based on the review questions and DRACC indicators, and inductive codes emerging during the reading and interpretation of the literature. These codes were grouped into higher-order categories, including: (1) climatic drivers (e.g., variability, El Niño, warming), (2) types of agricultural stress (e.g., soil degradation, water deficit), (3) adaptation practices (e.g., crop substitution, water harvesting, traditional knowledge), and (4) institutional dimensions (e.g., policy gaps, governance, knowledge transfer). Coding was managed manually using a structured matrix developed in Excel.

In addition, the synthesis incorporated spatial and temporal aggregation of findings to identify regional concentrations of evidence and temporal shifts in research focus. Studies were grouped by state or subnational region and by decade of publication to observe the evolution of academic attention and thematic priorities over time. This spatial-temporal mapping highlighted underrepresented areas such as the Orinoco Delta and parts of the Amazon basin, as well as periods with increased scholarly output, particularly post-2015, in connection with the Paris Agreement and national adaptation efforts.

2.8 Additional Data Sources and Visualizations

To support the interpretative depth of the narrative synthesis and enhance the contextual understanding of climatic and agricultural trends in Venezuela, this review integrated a set of visual and quantitative resources obtained from authoritative global and regional datasets (Table 2.2). These external sources provided high-resolution spatial and temporal data to complement the findings derived from the literature review.

Table 2.2 External data sources and visualizations used in the review

Source	Types of data/visualization	Period covered	Purpose of the review
World Bank—Climate Change Knowledge Portal (CCKP 2025)	(a) Current climatology (temperature and precipitation); 0.5° resolution (CRU) (b) Historical trends and variability (ERA5) (c) Climate projections (CMIP6, SSPs); 0.25° resolution	(a) 1991–2020 (b) 30, 50, 70 years (c) 2020–2100 (projected)	Establish historical climate baselines, visualize trends, and assess future risks under emission scenarios
MapBiomas Venezuela (2025)	(a) Land-use and land cover maps (raster, 30 × 30 m) (b) Surface water and glacier dynamics	1985–2023	Analyze land-use change, deforestation, agricultural expansion, and hydrological alterations
World Resources Institute (2024)	(a) Crop area by irrigation method (b) Exposure to water-related and drought risk (c) Food production vs. demand vs. trade balance	2020–2024	Assess agricultural vulnerability to water stress and food system dynamics
FAOSTAT (2024) and Climate Watch (2025)	(a) Agricultural emissions intensity (b) Agriculture share of GDP (c) Per capita N_2O emissions	1961–2021 (historical), 2021–2025 (latest available)	Contextualize environmental performance and economic contribution of agriculture
UNEP-WCMC (2025)	Index of Priority Rank % of Areas of global significance for biodiversity conservation, carbon storage, and water provision	2025	Map ecosystem resilience to climate change across Venezuelan biomes
Our World in Data (2025)	(a) Historical GDP (adjusted) (b) GHG emissions (total, per capita, by gas)	GDP: 1820–2022 GHG: 1850–2023	Analyze long-term economic and emission trends in national and regional contexts

2.9 Methodological Limitations

Despite the rigorous and systematic approach adopted in this review, several limitations must be acknowledged, as they may influence the interpretation and generalizability of the findings. First, although a wide range of databases and grey literature sources were consulted, the availability and accessibility of high-quality, Venezuela-specific studies remain limited, particularly for certain subregions (e.g., Guayana Shield, Orinoco Delta) and sectors (e.g., livestock, coastal fisheries). This spatial and thematic asymmetry could introduce coverage bias in the synthesized results.

Second, the review was constrained by the quality and transparency of the primary literature. Many studies lacked detailed methodological descriptions or failed to report key variables, such as uncertainty margins, sample sizes, or timeframes, which limited the comparability and depth of synthesis. Furthermore, language bias may have affected the inclusion of some relevant documents published in languages other than Spanish or English, though this is expected to be minimal.

Third, while the integration of grey literature allowed for the inclusion of policy-relevant and field-based information, the absence of peer review and standard methodological reporting in many of these sources introduces a degree of uncertainty in assessing their reliability. To mitigate this, quality assessment and triangulation procedures were applied, but limitations in metadata and documentation persist.

One important limitation to note is that some of the studies included under the mitigation and adaptation categories were conducted primarily from the perspective of agricultural practices aimed at increasing productivity, without explicitly linking these practices to climate change adaptation or mitigation frameworks. Although such studies did not frame their objectives within the climate change discourse, they were nonetheless valuable for this review. Their findings were incorporated and contextualized within the scope of this book, as they provide relevant evidence on agricultural management practices that, despite not being originally conceived as climate responses, have implications for enhancing resilience, reducing emissions, and promoting sustainability within Venezuelan agriculture.

Finally, data extracted from external platforms (e.g., CCKP, MapBiomas, Climate Watch), while valuable, are constrained by their modeling assumptions, resolution limits, and update cycles. These factors should be considered when interpreting projections or aggregated indicators.

Although this review was designed to address the specific context of Venezuelan agriculture, its methodological framework, analytical processes, and synthesized findings are also applicable to other intertropical regions. The climatic conditions, agricultural systems, and adaptation challenges examined here share strong parallels with those faced by countries across the intertropical belt, underscoring the broader relevance and potential for regional knowledge transfer.

2.10 Conclusions

This chapter describes the methodological foundations of a systematic and narrative review aimed at synthesizing current knowledge on the vulnerabilities and adaptation responses of Venezuelan agriculture to climate change. Grounded in internationally accepted guidelines such as PRISMA and the Cochrane Handbook, and informed by the DRACC initiative's objectives, the review employed transparent procedures for study identification, screening, data extraction, and synthesis.

By combining academic and grey literature, integrating quantitative datasets, and applying thematic and spatial-temporal analytical tools, the methodology ensured a comprehensive and multidimensional assessment of the topic. The use of thematic coding, comparative tables, and external visual data sources allowed for a nuanced understanding of regional disparities, sectoral vulnerabilities, and evolving adaptation strategies across Venezuela.

An important achievement stemming from this review process was the realization of the critical lack of regionalized information on climate change impacts in Venezuela. This gap led the team to adopt a regionalization approach that required generating new data and analyses to inform the book. As a direct result, two scientific papers were produced: *Mapping projected variations of temperature and precipitation due to climate change in Venezuela* (Viloria et al. 2023), which provided national-scale climate projections, and *How climate change will affect the ability of land in western Venezuela's plains to grow maize (Zea mays L.)* (currently under review), which assessed projected impacts on maize land suitability in the western Llanos. These studies represent key outputs that originated from the systematic efforts undertaken in this work and are among the valuable contributions harvested in this book.

Despite the challenges posed by data gaps and uneven evidence quality, the methodological approach adopted here contributes to strengthening the scientific basis for national adaptation planning and supports evidence-informed dialogue among researchers, policymakers, and civil society. The transparency and replicability of this review process provide a platform for future updates and expansions, in line with the dynamic nature of climate change research and policy.

References

Booth A, Sutton A, Papaioannou D (2016) Systematic approaches to a successful literature review, 2nd edn. SAGE Publications

Cumpston M, Li T, Page MJ, Chandler J, Welch VA, Higgins JP, Thomas J (2019) Updated guidance for trusted systematic reviews: a new edition of the Cochrane Handbook for Systematic Reviews of Interventions. Cochrane Database Syst Rev 3(10):ED000142. https://doi.org/10.1002/14651858.ED000142

Egger M, Smith GD, Schneider M, Minder C (1997) Bias in meta-analysis detected by a simple, graphical test. BMJ 315(7109), 629–634. https://doi.org/10.1136/bmj.315.7109.629

FAOSTAT (2024) Macro indicators. License: CC BY-NC-SA 3.0 IGO. Accessed 29 Apr 2024

Godin K, Stapleton J, Kirkpatrick SI, Hanning RM, Leatherdale ST (2015) Applying systematic review search methods to the grey literature: a case study examining guidelines for school-based breakfast programs in Canada. Syst Rev 4:138. https://doi.org/10.1186/s13643-015-0125-0

Haddaway NR, Collins AM, Coughlin D, Kirk S (2015) The role of Google Scholar in evidence reviews and its applicability to grey literature searching. PLoS One 10(9):e0138237. https://doi.org/10.1371/journal.pone.0138237

Lescher-Soto I, Villamizar A, Olivares BO, Gutiérrez ME, Nagy GJ (2024) Navigating the uncertain terrain: Venezuela's future using the shared socioeconomic pathways framework: a systematic review. Climate 12(7):98. https://doi.org/10.3390/cli12070098

Liberati A, Altman DG, Tetzlaff J, Mulrow C, Gøtzsche PC, Ioannidis JP et al (2009) The PRISMA statement for reporting systematic reviews and meta-analyses of studies that evaluate health care interventions: explanation and elaboration. PLoS Med 6(7): e1000100. https://doi.org/10.1371/journal.pmed.1000100

MapBiomas (2025) Collection 2 of the annual series of Land Cover and Surface Water Bodies, Maps of Venezuela, Provita. Consulted on May 17, 2025. https://plataforma.venezuela.mapbiomas.org/

Moher D, Liberati A, Tetzlaff J, Altman DG (2015) Preferred reporting items for systematic reviews and meta-analyses: the PRISMA statement. PLoS Med 6(7): e1000097. https://doi.org/10.1371/journal.pmed.1000097

Olivares BO, Lobo D, Rey JC, Vega A, Rueda MA (2023) Relationships between the visual evaluation of soil structure (VESS) and soil properties in agriculture: a meta-analysis. Sci Agropecuaria 14(1):67–78. https://doi.org/10.17268/sci.agropecu.2023.007

Our World in Data (2025) Our world in data. https://ourworldindata.org. Accessed 3 Jul 2025

Page MJ, McKenzie JE, Bossuyt PM, Boutron I, Hoffmann TC, Mulrow CD (2021) The PRISMA 2020 statement: an updated guideline for reporting systematic reviews. PLoS Med 18(3): e1003583. https://doi.org/10.1371/journal.pmed.1003583

Sánchez-Flores E, Olivares BO, Vinueza Chérrez R, Serrano Rodríguez R (2024) Are we doing enough to understand our relationship with the environment? A bibliometric analysis of the instruments that evaluate pro-environmental behavior. Sustain Dev 32(6):6555–6578. https://doi.org/10.1002/sd.3023

UNEP-WCMC (United Nations Environment Programme World Conservation Monitoring Centre) (2025) Index of Priority Rank % of Areas of global significance for biodiversity conservation, carbon storage, and water provision. [Map]. UN Biodiversity Lab. https://n9.cl/tpwxgu

Viloria JA, Olivares BO, García P, Paredes-Trejo F, Rosales A (2023) Mapping projected variations of temperature and precipitation due to climate change in Venezuela. Hydrology 10(4):96. https://doi.org/10.3390/hydrology10040096

Wohlin C (2014) Guidelines for snowballing in systematic literature studies and a replication in software engineering. In: Proceedings of the 18th international conference on evaluation and assessment in software engineering. ACM, pp 1–10. https://doi.org/10.1145/2601248.2601268

World Bank, Climate Change Knowledge Portal (2025). https://climateknowledgeportal.worldbank.org/. Accessed 5 Apr 2025

World Resources Institute (2025) Climate watch historical GHG emissions. World Resources Institute, Washington, DC. https://www.climatewatchdata.org/ghg-emissions. Accessed 7 May 2025

World Resources Institute (WRI) (2024) Aqueduct food: water risk and agricultural production platform—Venezuela. https://n9.cl/7et9nz

Chapter 3
Climate Variability and Its Effects on Agriculture

3.1 Introduction

The study of climate variability in Venezuela is highly relevant because of the country's geographical and climatic diversity, which renders it particularly vulnerable to the consequences of changing weather patterns. As a country rich in natural resources and reliant on agriculture, comprehending the effects of climate variability is crucial for ensuring sustainable agricultural practices (Barreto et al. 2024), safeguarding food security (Rodríguez García 2021), and maintaining ecological balance (Paredes-Trejo et al. 2023). In recent years, Venezuela has experienced shifts in precipitation patterns, rising temperatures, and an increased frequency of extreme weather events, all of which pose direct threats to local agricultural systems and natural ecosystems (Paredes-Trejo and Olivares 2018). Therefore, investigating these phenomena through scientific research is essential for developing informed adaptation strategies that can help mitigate the adverse effects of climate change.

The necessity of this type of study becomes increasingly evident when we consider the interconnectedness of climate patterns with key economic sectors in Venezuela, particularly agriculture, which remains a vital component of the nation's economy. Climate variability directly affects crop yields (Campos 2023), pest dynamics (Olivares et al. 2021a), soil fertility (Viloria et al. 2023), and water resources (Moreno-Pizani 2021), all of which are critical for maintaining agricultural productivity (Lescher Soto et al. 2024). Furthermore, the lack of sufficient scientific knowledge about how specific climatic shifts influence regional agricultural practices exacerbates the vulnerability of farmers, making it increasingly difficult to adapt to unpredictable conditions. Research on climate variability provides the tools necessary to predict and manage these challenges, helping to build resilience in local farming systems and improve long-term sustainability.

However, despite the critical importance of understanding climate variability, there appears to be a gap in the scientific literature regarding its specific effects in

A. Rosales Hernández et al., *Agriculture Under Climate Change*, The Latin American Studies Book Series, https://doi.org/10.1007/978-3-032-06368-7_3

Venezuela. While global studies on climate change and variability are abundant, localized research focused on the specific effects of climate variability in various Venezuelan regions remains sparse. This gap in the literature presents a significant challenge for effective decision-making and policy formulation in the country. Without comprehensive studies that address regional differences in climate variability and its effects on agriculture, designing tailored interventions to meet the unique needs of Venezuelan farmers and communities becomes increasingly difficult.

The objective of this chapter is to explore how climate variability manifests in Venezuela, its regional implications, and its impact on agriculture, with a focus on the various elements and characteristics that contribute to these shifts. This exploration will highlight the specific patterns of temperature and precipitation changes, their direct consequences for local agricultural systems, and the broader implications for food security and ecological sustainability.

Given the increasing urgency of adapting to climate change, this chapter aims to investigate whether there is sufficient scientific literature that addresses the effects of climate variability in Venezuela. It will explore the extent to which existing research covers key areas such as temperature and precipitation changes, their implications for agriculture, and the vulnerability of specific regions to climate-related risks. The central question to be explored is: Is there sufficient scientific literature that addresses climate variability in Venezuela and how can this knowledge be used to enhance agricultural resilience and support climate adaptation efforts? This inquiry is crucial for advancing climate research and ensuring that appropriate, evidence-based solutions are developed to protect Venezuela's agricultural sector and natural resources.

3.2 Current Climate

The climate distribution in Venezuela is diverse because of its geographical location, varied topography, and the influence of global climatic phenomena. Figure 3.1a shows the climate classification of Venezuela according to the Köppen-Geiger system, corresponding to the period 1991–2020 (Rubel and Kottek 2010; Beck et al. 2023). In the northern part of the country, tropical climates dominate, with classifications such as Af (tropical humid climate), found in coastal areas and near the plains. In contrast, the higher elevations of the Andes Mountain range are dominated by cooler climates, such as Cfb (temperate humid climate). In Venezuela's interior and southern regions, particularly in the Amazon region, there is a strong presence of Af and Am (tropical monsoon) climates, where precipitation is constant and temperatures remain high throughout the year.

In the Central and Eastern Llanos of Venezuela, the predominant climate is classified as As/Aw (tropical savanna climate with a defined dry season) (Fig. 3.1b). This classification is characterized by marked seasonality, with a rainy season concentrated from May to November and a dry season from December to April.

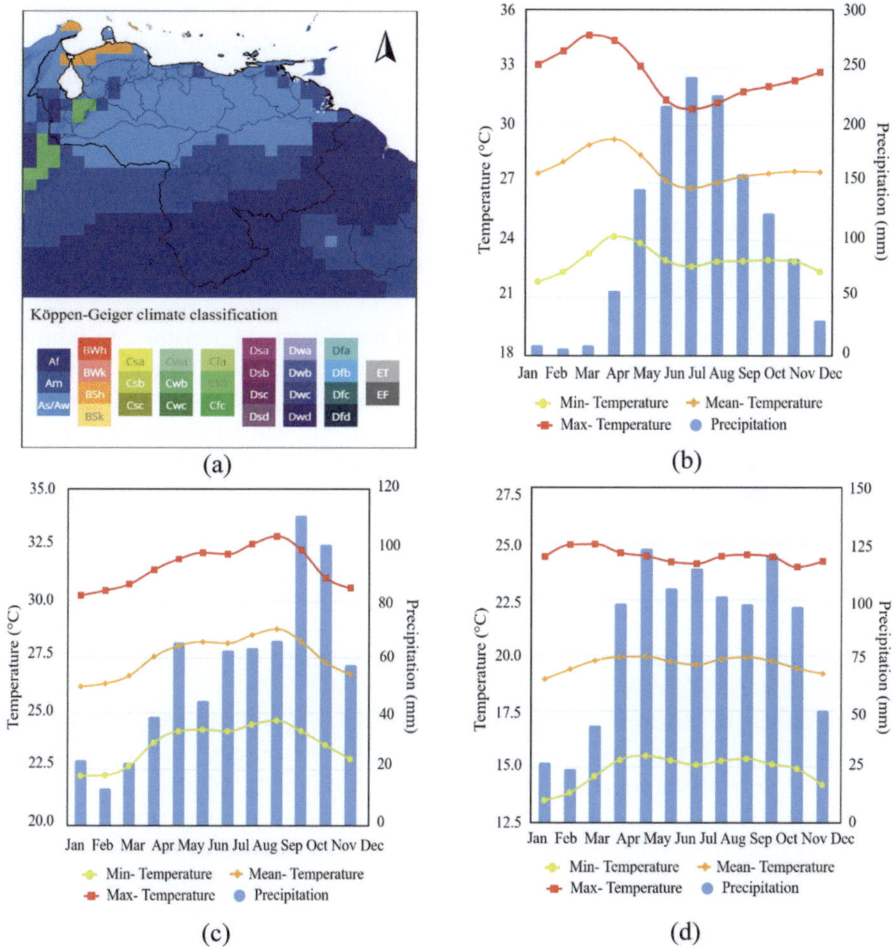

Fig. 3.1 (**a**) Köppen-Geiger climate classification system according to the World Bank, Climate Change Knowledge Portal (2025); monthly climatology of temperature means, minimum, maximum, and precipitation (1991–2023) in the states, (**b**) Guárico, (**c**) Falcón, (**d**) Mérida, according to the data source from Harris et al. (2020) and available in World Bank, Climate Change Knowledge Portal (2025)

Temperatures remain high throughout the year, ranging between 24 and 30 °C, contributing to a warm and humid environment during the rainy season. During the dry season, precipitation decreases significantly, which can lead to periods of water stress in both natural ecosystems and agriculture. This climate particularly affects agricultural production, crops such as maize, rice, and sorghum depend on water availability during the rainy season. However, higher-altitude areas in the region may experience slight thermal variations and higher humidity during the rainy season.

In the more arid and semi-arid regions of the west and north of the country, BWh and BSh climates are found, corresponding to desert and warm steppe areas, where precipitation is very low and temperatures are high (Fig. 3.1c). As one ascends in the mountains, especially in the Andean regions, there is a transition to cooler and colder climates, with classifications such as Csb (temperate mountain climate) (Fig. 3.1d).

3.3 Climate Variability and Scientific Production: Effects, Trends, and Agricultural Relevance

Climate trends (past, present, and future) must be understood in natural variability. The climate variability discussed in this section refers to how climatic conditions (e.g., air temperature and precipitation) fluctuate from year to year within their respective typical variability range. The cause of this natural variability may be because of the quasi-random internal variability of the coupled atmosphere-ocean-land-cryosphere system (since climate variability lasts for many years). A good example of a cause of this variability is that induced by the El Niño-Southern Oscillation (ENSO) (Álzate et al. 2015). This internal climate variability is always present, sometimes a little more exaggerated, sometimes a little less. The variability can be very large from year to year (e.g., at high latitudes) (Veettil et al. 2017), and in some places, and for specific variables, it can be small (e.g., temperatures in the tropics) (Vincenti-González et al. 2018).

The effects of climate variability in Venezuela are significant and far-reaching, particularly in sectors such as agriculture, water resources, and biodiversity. Variations in temperature and precipitation patterns directly influence crop yields, water availability, and the distribution of pests and diseases. For example, extreme weather events like droughts and floods, which are exacerbated by El Niño (ENSO) and La Niña phenomena, can devastate crops in rainfed agriculture regions. In recent years, the agricultural sector has experienced increasing uncertainty, as erratic rainfall patterns and temperature fluctuations threaten food security. Furthermore, the increased frequency of extreme events is associated with long-term shifts in climatic conditions, which challenge the resilience of local communities and ecosystems.

The number of scientific publications available to assess the effects of climate variability on agriculture was ($n = 64$), with the proportion of publications related to meteorological drought and the effects of water deficit in areas of agricultural interest in Venezuela being particularly notable. Publications on climate variability and change in agriculture have also increased, although they represent a small fraction of the total publications generated in the country.

Figure 3.2 shows that the highest proportion of scientific publications used monthly and annual precipitation data ($n = 21$) on a year-on-year scale at the level of the following states: Anzoátegui, Aragua, Bolívar, Carabobo, Guárico, Mérida, Monagas, and Sucre; followed by the Standardized Precipitation Index (SPI) for the characterization of meteorological drought ($n = 15$) in rainfed agriculture regions of

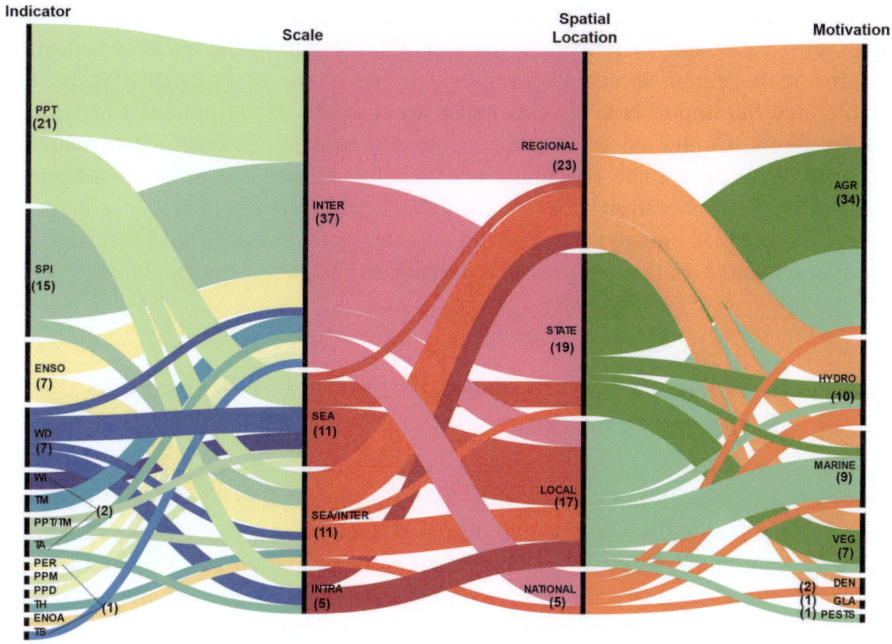

Fig. 3.2 Alluvial diagram of scientific publications on climate variability in Venezuela ($n = 64$). It shows correlations between categorical dimensions represented as flows (indicator of variability, scale, spatial location, and motivation), visually linking categories with shared elements. Each rectangle represents a unique value in the selected dimension; its height is proportional to its value. Note = Indicator: monthly precipitation (PPM), annual precipitation (PPT), Decadian precipitation (PPD), soil temperature (TS), air temperature (TA) and sea temperature (TM), El Niño-Southern Oscillation (ENSO), El Niño-Southern Oscillation (ENSO), North Pacific Oscillation (PDO), water balance water index (WI), water deficit (WD), Standardized Precipitation Index (SPI), and Community perception (PER). Scale = Interannual (INTER), seasonal (SEA), intrannual (INTRA); Spatial location = Regional (REG), State, Local, National; Motivation = Dryland agriculture (AGR), Hydrographic basin (HYDRO), Marine Ecosystem (MARINE), Vegetation (VEG), Dengue (DEN), Glacier (GLA), and Pests (PEST)

the Andes, Amazonas, central region, the Llanos, Falcón-Lara depression, insular region, and eastern Venezuela; ENSO/ENSO ($n = 8$) on a seasonal/interannual scale at the regional level such as the Andes, central region, and specifically in the states of Barinas, Delta Amacuro, and Carabobo, focused on watersheds, vegetation, and rainfed agriculture. The rest of the publications refer to the water index and the water deficit from the hydrological balance and its influence on rainfed agriculture, as well as studies of sea temperature at the level of the marine ecosystem. Finally, scientists are studying the influence of extreme weather events on agriculture and looking for solutions to minimize their impact.

Scientific research plays a pivotal role in addressing the challenges posed by climate variability in Venezuela. The growing body of literature on climate variability and its effects on agriculture provides valuable insights that can guide policy and decision-making. For example, publications analyzing the SPI have helped identify areas most vulnerable to meteorological drought, thus informing resource allocation and risk management strategies. Similarly, studies on the water balance index have highlighted the importance of sustainable water management practices in agricultural regions. Continued investment in climate research is essential to enhance understanding of climate dynamics and develop effective adaptation measures. Furthermore, collaboration between researchers, policymakers, and local communities is necessary to ensure that scientific knowledge is translated into practical, on-the-ground solutions that address both immediate and long-term challenges.

3.3.1 Significant Changes in the Water Regime in the Face of Natural Variability

Interpreting the results of the research reported in Table 3.1, in the state of Carabobo, recent meteorological studies have highlighted the importance of climate variability in rainfed agricultural systems. Fluctuations in precipitation and temperature significantly affect the productivity of crops such as maize, potato, tomato, and onion. In the Lake Valencia basin, decreased rainfall in some seasons, such as Yuma Caserío and Colonia El Trompillo, has resulted in water deficits that severely affect crop yields, especially in years where drought episodes have been more pronounced.

Excess water in some areas has led to logistical challenges, affecting both the harvesting process and the storage of agricultural products. The identification of suitable and moderately suitable areas for agriculture, based on the combination of meteorological and soil factors, has been essential to adjust planting dates and ensure better use of available resources.

These studies also highlight the influence of phenomena such as El Niño and La Niña on the occurrence of rainfall extremes, which impacts both production and the sustainability of agroecosystems (Table 3.1). The lack of access to irrigation water in drought-affected areas, coupled with the need for more efficient management of water resources, poses significant challenges for agricultural planning in the region. Through the use of historical data and multivariate analyses, critical areas have been identified where it is a priority to implement adaptation strategies that can mitigate the effects of climate variability. In this sense, agroclimatic zoning efforts provide essential guidance to improve the resilience of local agricultural systems, ensuring the long-term sustainability of crops in Carabobo.

To complement these efforts, Fig. 3.3a shows precipitation trends per decade for the period 1971–2020 in Venezuela, revealing significant spatial variability in rainfall patterns across the country, with marked differences between regions. In the western and northern parts of Venezuela, including Zulia and Falcón states, the precipitation levels have decreased, represented by the red and brown shades on the map. These areas experience annual precipitation deficits ranging from −145 to −39.4 mm, indicating a clear trend of drying conditions over the decades. These regions are vulnerable to increasing aridification, which can have severe

Table 3.1 Effects of climate variability on ecosystems and agricultural systems in Venezuela

Phenomenon	Region	Study period	Observed data/evidence	References
Interannual variability of precipitation	Carabobo State	1969–1999	The rainiest year was 1969 with 1422.0 mm, and the driest was 1989 with 862.8 mm. Annual average: 1096.1 mm	Olivares and Hernández (2019a)
Influence of El Niño and La Niña on seasonal precipitation	Lake Valencia Basin	1934–2005/1966–1992	No long-term trend in seasonal precipitation was found. El Niño was identified as being linked to drier episodes and La Niña to wetter episodes in some seasons, but not consistently	Paredes-Trejo et al. (2015)
Precipitation variability and agricultural land suitability	Carabobo State	1970–1999	Through multivariate analysis, areas with high agricultural suitability for different crops in Carabobo were identified. Libertador and Carlos Arvelo municipalities stand out with the best areas for corn, onion, tomato, and potato	Olivares and Hernández (2020)
Precipitation variation in maize production	Carabobo State	1970–2000	59,000 ha of maize-grade farmland in fertile, well-drained soils. Optimal planting dates vary between May and August	Olivares et al. (2018a)
Precipitation variation in tomato production	Carabobo State	1970–2000	Identification of 34,800 ha of suitable land and 32,600 ha of moderately suitable land. Variation in yields was observed according to the optimal planting date, such as in the second decade of May	Olivares et al. (2020)
Precipitation variation in potato production	Carabobo State	1970–2000	33,700 ha of suitable soils and 30,400 ha of moderately suitable soils were identified. The recommended planting dates are between May and August	Olivares and Hernández (2019b)
Precipitation variation in onion production	Carabobo State	1970–2000	78,700 ha of suitable land and 42,000 ha of moderately suitable land. The best planting dates are between the second decade of May and the first decade of August	Olivares et al. (2018b)
Water variability (water deficit/excess)	Carabobo State	1969–1999	Four regions were identified in Carabobo, with a water deficit ranging from −2.16 to −18.96, and annual rainfall between 921 and 1063 mm	Olivares et al. (2018c)
Climate variability (water deficit and excess)	Carabobo State	1969–1999	In agricultural areas of the Dry Sub-humid (C1) climate type, which require complementary irrigation, it is possible to combine irrigation and rainfed management to reach two harvests per year in short-cycle crops	Olivares et al. (2018d)

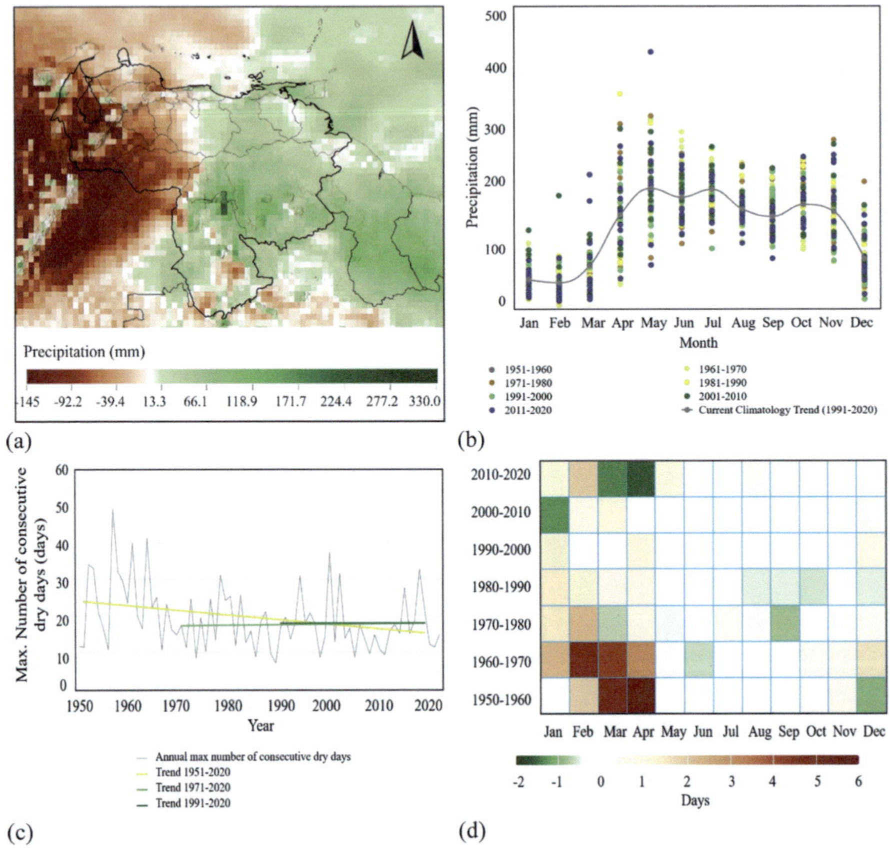

Fig. 3.3 (**a**) Precipitation trend per decade (annual) (1971–2020) in Venezuela; (**b**) variability and trends of precipitation across seasonal cycle (1951–2020) in Carabobo; (**c**) Maximum number of consecutive dry days annual trends with significance of trend per decade (1951–2023) in Carabobo; (**d**) maximum number of consecutive dry days monthly trends with significance of trend per decade (1951–2020); in Carabobo according to the data source from Harris et al. (2020) and available in World Bank, Climate Change Knowledge Portal (2025)

consequences for local agriculture, particularly in rainfed farming systems, and can stress water resources crucial for both human consumption and ecosystems.

In contrast, the central and eastern regions of the country, as well as the Amazon basin, show higher precipitation levels, reflected in the green to dark green areas on the map. Annual precipitation in these regions ranges from 13.3 to 330 mm, with areas experiencing higher rainfall levels, such as the Guayana region and the Andes, showing an increase in precipitation over the decades. While the central and eastern areas have remained humid, the observed increases in rainfall (in some regions up to 330 mm) may still be subject to seasonal and interannual variability, highlighting the need for careful management of water resources and agricultural practices to adapt to these changing patterns.

The disparities in precipitation trends across the country underscore the importance of region-specific climate studies and adaptive strategies. Areas with negative trends in precipitation, such as those with deficits of up to −145 mm, should prioritize water conservation measures, drought-resistant crops, and sustainable land management practices. On the other hand, regions with high precipitation, ranging from 118.9 to 277.2 mm, may face challenges such as flooding and soil erosion.

Figure 3.3b depicting the variability and trends of precipitation across the seasonal cycle from 1951 to 2020 in Carabobo, Venezuela, highlights notable fluctuations in precipitation throughout the year. The data reveals distinct peaks and valleys in monthly precipitation, with a significant maximum observed in May 2011 at 433.2 mm. This high value contrasts sharply with the minimum precipitation levels recorded in May 2020 (73.8 mm) and August 2019 (111.05 mm). The trend over the decades indicates a decline in precipitation by approximately −18.97 mm per decade, reflecting an overall decrease in rainfall intensity over the past 70 years. This trend, particularly during the critical rainy season months, suggests a changing climate pattern with potential effects on water resources and agricultural productivity.

The decreasing precipitation trend in Carabobo has had direct implications for agricultural production in the region, particularly in crops dependent on consistent rainfall, such as maize and rice. The marked decline in rainfall during key months of the rainy season, such as May and August, can lead to water stress in crops, which affects yields. In 2020, the low precipitation levels exacerbated challenges for farmers, with reduced soil moisture leading to a decrease in crop productivity.

Figure 3.3c illustrates the variability and trends in the frequency of dry periods across the years. The data highlights a significant trend between 1951 and 2020, with a decrease of −0.99 days per decade, which is statistically significant at the 96% confidence level. This suggests that over the past several decades, the length of consecutive dry days has decreased, meaning that droughts of extended duration have become less frequent. This trend could indicate changes in regional weather patterns, possibly associated with shifts in precipitation seasonality or intensity, affecting agricultural practices that rely on consistent rainfall patterns.

In contrast, the trends for the periods between 1971–2020 and 1991–2020 show much less significance. The 1971–2020 trend of −0.15 days per decade is not significant, with only a 15% confidence level, indicating that the decrease in dry days is not strong enough to be considered a reliable shift during this period. Similarly, the 1991–2020 trend, with a 0.05 days per decade increase, shows no significant change in consecutive dry days. This suggests that, although some periods in recent decades may have had shorter or longer dry spells, there has been no consistent or statistically significant change over the last three decades.

In the most recent decade (2010–2020) (Fig. 3.3d), significant decreases in the number of consecutive dry days were observed, particularly in March and April. In March, the trend showed a decrease of −1.37 days, while in April, it decreased by −1.59 days. These negative values indicate a reduction in the length of consecutive dry spells, suggesting that the dry periods in these months became shorter during this decade. This shift could have implications for agricultural practices, as it suggests more frequent rainfall during the traditionally drier months of the year, potentially benefiting crops that rely on consistent moisture levels.

In contrast, Table 3.2 includes the latest studies on the Andean region of Venezuela; the growing impact of climate change on the spatial dynamics of ecosystems is highlighted, particularly in the Cordillera de Mérida. The effects of global warming have resulted in an upward shift of the montane forest boundary toward the páramo, with a movement of approximately 12.8 m per decade. This phenomenon has affected the distribution of key ecosystems, such as the páramo and montane forest, transforming their structure and density, besides increasing the cover of woody vegetation at higher altitudes. The studies also highlight the progressive conversion of páramo lands into pastures and agricultural areas, which aggravates the loss of biodiversity in these areas. This change is driven by both the increase in temperature and the transformation of land use, affecting the resilience of these ecosystems to extreme weather events.

In addition, research on crops and agricultural productivity in regions such as Carabobo, Aragua, and Mérida shows how climate variability is affecting rainfed agriculture, especially on crops such as corn, potatoes, and bananas (Table 3.2). Water availability and the response of plants to water stress and high temperatures are altering the yield and quality of products. In the Venezuelan Andes, the densification of páramo forests and the colonization of higher elevations by woody species are also because of temperature fluctuations and changes in precipitation patterns. These studies highlight the need to implement adaptation strategies to mitigate the adverse effects of climate change on Andean ecosystems and agricultural production in the country.

Figure 3.4a shows the annual precipitation trends in the state of Aragua, highlighting a minimum in May 2011, with 314.9 mm of precipitation, and a more average behavior during the decade of 2011–2020, where precipitation levels remained close to the normal values. The overall trend per decade is −14.88 mm, indicating a sustained decrease in annual precipitation over time. Mérida highlights significant variations in precipitation throughout the year (−82.89 mm per decade) (Fig. 3.4b). In particular, July 2019 and August 2018 stand out as months with relatively low precipitation, recording 131.2 mm and 130 mm, respectively. These values are considered minimums for the region, representing a notable decrease compared to other years in the same months. The current climatology trend (1991–2020) indicates a general fluctuation with no drastic shifts, but it does show a trend of moderation in rainfall intensity during the rainy season.

These low precipitation values in 2018 and 2019 could have posed challenges for agriculture in Mérida, particularly for crops that depend heavily on consistent rainfall, such as coffee and vegetables. The decreased rainfall during these months could have led to water stress in rainfed agricultural systems, affecting crop yields and requiring additional irrigation measures.

In Aragua, Fig. 3.4c shows the variability in dry spells over time. The trend for the decade 1951–2020 indicates a decrease of −0.81 days per decade, which is not statistically significant, as reflected by the 92% confidence level. Similarly, the trends for the decades 1971–2020 and 1991–2020 show decreases of −0.57 days per

Table 3.2 Effects of climate variability on ecosystems and agricultural systems in Venezuela

Phenomenon	Region	Study period	Observed data/evidence	References
Seasonal precipitation variability	Lake Valencia Basin, Carabobo State	1969–1999	A reduction in annual rainfall was observed in stations such as Yuma Caserío, Colonia El Trompillo, Agua Blanca, and Las Dos Bocas, affecting rainfed crops such as sorghum, corn, and fruits	Olivares (2018)
Spatio-temporal variability of precipitation	Guárico State	1970–2000	The rainy period occurs from May to October, with August being the wettest month (240.9 mm). 58% of the years evaluated were below the average annual rainfall (1035 mm)	Cortez et al. (2016)
Seasonal precipitation variability	Anzoátegui State	1970–2000	Five groups of climatic stations with homogeneous precipitation patterns were identified. Annual precipitation values vary between 530 and 1400 mm	Olivares et al. (2017a)
Meteorological drought	Various locations in Venezuela	1969–1999	The municipalities most affected by the drought include Turén, Francisco de Miranda, and Catatumbo, with high rates of agricultural vulnerability	Olivares et al. (2017b)
Variability of precipitation and potential evapotranspiration	Aragua State	2016–2017	A positive correlation was observed between precipitation and NDVI, with a 1-month delay in banana response to rainy conditions. NDVI values ranged from 0.296 to 0.924	Olivares et al. (2021b)
Impact of climate change on precipitation and temperature	Central Region (Aragua/ Carabobo)	2060–2079 (projection)	Rainfall is expected to decrease from 1600–2000 mm/year to 800–1200 mm/ year, with a significant increase in water deficit, affecting 4,700,000 ha	Olivares et al. (2019)
Change in precipitation and evapotranspiration	Middle basin of the Pao River, Central Region	1976–1992 (current climate) and 2026–2042 (future scenarios)	Projections under RCP 4.5 and RCP 8.5 scenarios show a reduction of up to 34.6% in excess precipitation and an 8.2% increase in potential evapotranspiration	Silva and Tobón (2019)

(continued)

Table 3.2 (continued)

Phenomenon	Region	Study period	Observed data/evidence	References
Water stress and high temperatures	Aragua State	2013–2015	The Cleopatra-Dancy and Volk-Dancy (Citrus) combinations showed the best performance in terms of gas exchange under drought and high temperature conditions	Pérez-Macias et al. (2022)
Climate variability (precipitation, humidity, and temperature)	Eastern, Central, and Western Regions	Not specified (current study conditions)	Clones grown in the tropical rainforest had a higher production of phenolic compounds and lignin, while those in the tropical dry forest accumulated more non-structural carbohydrates	Álvarez-Barreto et al. (2018)
Altitudinal climate variability	Mérida State	1981–1995	Humid, sub-humid, and rainfall bioclimates were identified, with annual rainfall between 750 and 2500 mm, and temperatures between −0.2 and 27.4 °C	Chacón-Moreno and Suárez del Moral (2020)
Climate change (global warming) and the tree line shift	Mérida State	1952–2009	A displacement of 72.7 m was observed in 57 years. The change has been gradual, affecting the structure and density of the vegetation in the transition zone between the páramo and the montane forest	Chacón-Moreno et al. (2021)

decade and −1.22 days per decade, respectively, both of which are also not significant, with the confidence levels at 56% and 59%.

These trends suggest that while there has been a general reduction in the number of consecutive dry days over the decades, the changes are not statistically significant, indicating that the variability in dry periods remains within a natural range. The lack of significant trends implies that, despite some fluctuations, Aragua has not experienced consistent shifts in the frequency of dry spells over the decades. This could suggest that agricultural practices in the region may not need to make immediate adjustments for longer-term drought trends, but should still consider the possibility of irregular dry spells that can affect water availability and crop productivity in specific years.

The trend for the decade 1951–2020 in Mérida (Fig. 3.4d) indicates a decrease of −4.26 days per decade, which is statistically significant at the 100% confidence level, suggesting a reliable decline in the number of consecutive wet days over the decades. Similarly, the trend for the decade 1971–2020 shows a decrease of −5.25 days per decade, which is also statistically significant at the 99% confidence

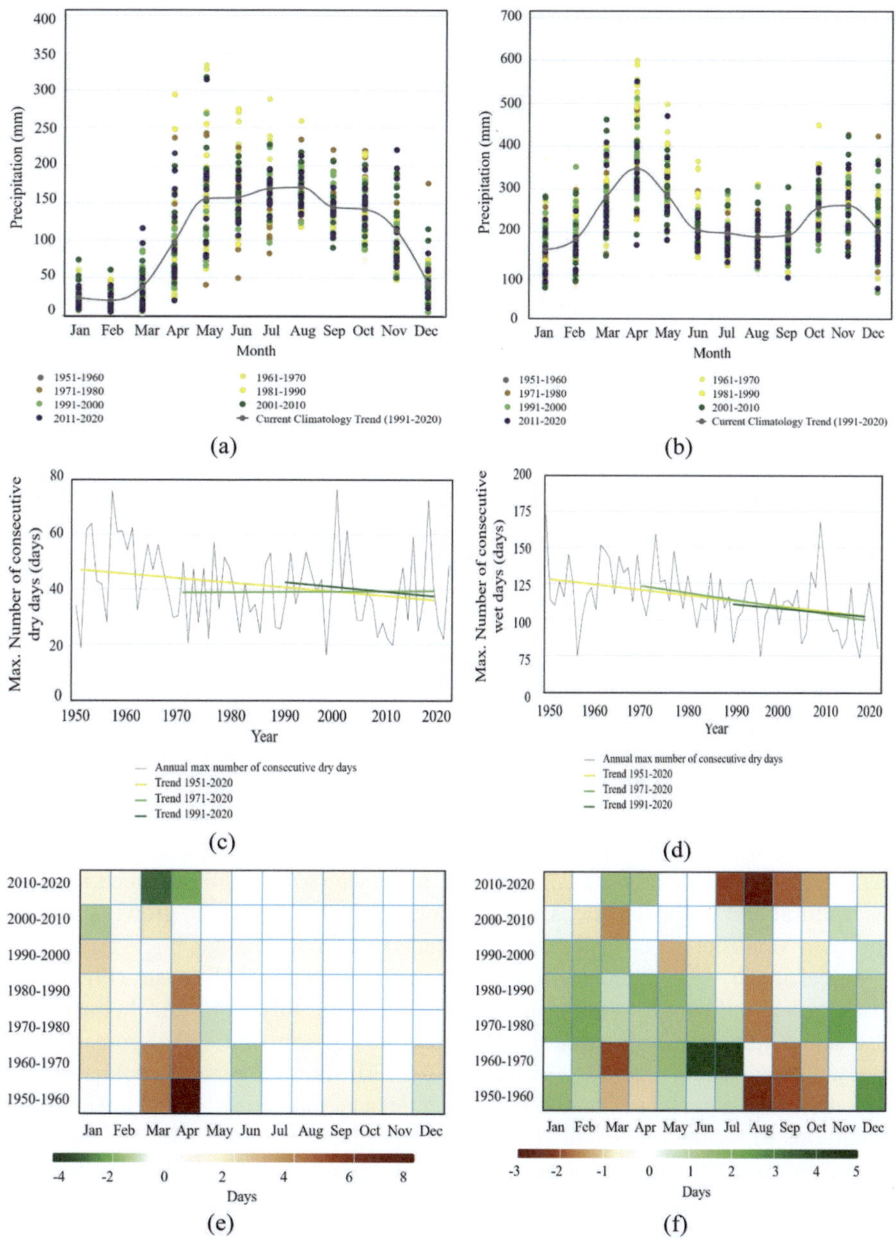

Fig. 3.4 (**a**) Variability and trends of precipitation across seasonal cycle (1951–2020) in Aragua and (**b**) Mérida; (**c**) Maximum number of consecutive dry days annual trends with significance of trend per decade (1951–2023) in Aragua and (**d**) wet days in Mérida; (**e**) maximum number of consecutive dry days monthly trends with significance of trend per decade (1951–2020); in Aragua and (**f**) wet days in Mérida according to the data source from Harris et al. (2020) and available in World Bank, Climate Change Knowledge Portal (2025)

level, further reinforcing the pattern of reduced consecutive wet days in recent decades.

Figure 3.4e shows that the decades with the largest increases in consecutive dry days are marked by brown color, indicating a significant extension of dry periods. Specifically, in March, the 1950–1960 decade experienced a maximum increase of +6 days of consecutive dry days, while in April, the 1970–1980 decade showed an even higher increase of +8 days. These notable increases in dry days suggest that during these decades, both March and April had substantially prolonged dry spells, which could have had adverse effects on agricultural production, particularly for crops that rely on timely rainfall.

In the last decade (2010–2020) in Mérida (Fig. 3.4f), there was a noticeable decrease in consecutive wet days from July to October, with values becoming negative. Specifically, August saw a decrease of −2.63 days, and July and September also exhibited negative trends. These reductions indicate shorter wet periods during these months compared to previous decades. The decrease in consecutive wet days during the mid-rainy season suggests a shift toward drier conditions during what is typically the wettest time of the year, which could have implications for water availability and agricultural productivity in Mérida, particularly for crops dependent on consistent rainfall.

Table 3.3 shows the last eight studies reviewed. In the tropical Andes, the studies pointed to the retreat of glaciers and the influence of climate variability on páramo ecosystems and watersheds, affecting water dynamics and crop distribution under climate change scenarios. In the Tinaco River basin, Cojedes state, the spatio-temporal analysis of the surface temperature of the soil showed a high variability, with elevated temperatures during dry periods, which affects water availability and agricultural development. In Lara state, the relationship between evaporation and climatic elements such as temperature and wind proved to be crucial for irrigation management in semi-arid areas. Similarly, these works highlight the need to adapt agricultural and water resources management strategies, given the growing vulnerability of Venezuelan ecosystems to the effects of climate change.

In particular, May 2020 recorded a low of 126 mm in Zulia state (Fig. 3.5a), marking a significant minimum in recent years, while July 2019 and August 2018 also saw relatively low precipitation values of 134.19 mm and 150.45 mm, respectively. The current climatology trend (1991–2020) shows a general decline in precipitation with a decrease of −82.98 mm per decade, indicating a trend toward drier conditions in recent years. This reduction in rainfall, particularly during the rainy season months, has had substantial implications for agricultural production in Zulia.

The decreasing precipitation trend, especially during critical months like May, July, and August, has likely exacerbated water stress in the region's agriculture, which is heavily dependent on consistent rainfall. Zulia, known for its crop production, particularly in the agricultural zones of the Llanos and surrounding areas, could face challenges such as lower crop yields and increased reliance on irrigation. This is particularly true for the banana-growing regions in the Sur del Lago de Maracaibo, including Colón, Francisco Javier Pulgar, Jesús Enrique Lossada, and

Table 3.3 Effects of climate variability on ecosystems and agricultural systems in Venezuela

Phenomenon	Region	Study period	Observed data/evidence	References
Glacial retreat because of global warming	Tropical Andes (Venezuela)	Recent decades (1975–2016)	An accelerated retreat of glaciers was observed, especially those at lower altitudes. Glaciers in Venezuela and Colombia are the most vulnerable to rising temperatures	Veettil et al. (2017)
Climate variability (El Niño/La Niña)	Miguaguó Micro-basin, Venezuelan Andes	1971–2017	Significant changes in wetland moisture were observed, with increased recovery during La Niña events and high wetland resilience	Mejía et al. (2022)
Climate Variability and Climate Change (ENSO)	Zulia and Pamplonita River basins, Norte de Santander, Colombia	1981–2010	A generalized increase in temperature and precipitation was observed in both basins, with a higher incidence of extreme events associated with the ENSO cycle	Álzate et al. (2015)
The temperature increases because of climate change	Cordillera de la Costa bioregion	1972–2004	Maximum temperatures in some stations reached 36.2 °C (Cúa), while Colonia Tovar had a minimum value of 8.8 °C. The ARIMA analysis showed significant trends of increase in the stations of San Felipe and Maracay	Romero and Millano (2021)
Climate variability of precipitation	Cordillera de la Costa bioregion,	1970–1999	Only two stations, San Felipe and Nirgua in the state of Yaracuy, present, according to the Mann-Kendall test, a significant growing trend. The El Torito, La Entrada, and Cúa-Tovar stations show a tendency not to a significant decrease	Millano and Romero (2021)
Variability of precipitation and biomass production	South of Lake Maracaibo	2020	During the rainy season, a higher biomass yield (up to 14,063 kg/ha) and higher crude protein content (8.6%) were recorded. In the dry season, yield and quality decreased considerably	Uzcátegui-Varela et al. (2022)
Climate change and crop distribution	Mérida State	Projections for 2035, 2065, 2100	The potential area for potato cultivation will decrease by 69% by 2100, while the area of cocoa will lose up to 63.8% of its surface	Paredes and Chacón-Moreno (2021)

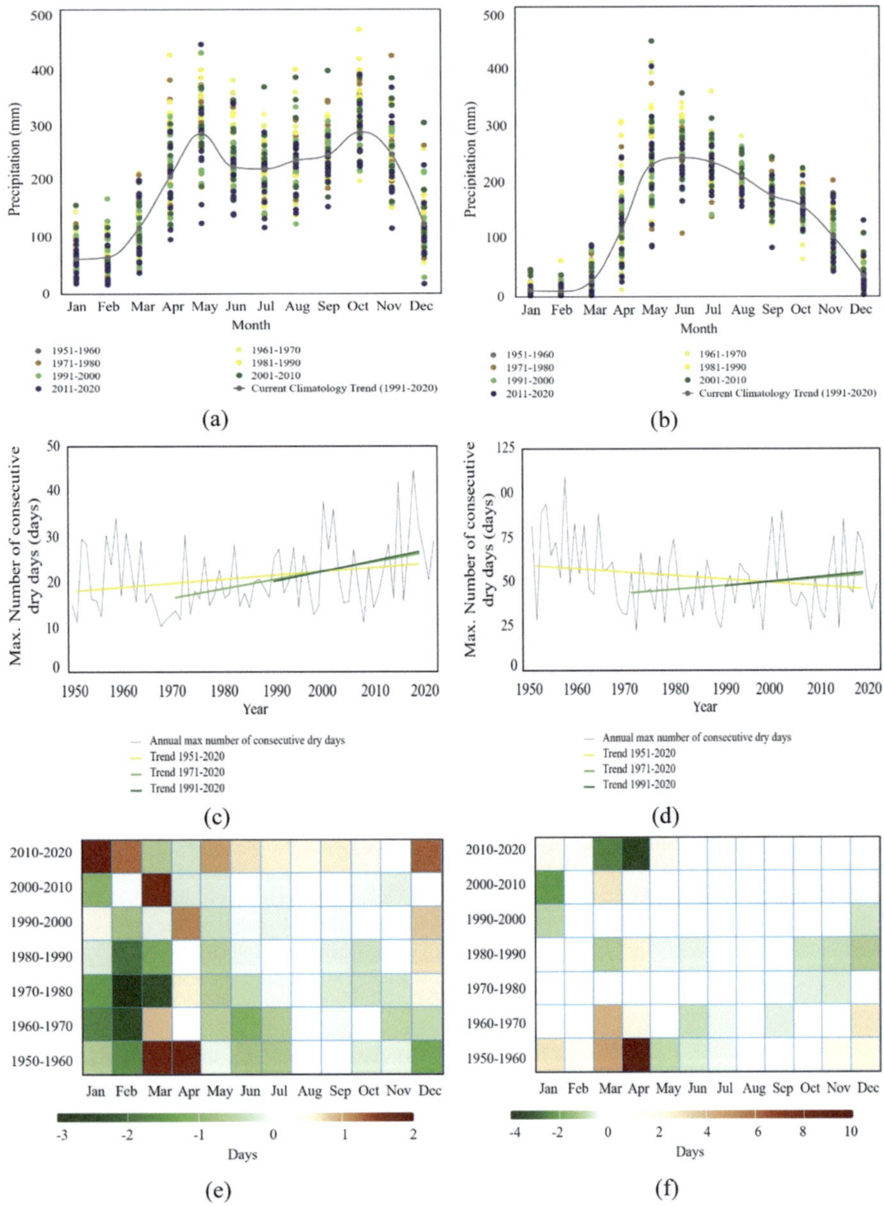

Fig. 3.5 (**a**) Variability and trends of precipitation across seasonal cycle (1951–2020) in Zulia and (**b**) Cojedes; (**c**) maximum number of consecutive dry days annual trends with significance of trend per decade (1951–2023) in Zulia and (**d**) in Cojedes; (**e**) maximum number of consecutive dry days monthly trends with significance of trend per decade (1951–2020); in Zulia and (**f**) in Cojedes according to the data source from Harris et al. (2020) and available in World Bank, Climate Change Knowledge Portal (2025)

Sucre municipalities, where the cultivation of bananas is a primary economic activity. These areas, along with municipalities like Mara and La Cañada de Urdaneta, depend on sufficient rainfall for irrigation, and the observed decline could lead to increased pressure on water resources.

Figure 3.5b reveals notable fluctuations across the seasonal cycle. In particular, May 2020 recorded a minimum of 87.81 mm, and September 2015 had a similarly low value of 85.85 mm, highlighting significant dry spells during these months. The current climatology trend (1991–2020) shows a general decline in precipitation, with a decrease of −36.64 mm per decade, suggesting a trend toward drier conditions over the years. This reduction in rainfall, especially during critical months like May and September, has likely exacerbated water stress in Cojedes, a region that depends heavily on consistent rainfall for its agricultural production, including maize, rice, and sugarcane.

The trend for the decade 1951–2020 shows a slight increase of +0.77 days in Zulia state (Fig. 3.5c), but this trend is not statistically significant (92% confidence), indicating that the changes in consecutive wet days during this period were not substantial enough to be considered a reliable shift. However, the trend for the decade 1971–2020 indicates a more significant increase of +1.67 days, with a 98% confidence level, suggesting that consecutive wet days have increased notably during this period. In contrast, the trend for the decade 1991–2020 shows an increase of +1.26 days, but this trend is not significant (66% confidence), indicating that while there was some increase in consecutive wet days, it was not consistent enough to be considered a definitive shift.

The trend for the decade 1991–2020 in Cojedes shows a smaller increase of +1.39 days (Fig. 3.5d), but this is not significant either (14% confidence), reinforcing the variability of the wet periods in recent years. These findings suggest that the length of consecutive wet days in Cojedes has fluctuated over the decades without clear or consistent trends. While there have been periods of both decrease and increase in consecutive wet days, these changes have not been statistically significant, indicating that other factors, such as natural climate variability, may be influencing the region's rainfall patterns.

Figure 3.5e indicates significant changes in dry periods in Zulia, especially during the last decade (2010–2020). In this period, the maximum value of consecutive dry days occurred in January with an increase of +2 days (indicated by dark brown), reflecting a longer dry spell during this month. Cojedes highlights significant changes in the dry spells across the decades (Fig. 3.5f). In the last decade (2010–2020), the maximum reduction in consecutive dry days occurred in March, with a decrease of −2.42 days (shown in brown), indicating that the length of dry spells in March became shorter. Similarly, April also experienced a significant reduction of −3.35 days (shown in brown), suggesting that the dry periods in this month were reduced compared to earlier decades.

3.3.2 *Influence of Meteorological Droughts in Agricultural Areas*

Meteorological drought refers to a prolonged period of dry weather caused by a lack of rainfall in a region. It is characterized by a scarcity of rainfall over a significant period, resulting in a decrease in soil moisture and a deficit of water in the atmosphere. Meteorological drought is assessed by calculating specific indices of precipitation and moisture demand from the atmosphere. Meteorological droughts are occurring with greater frequency and duration in agricultural areas which had led to a reduction in the production of cereals, fruit trees, and vegetables, among others, this influence being difficult to quantify. Figure 3.6 shows the number of publications that address drought and meteorology in the national territory ($n = 18$). Within these meteorological drought studies, those carried out by Parra et al. (2018) stand out, where the most relevant maximum magnitude was the one recorded in Bramón for the year 1993, whose magnitude value reached 11.56. Followed by the results of

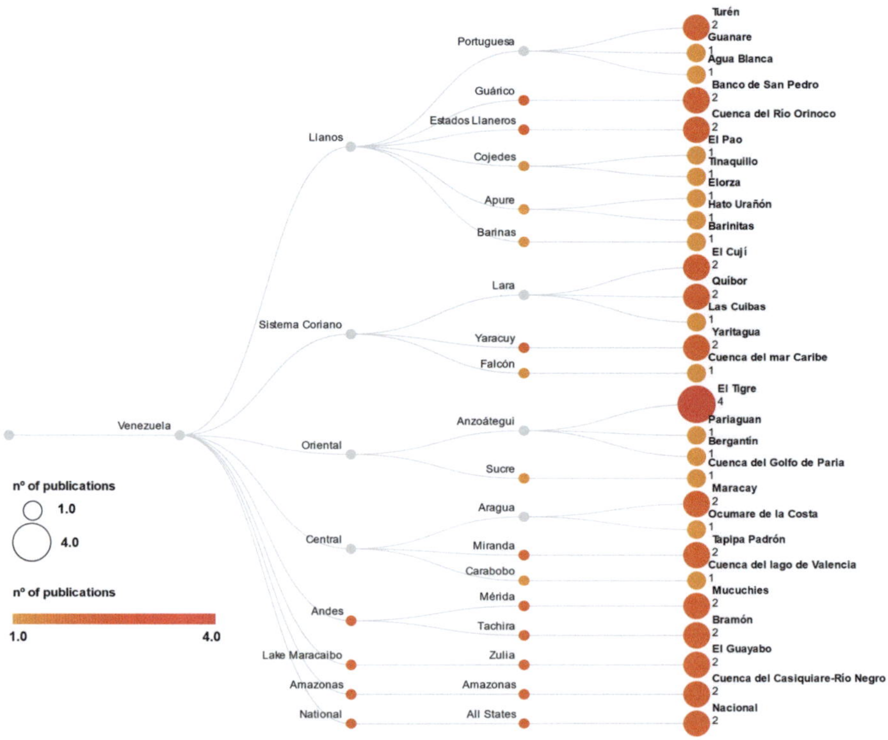

Fig. 3.6 Number of publications on the subject of drought and meteorology in the national territory ($n = 18$). Hierarchically structured data is displayed with a tree structure, where the root node is on the left (Venezuela) and the leaves are on the right (regions and states of Venezuela). The size of the nodes encodes the additional quantitative dimension represented by the number of related publications. (Source: Authors' elaboration based on data collected by the authors)

Cortez et al. (2018), where the most relevant maximum magnitudes were recorded at the Tapipa-Padrón station in 2001 with a magnitude of 9.70 and in 2003 with a magnitude of 9.50; the CENIAP station (Aragua) also reported a maximum magnitude of 9.20 during 1982. According to Olivares and Zingaretti (2018), the percentage of agricultural areas affected by the drought that occurred in 1991 in Venezuela was around 40–50%, with the towns of El Tigre and Tapipa-Padrón being the most affected.

Olivares et al. (2016b) and Villegas et al. (2021) established that the most relevant maximum drought magnitude in El Tigre was 10.15 during 2014. Also, the drought event that occurred from the beginning of the pre-humid period (April) of 2001 to the beginning (February and March) of 2002 was one of the most intense that has been known in most of the national territory throughout the historical record. On the other hand, Maríñez et al. (2021) observed extreme drought events of 32.1% in El Tigre, Anzoátegui state, seeing this behavior for the years 1974, 1984, 1994, 2001, and 2014.

It is interesting to cite the study by Olivares et al. (2016a) in which they point out that there is a generalized perception of agricultural producers in Anzoátegui about climate variability in recent years, which was explained through the occurrence of natural phenomena, such as meteorological drought, excess water, and water erosion coinciding with the rain and drought events reported in Olivares et al. (2013).

Also, the findings of Paredes-Trejo et al. (2018) establish that it is possible to hypothesize that meteorological droughts in the basins of the Caribbean Sea and Lake Maracaibo have more potential to trigger hydrological droughts than in the rest of the regions, such as the Orinoco River basin with 643,875 km^2 (area in Venezuelan territory). Likewise, Olivares and Zingaretti (2019) reported that localities such as CENIAP in Aragua state, El Cují Lara state, Yaritagua Yaracuy state, Turén Portuguesa state, Quíbor Lara state, Mucuchíes Mérida state and Bramón Táchira state presented in their historical rainfall record (1980–2014) with wet years, intermediate years, and years with significant water deficit mostly in the period 2008 to 2013. Also, the meteorological drought showed temporal variations in intensity and magnitude in the localities of the Coriano System. The most relevant maximum magnitudes were those recorded in Quíbor in 2001 with 11.29 and El Cují during 2005 with a value of 11.18 (Olivares et al. 2016c). According to the report by Paredes-Trejo and Olivares (2018), drought as a climate threat has gained media notoriety because of the severe economic and social effects caused by the 2002/2003, 2009/2010, and 2013/2016 episodes in Venezuela.

In southern Venezuela, it is worth mentioning that the most recent droughts in the Venezuelan Amazon have been attributed to a combination of central El Niño (summer) and warm conditions in the northern tropical Atlantic (autumn, winter) (Enciso et al. 2019). At the international level, the study by Zhang et al. (2022) established that in 1982 and 2015, the drought hotspots according to their frequency were located in the western United States, Venezuela, Colombia, eastern Brazil, and southern Argentina in South America, with the average frequency of droughts in these regions being more than nine events in that period.

The results of Rivas et al. (2021) indicated that the level of extreme drought intensity in the locality during the period 1982–2015 was very low in Guanare, Portuguesa state, not exceeding 2% and being incongruent with extremely strong magnitudes. These meteorological drought events occurred coincidentally for the period evaluated between December and March, while the most significant humidity was between May, June, and July.

The results of Tezara et al. (2020) noted that the best performance of cacao plants during drought was shown by the Criollos Moderno's (genetically improved types of the original criollo cacao) with the highest instantaneous efficiency of water use, while five Trinitarios clones (hybrid groups that come from the mixture between criollo cacao and forastero cacao) appeared to be less sensitive to drought, since neither chlorophyll, total soluble protein concentration, nor stomatal conductance changed with drought, indicating that these Trinitarian clones, with a lower net photosynthetic rate, have a conservative use of water.

The findings of Paredes-Trejo et al. (2023) establish that El Niño events worsen the drought conditions of the Orinoco River basin, which covers a total area of 994,368 km^2, including Colombia, while Atlantic Sea surface temperature (SST) variability has less influence. The study also notes that long-term surface warming trends increase the demand for evapotranspiration, which exacerbates drought conditions. However, vegetation greenness remains high despite high surface temperatures, suggesting that temperature alone does not significantly affect vegetation phenology or root-zone water reserves. In summary, droughts seriously disrupt the infiltration-percolation process, affecting the long-term recharge rate of aquifers and their contributions to ecological flows.

Concluding with the drought studies, the results of Silva and Mendoza (2021) show a high variability of drought events between seasons, models, and climate change scenarios but with a common increase in the magnitude and duration of drought events for all models in Barinitas, Barinas State, and Pariaguán Anzoátegui State, mainly under the RCP 8.5 scenario. Drought reduced leaf N content, chlorophyll content, and photochemical activity, suggesting inhibited metabolism.

Figure 3.7 indicates that, over the decades, the distribution of precipitation has shifted, with a more variable and slightly higher annual rainfall observed in the last decades (1991–2020) in Anzoátegui (Fig. 3.7a) and Lara (Fig. 3.7c) which may have implications for water management, agriculture, and ecosystem stability in the region. Also, Fig. 3.7 shows an increase in the frequency and intensity of extreme precipitation events over the decades in Anzoátegui (Fig. 3.7b) and Lara (Fig. 3.7d), particularly in recent years. The green circles, indicating events with an intensity greater than 2.5 standard deviations (SD), become more prevalent toward the end of the study period, suggesting a rise in the magnitude of precipitation events compared to the historical average. The shift in precipitation distribution has significant implications for regional agriculture; flooding, soil erosion, and waterlogging may severely affect crops sensitive to excess water, such as maize and beans. While these events may provide short-term benefits to rainfed farming systems, the long-term effects of extreme rainfall, such as crop loss and damage to irrigation infrastructure, could pose greater challenges.

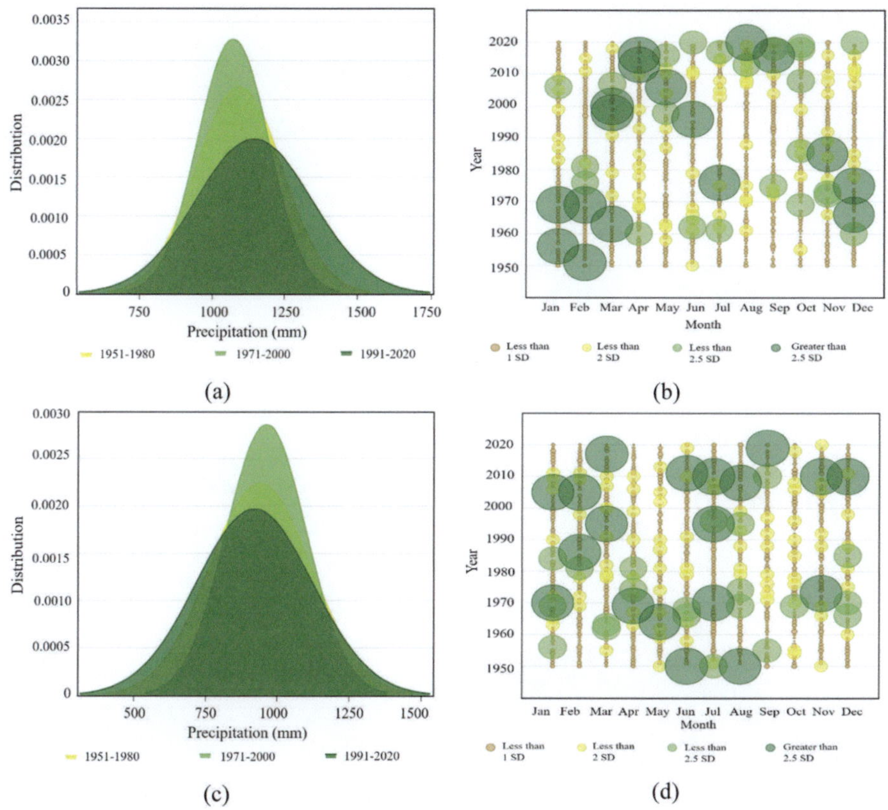

Fig. 3.7 (**a**) Change in distribution of precipitation (1951–2020) in Anzoátegui; (**b**) change in event intensity of average largest 5-day cumulative precipitation (1951–2020) in Anzoátegui; (**c**) change in distribution of precipitation (1951–2020) in Lara; (**d**) change in event intensity of average largest 5-day cumulative precipitation (1951–2020) in Lara according to the data source from Harris et al. (2020) and available in World Bank, Climate Change Knowledge Portal (2025)

Figure 3.8a illustrates the change in distribution of precipitation in Portuguesa (1951–2020), revealing significant shifts in precipitation patterns over the decades. During the period from 1951 to 1980, precipitation was concentrated around the 1500 mm mark, as shown by the narrow peak in the yellow area, indicating a more consistent and predictable rainfall pattern. However, in the following decades, particularly from 1971 to 2000 and 1991 to 2020, there was a noticeable broadening of the distribution, with precipitation levels becoming more spread out, as shown by the green areas. This indicates an increase in variability in precipitation, with both lower and higher precipitation values observed, especially during the 1991–2020 period, where the distribution peak shifted slightly toward higher values.

Figure 3.8c shows a clear shift in precipitation patterns over the decades. From 1951 to 1980, a narrow peak in the yellow area shows precipitation concentrated around 2500 mm, indicating more consistent rainfall levels then. However, in the

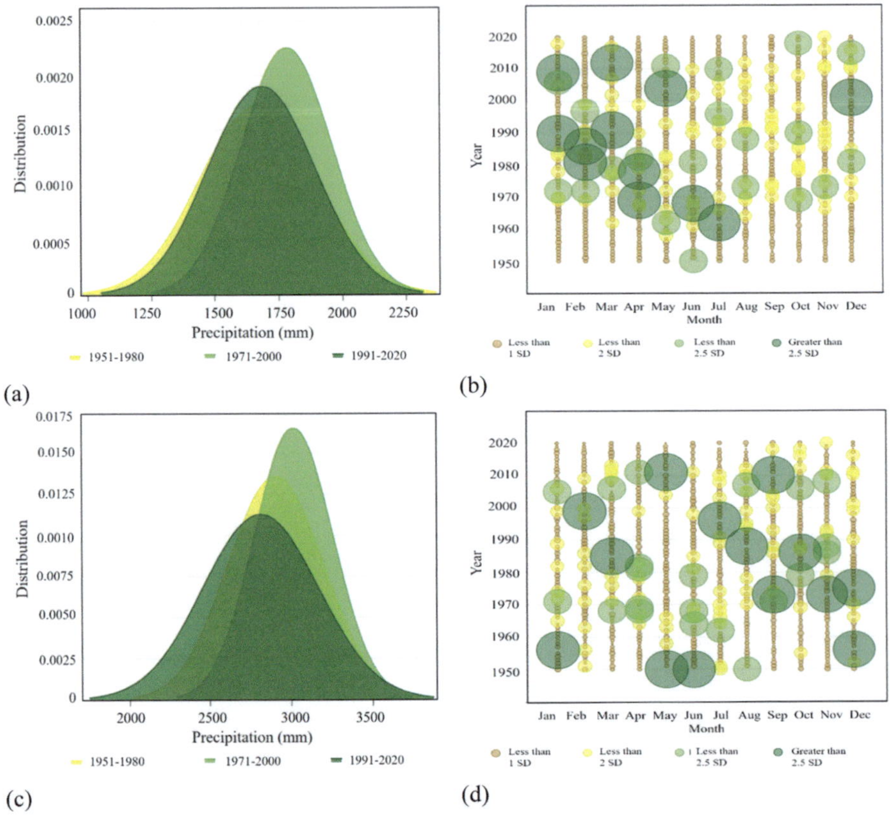

Fig. 3.8 (**a**) Change in distribution of precipitation in Portuguesa (1951–2020); (**b**) change in event intensity of precipitation in Portuguesa (1951–2020); (**c**) change in distribution of precipitation in Tachira (1951–2020); (**d**) change in event intensity of precipitation in Tachira (1951–2020) according to the data source from Harris et al. (2020) and available in World Bank, Climate Change Knowledge Portal (2025)

subsequent decades, especially between 1971–2000 and 1991–2020, there is a broadening of the distribution, particularly in the green areas, which reflects an increase in the variability of precipitation. The peak of precipitation in 1991–2020 shifts slightly toward 3000 mm, suggesting that the region has seen higher rainfall totals on average in recent years, along with increased variation in precipitation amounts. This shift may have important implications for agriculture and water resource management in Táchira, particularly in planning for more extreme weather events and adapting to higher precipitation variability.

Figure 3.8 illustrates the change in event intensity of precipitation for Portuguesa (Fig. 3.8b) and Táchira (Fig. 3.8d) from 1951 to 2020. Both regions show a shift toward more intense precipitation events over the years, reflected by the increasing frequency of larger circles (representing larger precipitation events) as the decades progress. In Portuguesa (Fig. 3.8b), from 1951 to 2020, the intensity of precipitation events has increased, with more frequent green circles (indicating events greater than 2.5 standard deviations, SD) in recent decades. The 2000s and 2010s show a marked increase in extreme precipitation events, especially in May, June, and October, where the largest circles (greater than 2.5 SD) are increasingly common. These changes suggest a rising frequency of extreme rainfall events, which could have significant implications for agricultural activities, potentially increasing the risk of flooding, soil erosion and landslides, and waterlogging, particularly for crops sensitive to excessive water.

In Táchira (Fig. 3.8d), similar trends are observed, with a noticeable increase in extreme precipitation events over time. The 2010s stand out with larger green circles, especially in April, May, and October, indicating more frequent intense rainfall events during these months. This trend suggests that the region has experienced a higher frequency of extreme precipitation, which could affect agricultural practices by increasing the likelihood of extreme weather events such as flooding or droughts, depending on the timing of the rains.

3.3.3 Relationships of Climate Variability with Pests and Diseases

Climate variability has had a significant impact on the incidence of plant pests and diseases in Venezuela, with changes in growth patterns and distribution of species harmful to agricultural production. There are scientific reports that show the relationship between climate variability and the appearance of pests and diseases, such as Bertorelli and Olivares (2020) who point out a clear relationship between these factors of population fluctuation of the insect *Spodoptera frugiperda* (JE Smith) (Fall armyworm) in El Tigre, Anzoátegui state, and the climatic conditions in the days after planting; a greater pest attack occurred in the vegetative period when there were low amounts of daily precipitation and high temperatures.

Figure 3.9a shows the annual trends of average maximum surface air temperature in Anzoátegui from 1951 to 2023, with significant increases in temperature over the decades. The trend for 1951–2020 shows an average increase of 0.26 °C, while the trends for 1971–2020 and 1991–2020 show increases of 0.33 °C and 0.4 °C, respectively. These trends are statistically significant at 100% confidence, indicating a clear and consistent rise in maximum surface temperatures across the decades. This increase in temperature has significant implications for agricultural pests, particularly the *Spodoptera* genus, which includes major pest species like *Spodoptera frugiperda* (fall armyworm).

Warmer temperatures can accelerate the development and reproduction of these pests, leading to higher infestation rates and potentially greater damage to crops. The trend of rising temperatures, combined with changing precipitation patterns, may enhance the survival and spread of such pests, necessitating more effective pest management strategies in agriculture to mitigate the impact on crop yields.

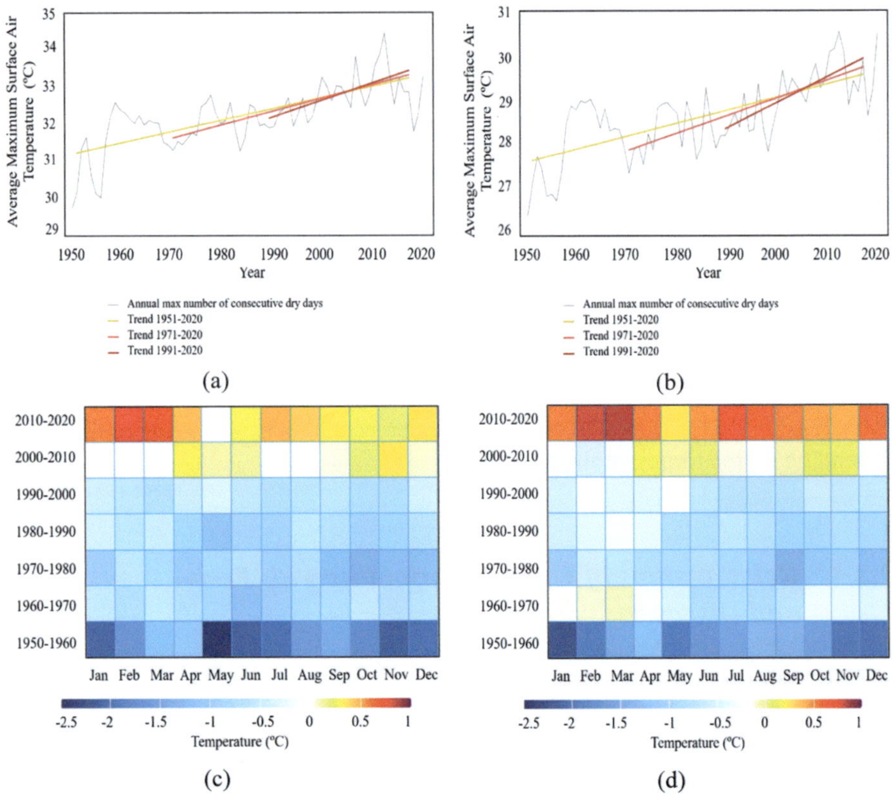

Fig. 3.9 (**a**) Average maximum surface air temperature annual trends with significance of trend per decade (1951–2023) in Anzoátegui and (**b**) Miranda; (**c**) average maximum surface air temperature monthly trends (1951–2023) in Anzoátegui and (**d**) Miranda according to the data source from Harris et al. (2020) and available in World Bank, Climate Change Knowledge Portal (2025)

On the other hand, Vincenti-González et al. (2018) found a strong and significant association between the interannual cycles of dengue and ENSO, suggesting that El Niño events have been partly responsible for periodic dengue outbreaks in northern Venezuela. The proposal of Fernández and Sáez (2018) seeks to generate a spatial model that allows projecting, based on environmental variables, the behavior of the disease in the geographical context. Elevated temperatures during El Niño years can shorten the mosquito's incubation period, increasing the transmission rate of the virus. Therefore, the rise in average temperatures, coupled with ENSO-driven climatic changes, creates a favorable environment for both vector-borne diseases like dengue and agricultural pests.

Also, Moreno et al. (2020) indicated that the diversity of insects with nocturnal activity in two rice fields in the Northwestern Plains of Venezuela could respond positively to rainy conditions and an abundance of high-quality food resources (mature rice plants). On the other hand, the results of Marval-Rodríguez et al. (2019)

indicated that the ichthyofauna of the CañoManamo region, in the Orinoco River Delta, is composed of characteristic elements of an estuary, with the climatic seasons being the most important factors for the evaluated group and this reflects the seasonality of this community. Likewise, Zamora-Figueroa et al. (2021) established that the complex microbial interactions in the floodplain of the Mapire River south of Anzoátegui regulate biogeochemical processes, interactions dependent on physicochemical changes in the soil throughout the hydroperiod caused by spatial differences along the gradient.

Figure 3.9b shows that from 1951 to 2020, the temperature rose by 0.28 °C in Miranda, while the increase was more pronounced in the subsequent decades: 0.38 °C from 1971 to 2020 and 0.57 °C from 1991 to 2020, with all trends being statistically significant at 100% confidence. This steady temperature rise has important implications for local ecosystems, particularly for aquatic species such as amphipods. Amphipods, being sensitive to temperature changes, may experience shifts in their abundance and distribution as temperatures rise. Warmer waters can accelerate the metabolic rates of these organisms, potentially leading to an increase in their abundance in certain areas where the temperature becomes more favorable for their survival. In the rocky coastlines of Miranda state, Hernández et al. (2020) mentioned that there is temporal variability of the total abundance of amphipod individuals in two coastal locations, with the highest density in August (618 ± 3 ind/100 cm^2) and the lowest in December (48 ± 2 ind/100 cm^2). Abundance patterns are related to physicochemical variables, sea temperature, and salinity, suggesting that the amphipod community may be sensitive to fluctuations in environmental conditions.

Figure 3.9c reveals a clear warming trend over the decades, particularly noticeable in the 2010–2020 period, where temperatures in months such as January, February, and March, exhibit a significant increase, with the average maximum temperature rising by more than 1 °C in these months. This warming trend, especially during the early part of the year, can have considerable implications for pests and diseases affecting local crops. In Miranda, Fig. 3.9d highlights a significant warming trend, particularly in February and March, with temperatures rising by 0.81 °C during this period. Additionally, the increased temperature may lead to the spread of plant diseases, as warmer conditions favor the growth of pathogens like Fusarium and Phytophthora, which affect key crops such as maize, rice, and vegetables.

In the state of Aragua, the study by Olivares et al. (2021) points out that banana cultivation is highly dependent on rainfall that has an important influence on the manifestation of symptoms of banana wilt generated by a fungus-bacterium complex; however, the low spatial resolution (250 m) of the Normalized Difference Vegetation Index (NDVI) images was not suitable for the correct identification of banana wilt disease. In this same study area, the research of León-Pacheco et al. (2021) pointed out that sweet potato plants (*Ipomoea batatas* L.) decreased their growth and leaf expansion in the face of drought conditions, partially closed the stomata to avoid water loss through transpiration, without affecting the variables of net rate of CO_2 assimilation, yield, and accumulation of biomass. The best clone was 64, with the highest transpiration, biomass accumulation, and yield of 18.3 t ha^{-1}. Irrigation throughout the crop generated the highest vegetative growth with 125 $plant^{-1}$ leaves.

3.3.4 Relationships of Climate Variability with Vegetation

Similarly, an important contribution to the study of the relationships of variability with the agroecosystem is represented by the research of Olivares and López (2019) where the largest proportion of land areas in Kashaama, Anzoátegui state, that went from little vegetation to dense vegetation correspond to the rainy months (June–September), this because the distribution of the water deficit covers the period from October to May. With a maximum of 212 mm in March, those crops that are sown after July are at risk of suffering from water deficit in the flowering phase.

An interesting contribution along the same lines is represented by the research of Vilanova et al. (2018) when they establish that humidity is a key driver of forest structure rotation and stem turnover in Venezuelan tropical forests, with longer dry seasons favoring higher rates of tree rotation and, therefore, lower biomass, which has important implications in climate change, given the increase in the frequency of droughts in many tropical forest areas. Regional variation in woody productivity among Venezuelan forests strongly reflects climate effects, with the highest woody productivity when precipitation and temperatures are high.

In Bolívar, Fig. 3.10a shows a consistent increase in temperature over the decades. From 1951 to 2020, the temperature rose by 0.16 °C, with a more pronounced increase in the following decades: 0.26 °C from 1971 to 2020, and 0.36 °C from 1991 to 2020. All trends are statistically significant at 100% confidence, indicating a clear and ongoing warming trend in the region. This temperature rise is particularly concerning for the tropical forest areas in the south of Venezuela, where higher temperatures can have significant ecological implications, especially from 2010 to 2020, with January, February, and March showing significant warming, particularly in March with temperatures rising over 0.5 °C (Fig. 3.10b). Warmer temperatures can exacerbate stress on forest ecosystems, potentially leading to changes in species composition, reduced biodiversity, and altered carbon sequestration rates. Additionally, increased temperatures can exacerbate drought conditions and lead to more frequent and intense forest fires, threatening the health of these crucial ecosystems.

In Mérida, the 1991–2020 period (shown in red) (Fig. 3.10c), the peak of the distribution has further shifted to temperatures around 22–23 °C. This shift indicates a consistent rise in average maximum temperatures over the last several decades. This warming trend in Mérida, particularly in the tropical mountain regions, has significant ecological implications. The increase in temperature could alter the functioning of the local ecosystems, particularly the Andean Forest

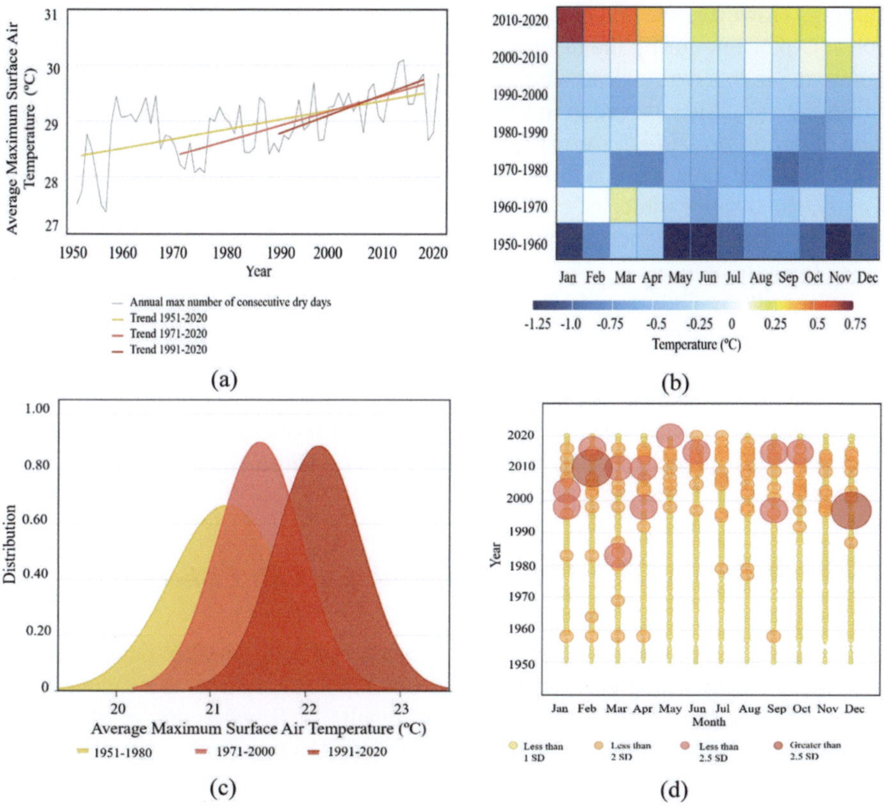

Fig. 3.10 (**a**) Average maximum surface air temperature annual trends with significance of trend per decade (1951–2023) in Bolivar; (**b**) average maximum surface air temperature monthly trends (1951–2023) in Bolivar; (**c**) change in distribution of average maximum surface air temperature (1951–2020) in Mérida; (**d**) change in event intensity of maximum of daily max-temperature (1951–2020) in Bolivar according to the data source from Harris et al. (2020) and available in World Bank, Climate Change Knowledge Portal (2025)

ecosystems (Vilanova et al. 2018). Warmer temperatures could lead to shifts in species distributions, with species that are adapted to cooler conditions potentially migrating to higher altitudes or facing increased stress.

Also, Fig. 3.10d highlights a clear upward trend in the intensity of temperature maxima, particularly from the 2010s onward. The presence of such intense temperature extremes in the most recent decade suggests that Mérida is experiencing more frequent and severe temperature anomalies, with implications for both the environment and agriculture.

On the other hand, in the Orinoco plains, in the study by Leal et al. (2019), it was established that the different climatic phases recorded in El Cedral (Western Plains of Venezuela) coincide with those described in the Cariaco Basin, which indicates that the variability of the El Niño Southern Oscillation was the main climatic forcing. However, the intensification of anthropic activities limited the expected forest expansion during the last two millennia, evidencing the domestication of the landscape in this region. Likewise, Montoya et al. (2019) described that the increase in ENSO variability recorded in Cariaco may have played a key role in the expansion of palm (*Mauritia flexuosa*) in the Orinoco Delta, where potential human populations in that area were influenced differently by these environmental changes. Firstly, the disappearance of coastal resources may have favored the abandonment of the land, while the increase in the abundance of palm may have influenced other crops from the interior.

The study by Galviz and Valerio (2021) showed that the plants collected from *Jacquinia armillaris* on La Tortuga Island (Federal Dependencies) showed characteristics that suggest adaptation to the saline and more arid conditions of the island compared to the plants of Turpialito (Sucre), which presented more leaf traits related to adaptations to water deficit. These results demonstrate the phenotypic plasticity of *J. armillaris* plants growing in two different xeric shrublands.

3.3.5 Relationships of Climate Variability with Fisheries

In the upper delta of the Orinoco River, climatic and water quality factors, associated with the hydrological cycle, played an important role in the regulation of artisanal fishing activity, according to the results of Silva-Acuña et al. (2012). Likewise, in the continental area adjacent to the Araya submarine valley, Sucre state, on the northern (Caribbean) coast of the Araya peninsula, a study by Quintero and Galina (2016) indicates that agricultural practices are limited in this area, although because of the nature of the plain, an attempt has been made to encourage aquaculture through the construction of ponds. The trade winds, which were characterized in the study, contribute to the phenomenon of upwelling, consisting of the ascent of deep and cold waters loaded with nutrients to the surface layer of the sea, this being the origin of the marine wealth of the area. For their part, Marrero and Rodríguez-Olarte (2017) establish that the rise in sea levels with climate change would increase saline intrusion into estuaries and deltas, changing the properties of the waters, affecting the mechanisms of operation of these environments and thereby affecting plant and animal communities; thus establishing that in the climate change scenarios (weather patterns and variations in sea levels), a displacement of some species and loss of wetlands could be generated although the resilience of these habitats would lead to a rearrangement in new areas. Research by Márquez-Rojas et al. (2021) indicates that the highest concentrations of chlorophyll (Cl-a) (0–26.49 mg m^{-3}) and phytoplankton abundance (37–1.247 cel. mL^{-1}) were reported in the rainy season (June–July 2009) in the eastern Gulf of Cariaco.

3.4 Scientific Production of the Effects of Climate Variability in the Different Regions of Venezuela

Figure 3.11 presents the number of publications that address the effects of climate variability in the different regions of Venezuela. The largest number of studies visited refers to significant changes in the water regime, as an expression of climate variability, and to meteorological drought in agricultural areas, with a proportion of 43.8% and 28.1% of the total works, respectively. One of the main effects of the change in the water regime is on the planting dates of crops, which is "felt" as very important by farmers. Meteorological droughts are becoming more frequent in the country's agricultural regions, with significant effects on crop productivity. The relatively higher number of publications on these two topics seems to reflect the attention that farmers are asking for to contribute to solving the incidence of climate variability in agricultural production in their regions.

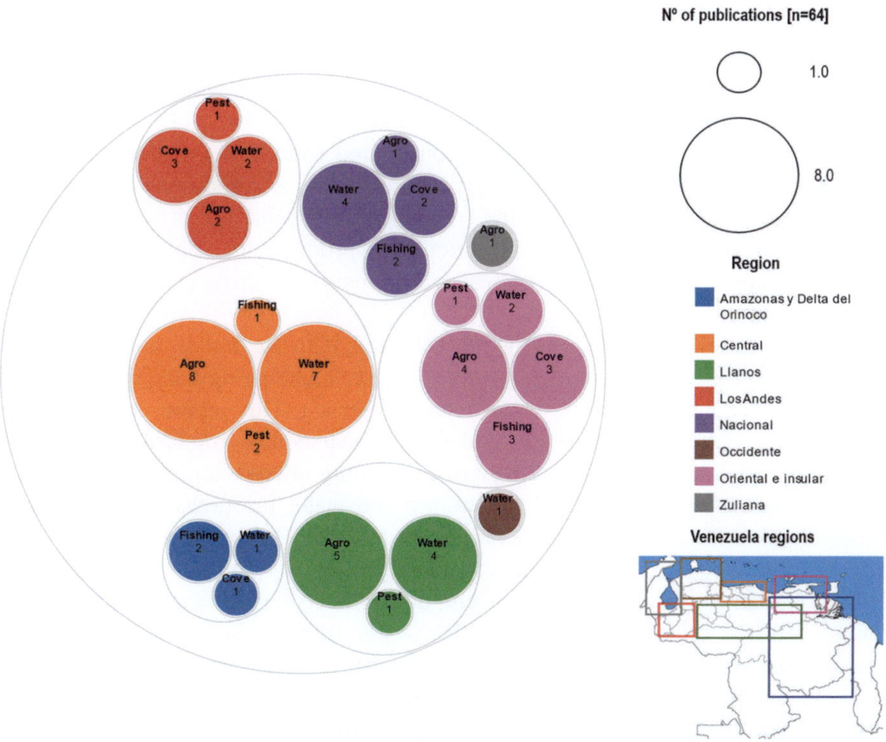

Fig. 3.11 Number of publications addressing the effects of climate variability in Venezuela ($n = 64$). Display values in a hierarchical structure by using circle areas. The hierarchical structure is represented by nested circles that represent the different effects of climate variability (impact on the water regime (Water), agricultural planning (Agro), the area or cover affected (Cove), the incidence of pests and diseases (Pest), and on fishing. It is an additional quantitative dimension with size (number of scientific publications) and a qualitative dimension in color by natural regions. (Source: Authors' elaboration based on data collected by the authors)

The largest proportion of studies seems to be concentrated in the central region of Venezuela ($n = 18$), followed by the eastern region ($n = 13$), the plains ($n = 10$), and the Andes ($n = 08$). Few studies address climate variability at the national level ($n = 09$). The proportion of publications that establish the effect of climate variability on agricultural planning significantly exceeds the rest of the effects ($n = 21$).

The analysis of the scientific production regarding the effects of climate variability across the different regions of Venezuela reveals notable patterns and significant gaps in research. The Central region also stands out with a significant number of studies, primarily focusing on agriculture and water resources. This region, home to key agricultural production areas such as Aragua, Carabobo, and Miranda, faces challenges related to the management of water resources and crop health. Given the increasing risks posed by climate change, there is a pressing need for more research on agricultural pests and disease dynamics, particularly about temperature increases and shifts in precipitation patterns. While some research is underway, more studies on climate-smart agricultural practices and integrated pest management are needed to build resilience against climate variability.

The Eastern and Insular region of Venezuela, despite being an area of significant agricultural and environmental importance, has received relatively limited attention in terms of scientific publications, with only 13 studies. These publications primarily focus on agriculture, water resources, and affected surfaces. The region, which includes states such as Sucre, Nueva Esparta, and Monagas, is vulnerable to the effects of climate variability, particularly regarding rising temperatures, altered precipitation patterns, and the increasing frequency of extreme weather events. These climatic changes can directly affect agricultural productivity, especially for crops like cocoa, corn, and tropical fruits, which are sensitive to shifts in seasonal rainfall and temperature.

The Llanos region emerges as the most studied, with the highest number of publications (8 on agriculture and 7 on water resources), indicating the region's central importance for agriculture and water-related research. This region, known for its extensive agricultural activities, particularly in livestock and crop farming, is highly susceptible to changes in climate patterns, including altered rainfall and increased temperatures. The dominance of agriculture and water resources as key research areas aligns with the region's vulnerability to droughts and floods, which can directly affect crop yields and water availability.

In contrast, the Amazon and the Delta del Orinoco region, with their predominantly forested areas, have fewer studies (only four publications). The limited research on the effects of climate variability in this region is a significant gap, especially given the potential impact of temperature increases and deforestation on biodiversity and hydrological cycles. The tropical forest ecosystems in Amazonia are highly sensitive to climate change, and the lack of comprehensive studies on the effects of climate variability in this region leaves a gap in understanding how these ecosystems may respond to future climate shifts.

The Los Andes region, despite its importance for agricultural production in Venezuela, has relatively few studies (eight publications), with a focus on agriculture and water resources. This mountainous region faces significant challenges from temperature increases, reduced precipitation, and changes in the timing of seasons, all of which can affect crop yields and water availability. The lack of detailed studies on pest outbreaks, crop disease, and the impact of rising temperatures on Andean ecosystems represents a critical gap in the scientific literature. More research is needed to assess the potential for climate-induced disruptions in agricultural cycles, particularly for crops like coffee and vegetables, which are highly sensitive to climatic changes.

The Zuliana region, which is particularly important for agriculture, especially in farming and fisheries, has minimal research (only one study) on the effects of climate variability. The lack of studies on pests and diseases, which can significantly affect the agriculture and fisheries sectors in this region, is a crucial research gap. Temperature rises and changes in rainfall patterns in Zuliana could exacerbate existing challenges, including water scarcity and reduced agricultural productivity, particularly for crops like plantains and rice, which are vulnerable to changing weather patterns.

While certain regions, such as the Llanos and Central, have extensive research coverage, there is a clear need for further investigation in areas like Amazonas, Zuliana, and Los Andes. Key topics that need more attention include the impact of rising temperatures on agricultural pests and crop diseases, especially in areas where climate change is expected to lead to longer and more intense dry spells or more frequent extreme weather events. The increasing frequency of El Niño events and the associated changes in temperature and precipitation could potentially worsen the situation, leading to crop failures, increased pest outbreaks, and water scarcity.

The lack of research in some of these vulnerable regions could have serious consequences for Venezuelan agriculture. With climate variability likely to become more extreme and unpredictable, the agricultural sector in Venezuela could face a significant decline in crop yields, particularly in the Llanos and Zuliana regions. Without further research, these areas may not adequately prepare for future climate-induced disruptions. Increasing the scientific understanding of how climate change will affect different regions of Venezuela is essential for developing effective climate adaptation strategies, including water management, crop diversification, and resilience-building measures to ensure the continued productivity of Venezuelan agriculture in the face of climate uncertainty.

However, in Venezuela, agricultural systems are not consolidated in terms of agrometeorological research, and there is no operational network of meteorological observations, so there is no greater access to forecasts and predictions in areas of agricultural interest (Torín et al. 2018; Landaeta 2018, 2019; Grassi et al. 2019). This makes it difficult for agricultural systems such as rice, corn, sorghum, coffee, sugarcane, and bananas not to have the necessary development of various models that allow simulating water availability, potential yields, crop growth simulation, as well as the response of these crops to various management alternatives. On the other hand, there are smallholder farmers who have little or no access to agroclimatic information. This is why the negative impact of climatic behavior in the country, particularly on the production and agricultural yield of various crops, makes it difficult to make decisions for the planting period and monitoring in the phenological cycle in the crops of small farmers or rural and indigenous economies.

Faced with climate variability and its impact on Venezuelan agriculture, there is an urgent need to embrace innovative techniques such as machine learning algorithms, artificial intelligence (AI), and big data analytics to move beyond conventional methods and address the obsolescence of the few research institutes in the country. These advanced technologies can significantly enhance the efficiency and accuracy of agricultural research by analyzing vast datasets related to climate patterns (Campos 2023), crop health (Olivares et al. 2022), pest outbreaks (Rodríguez-Yzquierdo et al. 2023), and water resources (Vivas et al. 2023).

For example, AI-driven predictive models could help forecast pest infestations, allowing farmers to take proactive measures, while machine learning could optimize irrigation systems by analyzing weather patterns and soil moisture levels in real time. Big data can enable more precise agricultural decision-making by integrating diverse sources of information, such as satellite imagery, local weather data, and crop yield statistics, thus improving crop management and resource allocation. By leveraging these technologies, Venezuelan research institutes can overcome their outdated infrastructure and provide farmers with data-driven insights that enhance productivity, reduce costs, and promote sustainable practices. This shift toward cutting-edge technologies would not only revitalize the research sector but also empower agricultural producers to make more informed decisions, ultimately improving resilience to climate change and boosting the country's food security.

3.5 Conclusions

Venezuela's climate distribution is a result of its unique geographical position, topography, and global climate phenomena, which shape distinct climate zones across the country. The diverse climatic conditions, from tropical humid in the north to temperate mountain climates in the Andes, highlight the need for region-specific agricultural practices and resource management strategies. Understanding this complexity is crucial for adapting agricultural systems to the specific challenges posed by each region's climate variability, ensuring the sustainability of local food production systems.

The significant effects of climate variability in Venezuela, especially with fluctuations in temperature and precipitation, have profound implications for agriculture. With increasing temperatures and changing rainfall patterns, vulnerable crops face greater risk from droughts, floods, and pest outbreaks. However, the scientific community has made strides in understanding these changes, with an increase in publications addressing the relationship between climate variability and agricultural productivity. Yet, the challenge remains in translating this knowledge into effective, localized strategies for farmers.

The variability in rainfall patterns, particularly the spatial and temporal shifts, calls for more effective water resource management. Regions like the Llanos and central states like Aragua, Carabobo, and Miranda have experienced significant changes in precipitation, exacerbating the challenges faced by rainfed agriculture. The decline in precipitation during crucial growing months is a clear indicator of the need for sustainable water management practices and irrigation systems to mitigate the adverse effects on crop yields and ensure food security.

The increasing frequency and intensity of meteorological droughts pose a serious threat to agricultural systems across Venezuela, particularly in regions dependent on rainfed farming. Studies have shown that droughts can severely reduce crop yields, especially in the Llanos, Zulia, and central regions, where water stress has become a recurring issue. Addressing this challenge requires the implementation of drought-resistant crops and improved water conservation techniques to safeguard agricultural productivity.

The changing patterns of precipitation across Venezuela, as seen in regions like Anzoátegui and Lara, reflect a shift toward more extreme events. While some areas are facing drier conditions, others experience more intense rainfall, which can lead to flooding and soil erosion. These shifts underscore the need for adaptive strategies in agricultural planning, including the use of weather data and forecasting models to help farmers optimize planting schedules and mitigate the effects of unpredictable weather events.

Rising temperatures have a direct impact on agricultural pests and diseases, with species benefiting from warmer conditions. As temperatures continue to rise, pest populations are likely to increase, leading to more frequent infestations and greater crop damage. Understanding the link between temperature trends and pest behavior is essential for developing pest management strategies and ensuring that agricultural systems can withstand the increased threat of pests and diseases under changing climate conditions.

Despite the growing body of scientific literature on climate variability in Venezuela, there are significant gaps in research, particularly in less-studied regions like Amazonas, Zulia, and the Andes. Research on the effects of climate change on pest dynamics, disease outbreaks, and crop resilience remains limited. Closing these gaps through targeted research can provide valuable insights into climate adaptation strategies, helping farmers make informed decisions and improving the overall resilience of the agricultural sector.

In light of the growing challenges posed by climate variability, the incorporation of advanced technologies such as machine learning, artificial intelligence, and big data analytics into agricultural research is crucial. These technologies can provide more accurate predictions of pest outbreaks, optimize irrigation systems, and help farmers manage resources more efficiently. By leveraging these technologies, Venezuelan agriculture can move beyond conventional methods, addressing the limitations of existing research infrastructure and improving the resilience of agricultural systems to climate change.

References

Álvarez-Barreto JM, Quintero D, Rodríguez M, Rea R, Sosa D (2018) Evaluation of phenolic compounds, lignin, amino acids, and carbohydrates in Theobroma cacao L. from three different climate regions in Venezuela. Emirates J Food Agric 30(6):522–530. https://doi.org/10.9755/ejfa.2018.v30.i6.1726

Álzate D, Rojas E, Mosquera J, Ramón J (2015) Cambio climático y variabilidad climática para el periodo 1981–2010 en las cuencas de losríos Zulia y Pamplonita, Norte de Santander, Colombia. Luna Azul 40:127–153. https://www.redalyc.org/pdf/3217/321733015010.pdf

Barreto W, García-Orellana Y, Torres-Rodríguez D (2024) From drought to sustainable agricultural production. Historical water resource management experiences in the Quíbor Valley (Venezuela). In: Corzo Perez GA, Hayde LG, Romero-López R, Jiménez Palacios A (eds) Aqua vitae, The Latin American studies book series. Springer, Cham. https://doi.org/10.1007/978-3-031-57731-4_11

Beck HE, McVicar TR, Vergopolan N, Berg A, Lutsko NJ, Dufour A et al (2023) High-resolution (1 km) Köppen-Geiger maps for 1901–2099 based on constrained CMIP6 projections. Sci Data 10(1):724. https://doi.org/10.1038/s41597-023-02549-6

Bertorelli M, Olivares BO (2020) Population fluctuation of *Spodoptera frugiperda* (J.E. Smith) (Lepidoptera: Noctuidae) in sorghum cultivation in Southern Anzoátegui, Venezuela. J Agric Univ Puerto Rico 104(1):1–16. https://n9.cl/igd6i. Accessed 17 Feb 2025

Chacón-Moreno E, Suárez del Moral P (2020) Mapa bioclimático de la Cordillera de Mérida. Ecotropicos 32:e0010. https://doi.org/10.53157/ecotropicos.32e0010

Chacón-Moreno E, Rodríguez-Morales M, Paredes D, Suárez del Moral P, Albarrán A (2021) Impacts of global change on the spatial dynamics of tree line in Venezuelan Andes. Front Ecol Evol 9:74. https://doi.org/10.3389/fevo.2021.615223

Cortez A, Rodríguez MF, Rey JC, Ovalles F, González W, Parra R, Olivares B, Marquina J (2016) Variabilidad espacio temporal de la precipitación en el estado Guárico, Venezuela. Rev Facult Agron (LUZ) 33(3):292–310. https://n9.cl/pmdck. Accessed 20 Mar 2025

Cortez A, Olivares BO, Parra RM, Lobo D, Rey JC, Rodríguez MF (2018) Descripción de los eventos de sequía meteorológica en localidades de la Cordillera Central, Venezuela. Ciencia, Ingenierías y Aplicaciones, I 1:23–45. https://doi.org/10.22206/cyap.2018.v1i1.pp23-45

Enciso C, Sulca J, Espinoza JC, Silva Y (2019) Pronóstico de sequías estacionales en la cuenca amazónica durante eventos ENOS. Boletín Técnico 6(12):9–12. http://hdl.handle.net/20.500.12816/4824. Accessed 17 Mar 2025

Fernández G, Sáez VS (2018) Distribución espacial del dengue en la Región Andina de Venezuela en distintos escenarios de cambio climático, una propuesta de investigación. Terra Nueva Etapa 34(55):281–293. https://n9.cl/1wniw. Accessed 7 Feb 2025

Galviz YC, Valerio R (2021) Leaf morphoanatomical traits of Jacquinia armillaris Jacq. (Theophrastoideae-Primulaceae) in two xeric shrublands from Venezuela. Neotrop Biodivers 7(1):364–375. https://doi.org/10.1080/23766808.2021.1964911

Grassi JEA, Polanco MR, Mantilla HAT, Hernández JYL (2019) Análisis espacio temporal de la homogeneidad de estaciones de precipitación en una zona árida y semiárida del Centro Occidente de Venezuela. Ciencia e Ingeniería 40(2):185–194. https://n9.cl/8qv3a. Accessed 7 Apr 2025

Harris I, Osborn TJ, Jones P et al (2020) Version 4 of the CRU TS monthly high-resolution gridded multivariate climate dataset. Sci Data 7:109. https://doi.org/10.1038/s41597-020-0453-3

Hernández V, Villamizar E, García M, Pereira C, Muñoz N (2020) Dinámica de la comunidad de anfípodosen en dos litorales rocosos del estado Miranda, Venezuela. Acta Biol Venezuelica 40(1):45–63. https://n9.cl/e165w. Accessed 27 Feb 2025

Landaeta LJ (2018) Análisis de las series de precipitación mensual en la región Capital de Venezuela, 1891–2006. Rev Geogr Venezolana 59(2):280–297. https://www.redalyc.org/articulo.oa?id=347760473004. Accessed 7 Apr 2025

Landaeta LJ (2019) Contribución al estudio de la pluviometría del estado Miranda (Venezuela) mediante Sistemas de Información Geográfica. Warisata-Revista de Educación 1(3):146–166. https://doi.org/10.33996/warisata.v1i3.72

Leal A, Gasson R, Behling H (2019) Human-made fires and forest clearance as evidence for late Holocene landscape domestication in the Orinoco Llanos (Venezuela). Veg Hist Archaeobotany 28:545–557. https://doi.org/10.1007/s00334-019-00713-w

León-Pacheco RI, Pérez-Macias M, Fuenmayor-Campos FC, Rodríguez-Izquierdo AJ, Rodríguez-Yzquierdo GA, Villagran-Munar EA (2021) Evaluación agronómica y fisiológica en clones de camote (*Ipomoea batatas*) sometidos a condiciones de estrés hídrico. Agronomía Mesoamericana 32(3):719–732. https://doi.org/10.15517/am.v32i3.42303

Lescher Soto I, Villamizar A, Olivares BO, Gutiérrez ME, Nagy GJ (2024) Navigating the uncertain terrain: Venezuela's future using the shared socioeconomic pathways framework—a systematic review. Climate 12(7):98. https://doi.org/10.3390/cli12070098

Maríñez AN, Palacios Sánchez JM, Osorio Carrera C et al (2021) Análisis comparativo de cuatro índices para la caracterización de sequías meteorológicas en la localidad El Tigre, Estado Anzoátegui, Venezuela. Rev Climatol 21:146–156. https://n9.cl/vicxs. Accessed 7 Mar 2025

Márquez-Rojas B, Troccoli L, Díaz-Ramos R (2021) Bloom algal y abundancia de copépodos en el golfo de Cariaco, Venezuela (Caribe Suroriental). Intropica 16(2):133–152. https://doi.org/10.21676/23897864.4010

Marrero C, Rodríguez-Olarte D (2017) Los humedales costeros venezolanos en los escenarios de cambios climáticos: vulnerabilidad, perspectivas y tendencias. In: Botello AV, Villanueva S, Gutiérrez J, Rojas Galaviz JL (eds) Vulnerabilidad de las zonas costeras de Latinoamérica al cambio climático. UJAT, UNAM, UAC, pp 461–476. 476 p

Marval-Rodríguez Á, Alió-Mingo J, Carvajal Y, Altuve D, Vizcaíno G (2019) Spatio-temporal variation of the ichthyofauna in the estuary from Manamo Channel, Orinoco River Delta, Venezuela. Biología Ciencia y Tecnología 12:45–48. https://doi.org/10.22201/fesi.20072082.2019.12.70409

Mejía JF, González JD, Albarrán A (2022) Influencia de la variabilidad climática en los pantanos altoandinos de la microcuenca Miguaguó en los Andes venezolanos. Revista de Ciencias Ambientales 56(2):38–62. https://doi.org/10.15359/rca.56-2.3

Millano J, Romero AJ (2021) Comportamiento de las precipitaciones en la Bioregión de la Cordillera de la Costa Central de Venezuela. Ciencia y Tecnología Agropecuaria 6(2):60–70. https://doi.org/10.24054/cyta.v6i2.2307

Montoya E, Pedra-Méndez J, Garcia-Falco E, Gomez-Paccard M, Giralt S, Vegas-Vilarrubia T, Rull V (2019) Long-term vegetation dynamics of a tropical megadelta: mid-Holocene paleoecology of the Orinoco Delta (NE Venezuela). Quat Sci Rev 221:105874. https://doi.org/10.1016/j.quascirev.2019.105874

Moreno A, Azofeifa Y, Dominguez C, Nassar JM (2020) Ensamblaje de insectos con actividad nocturna en dos arrozales de los Llanos Noroccidentales de Venezuela. Anartia 30:35–48. https://zenodo.org/records/3967863

Moreno-Pizani MA (2021) Water management in agricultural production, the economy, and Venezuelan society. Front Sustain Food Syst 4:624066. https://doi.org/10.3389/fsufs.2020.624066

Olivares BO (2018) Tropical conditions of seasonal rain in the dry-land agriculture of Carabobo, Venezuela. La Granja 27(1):86–102. https://doi.org/10.17163/lgr.n27.2018.07

Olivares BO (2023) Banana production in Venezuela: novel solutions to productivity and plant health, The Latin American studies book series. Springer Nature. https://doi.org/10.1007/978-3-031-34475-6

Olivares BO, Hernández R (2019a) Análisis regional de zonas homogéneas de precipitación en Carabobo, Venezuela. Revista Lasallista de Investigación 16(2):90–105. https://doi.org/10.22507/rli.v16n2a9

Olivares BO, Hernández RÁ (2019b) Sectorización ecoterritorial para la producción agrícola sostenible del cultivo de papa (*Solanum tuberosum* L.) en Carabobo, Venezuela. Ciencia y Tecnología Agropecuaria 20(2):323–354. https://doi.org/10.21930/rcta.vol20_num2_art:1462

Olivares BO, Hernández R (2020) Application of multivariate techniques in aptitude in Carabobo, Venezuela. Trop Subtrop Agroecosyst 23(2):1–12. https://doi.org/10.56369/tsaes.3233
Olivares BO, López M (2019) Normalized Difference Vegetation Index (NDVI) applied to the agricultural indigenous territory of Kashaama, Venezuela. UNED Res J 11(2):112–121. https://n9.cl/b2dww. Accessed 27 Feb 2025
Olivares BO, Zingaretti ML (2018) Analysis of the meteorological drought in four agricultural locations of Venezuela by the combination of multivariate methods. UNED Res J 10(1):192–203. https://doi.org/10.22458/urj.v10i1.2026
Olivares BO, Zingaretti ML (2019) Aplicación de métodos multivariados para la caracterización de periodos de sequía meteorológica en Venezuela. Revista Luna Azul 48:172–192. https://doi.org/10.17151/luaz.2019.48.10
Olivares BO, Caraballo L, Torrealba J (2013) Variabilidad del régimen de precipitación en el periodo 1990–2009 en la localidad de El Tigre estado Anzoátegui, Venezuela. Rev Fac Agron 30(1):19–32. https://n9.cl/lj6k2. Accessed 17 Mar 2025
Olivares BO, Zingaretti ML, Demey Zambrano JA, Demey JR (2016a) Tipificación de los sistemas de producción agrícola y la percepción de la variabilidad climática en Anzoátegui, Venezuela. Rev FAVE Ciencias Agrarias 15(2):39–50. https://doi.org/10.14409/fa.v15i2.6587
Olivares BO, Cortez A, Lobo D, Parra R, Rey JC, Rodríguez MF (2016b) Estudio de la Sequía Meteorológica en Localidades de los Llanos de Venezuela Mediante el Índice de Precipitación Estandarizado. Rev Acta Nova 7(3):266–283. https://n9.cl/j7k19. Accessed 7 Mar 2025
Olivares BO, Cortez A, Parra R, Lobo D, Rey J, Rodríguez M (2016c) Análisis temporal de la sequía meteorológica en localidades semiáridas de Venezuela. UG Ciencia 22(1):11–24. https://n9.cl/lo415. Accessed 17 Mar 2025
Olivares BO, Cortez A, Lobo D, Parra R, Rey J, Rodríguez M (2017a) Evaluation of agricultural vulnerability to drought weather in different locations of Venezuela. Rev Fac Agron 34(1):103–129. https://n9cl/d827w. Accessed 20 Mar 2025
Olivares BO, Parra R, Cortez A (2017b) Characterization of precipitation patterns in Anzoátegui state, Venezuela. Ería 3(3):353–365. https://doi.org/10.17811/er.3.2017.353-365
Olivares BO, Hernández R, Arias A, Molina JC, Pereira Y (2018a) Zonificación agroclimática del cultivo de maíz para la sostenibilidad de la producción agrícola en Carabobo, Venezuela. Rev Univ Geogr 27(2):139–159. https://n9.cl/r3ygl. Accessed 20 Mar 2025
Olivares BO, Hernández R, Arias A, Molina JC, Pereira Y (2018b) Identificación de zonas agroclimáticas potenciales para producción de cebolla (*Allium cepa* L.) en Carabobo, Venezuela. J Selva Andina Biosphere 6(2):70–82. https://n9.cl/ugm2e. Accessed 20 Mar 2025
Olivares BO, Hernández R, Coelho R, Molina JC, Pereira Y (2018c) Análisis espacial del índice hídrico: avance en la adopción de decisiones sostenibles en los territorios agrícolas de Carabobo, Venezuela. Rev Geogr Am Cent 60(1):277–299. https://doi.org/10.15359/rgac.60-1.10
Olivares BO, Hernández R, Coelho R, Molina JC, Pereira Y (2018d) Analysis of climate types: main strategies for sustainable decisions in agricultural areas of Carabobo, Venezuela. Scientia Agropecuária 9(3):359–369. https://doi.org/10.17268/sci.agropecu.2018.03.07
Olivares BO, Rey JC, Lobo D, Gómez JA, Landa BB (2019) El cambio climático en zonas bananeras de la Región Central de Venezuela: El futuro de los bananos con un escenario hídrico incierto. In: III Simposio Venezolano de Cambio Climático, Agricultura y Seguridad Alimentaria. 08 y 09 de octubre 2019. Universidad Metropolitana Caracas, Venezuela. https://n9.cl/l6kgk. Accessed 20 Mar 2025
Olivares BO, Hernández R, Arias A, Molina JC, Pereira Y (2020) Eco-territorial adaptability of tomato crops for sustainable agricultural production in Carabobo, Venezuela. Idesia 38(2):95–102. https://doi.org/10.4067/S0718-34292020000200095
Olivares BO, Rey JC, Lobo D, Navas-Cortés JA, Gómez JA, Landa BB (2021a) Fusarium Wilt of Bananas: A review of agro-environmental factors in the Venezuelan production system affecting its development. Agronomy 11(5) 986. https://doi.org/10.3390/agronomy11050986
Olivares BO, Paredes F, Rey J, Lobo D, Galvis-Causil S (2021b) The relationship between the normalized difference vegetation index, rainfall, and potential evapotranspiration in a banana plantation of Venezuela. Sains Tanah 18(1):58–64. https://doi.org/10.20961/stjssa.v18i1.50379

Olivares BO, Rey JC, Lobo D, Navas-Cortés JA, Gómez JA, Landa BB (2022) Machine learning and the new sustainable agriculture: applications in banana production systems of Venezuela. Agric Res U pdates 42:133–157

Paredes MY, Chacón-Moreno E (2021) Distribución potencial de cultivos agrícolas en el estado Mérida (Venezuela), en escenarios de cambio climático. Revista Geográfica Venezolana. Volumen 62 (2): 428–444. https://www.researchgate.net/publication/356970049

Paredes-Trejo F, Olivares B (2018) El desafío de la sequía en Venezuela. In: Núñez Cobo J, Verbist K (eds) Atlas de Sequía de América Latina y el Caribe. UNESCO/CAZALAC (Centro del Agua para Zonas Áridas y Semiáridas de América Latina y el Caribe), France, pp 127–136. https://unesdoc.unesco.org/ark:/48223/pf0000265894

Paredes-Trejo F, Guevara-Pérez E, Barbosa-Alves H, Uzcátegui-Briceño C (2015) Tendencia de la precipitación estacional e influencia de El Niño-Oscilación Austral sobre la ocurrencia de extremos pluviométricos en la cuenca del lago de Valencia, Venezuela. Tecnología y Ciencias del Agua 6(6):33–48. https://n9.cl/zkwp2. Accessed 20 Mar 2025

Paredes-Trejo F, Barbosa-Alves H, Moreno-Pizani MA, Farías-Ramírez A (2018) Influencia de las sequías sobre el régimen hídrico de los ríos en Venezuela. Ríos En Riesgo de Venezuela 2:171–184. https://n9.cl/3xs3d. Accessed 17 Mar 2025

Paredes-Trejo F, Olivares BO, Móvil-Fuentes Y, Arevalo-Groening J, Gil A (2023) Assessing the spatiotemporal patterns and impacts of droughts in the Orinoco River basin using Earth observation data and surface observations. Hydrology 10(10):195. https://doi.org/10.3390/hydrology10100195

Parra RM, Olivares BO, Cortez A, Lobo D, Rey JC, Rodríguez MF (2018) Características de la sequía meteorológica (1980–2014) en dos localidades agrícolas de los Andes Venezolanos. Rev Invest 42(95):38–55. https://n9.cl/mm6d. Accessed 7 Mar 2025

Pérez-Macias M, Soto E, León-Pacheco RI, Rodríguez-Yzquierdo GA, Villagrán-Munar EA (2022) Injertos de naranja (*Citrus sinensis*) y mandarina (*Citrus reticulata*) en fase de producción. Agronomía Mesoamericana 33(1):1–17. https://doi.org/10.15517/am.v33i1.45264

Quintero A, Galina T (2016) Evaluación de los parámetros del viento en los períodos de lluvia y sequía en el valle submarino de Araya, Venezuela. I Parte. Boletín del Instituto Oceanográfico de Venezuela 55(2):3–16. https://zenodo.org/records/3975473

Rivas DV, Mariñez AN, Pino MM, Valque RB, Cruz MT, Vásquez MG, Calderón LR (2021) Standardized Precipitation Index (SPI) for the characterization of meteorological droughts in the town of Guanare, Portuguesa State, Venezuela. Turk J Comput Mathemat Educ 12(7):2824–2830

Rodríguez García JJ (2021) Food security in Venezuela: from policies to facts. Front Sustain Food Syst 5:617907. https://doi.org/10.3389/fsufs.2021.617907

Rodríguez-Yzquierdo G, Olivares BO, Silva-Escobar O, González-Ulloa A, Soto-Suarez M, Betancourt-Vásquez M (2023) Mapping of the susceptibility of Colombian Musaceae lands to a deadly disease: *Fusarium oxysporum* f. sp. *cubense Tropical* Race 4. Horticulturae 9(7):757. https://doi.org/10.3390/horticulturae9070757

Romero AJ, Millano JL (2021) Comportamiento de las temperaturas en la Bioregión de la Cordillera de la Costa, Venezuela. Ciencia y Tecnología Agropecuaria 6(1):9–16. https://doi.org/10.24054/cyta.v6i1.1076

Rubel F, Kottek M (2010) Observed and projected climate shifts 1901–2100 depicted by world maps of the Köppen-Geiger climate classification. Meteorol Z 19:135–141. https://doi.org/10.1127/0941-2948/2010/0430

Silva MI, Mendoza NM (2021) Evaluación de la sequía meteorológica en locaciones agrícolas de Venezuela. Rev Fac Agron LUZ 38(3):548–572. https://doi.org/10.47280/RevFacAgron(LUZ).v38.n3.05

Silva O, Tobón C (2019) Evaluación de impactos ecohidrológicos del cambio climático en una cuenca de la región central de Venezuela. In: III Simposio Venezolano de Cambio Climático, Agricultura y Seguridad Alimentaria. 08 y 09 de octubre 2019. Universidad Metropolitana Caracas, Venezuela. https://doi.org/10.13140/RG.2.2.25317.96481. Accessed 20 Mar 2025

Silva-Acuña AS, Urbano T, Medina L, Moreno C, Figueroa M, Araújo D (2012) Relación entre la producción de peces y la variación ambiental en el delta superior del río Orinoco, Venezuela. Revista Científica UDO Agrícola 12(2):456–462. https://n9.cl/b46oj. Accessed 7 Feb 2025

Tezara W, Pereyra G, Ávila-Lovera E, Herrera A (2020) Variability in physiological responses of Venezuelan cacao to drought. Exp Agric 56(3):407–421. https://doi.org/10.1017/S0014479720000058

Torín CA, Monasterio PP, Marturet W (2018) Aplicación del control de calidad a series temporales de precipitación mensual de tres localidades del estado Yaracuy, Venezuela. Agron Trop 68(1–2):35–48. https://n9.cl/sgung. Accessed 7 Apr 2025

Uzcátegui-Varela JP, Chompre K, Castillo D, Rangel S, Briceño-Rangel A, Piña A (2022) Nutritional assessment of tropical pastures as a sustainability strategy in dual-purpose cattle ranching in the South of Lake Maracaibo, Venezuela. J Saudi Soc Agric Sci 21(7):432–439. https://doi.org/10.1016/j.jssas.2021.11.005

Veettil BK, Wang S, de Souza SF, Bremer UF, Simões JC (2017) Glacier monitoring and glacier-climate interactions in the tropical Andes: a review. J S Am Earth Sci 77:218–246. https://doi.org/10.1016/j.jsames.2017.04.009

Vilanova E, Ramírez-Angulo H, Torres-Lezama A, Aymard G, Gámez L, Durán C et al (2018) Environmental drivers of forest structure and stem turnover across Venezuelan tropical forests. PLoS One 13(6):e0198489. https://doi.org/10.1371/journal.pone.0198489

Villegas S, Ramírez RM, Camacho WR, Chamiquit CJ, Bazan ED, Campana ZG, Guerrero AP (2021) Detección de sequías meteorológicas en una región seco tropical de Venezuela, durante el período 1982–2015. Rev Climatol 21:157–166. https://n9.cl/2bx65. Accessed 7 Mar 2025

Viloria JA, Olivares BO, García P, Paredes-Trejo F, Rosales A (2023) Mapping projected variations of temperature and precipitation because of climate change in Venezuela. Hydrology 10(4):96. https://doi.org/10.3390/hydrology10040096

Vincenti-González MF, Tami A, Lizarazo EF, Grillet ME (2018) ENSO-driven climate variability promotes periodic major outbreaks of dengue in Venezuela. Sci Rep 8(1):1–11. https://doi.org/10.1038/s41598-018-24003-z

Vivas E, De Guenni LB, Allende-Cid H, Salas R (2023) Deep Lagged-Wavelet for monthly rainfall forecasting in a tropical region. Stoch Environ Res Risk Assess 37(3):831–848. https://doi.org/10.1007/s00477-022-02323-x

World Bank, Climate Change Knowledge Portal (2025). https://climateknowledgeportal.worldbank.org/. Accessed 5 Apr 2025

Zamora-Figueroa A, Malaver N, Hernández-Valencia I (2021) Diversidad functional microbiana y biogeoquímica en la planicie de inundación del río Mapire (estado Anzoátegui, Venezuela). Acta Biol Venezuelica 41(1):97–106. https://n9.cl/pftfo. Accessed 27 Feb 2025

Zhang Y, Fu B, Feng X, Pan N (2022) Response of ecohydrological variables to meteorological drought under climate change. Remote Sens 14(8):1920. https://doi.org/10.3390/rs14081920

Chapter 4
Impact of Climate Change in Venezuelan Agricultural Landscapes

4.1 Introduction

It is foreseen that the changes in the temperature and precipitation variables traced from climate change will increase the pressure on agricultural lands and the natural resources of the country's ecosystems. Such pressure will affect both mountain and lowland landacapes geographically distributed in the country. In a country where agricultural production depends mainly on rainfall, climate change becomes a phenomenon with severe consequences for rainfed agriculture. Trends and prognosis suggest that climate-related pressures will increase due to climate change, requiring changes in land use, production systems, local knowledge systems, managing mechanisms, and farmers' survival strategies. However, the perturbing threat of climate change places the country in an unfavorable condition, due to the limited research and production of information that has been generated to guide agricultural producers and agricultural public and private policies in facing climate change (García 2022).

This information breach constitutes a serious limitation when making decisions for the planning of agricultural production, for the design and implementation of mitigation and adaptation measures, and finally for the country not to be vulnerable in terms of its food security. The output of information will make it possible to decipher the different responses of the country's agricultural regions to climate change, as a result of the variability of relief, soils, climate, and the different agricultural practices applied by producers, among other factors. In the last 50 years, an increase in the altitudinal floor has been observed at the upper limit of high mountain agriculture. The impacts of climate change in the form of drought or more severe rainfall, the expansion of the agricultural frontier toward higher elevations, and the intensification of agriculture would generate production systems that intensify soil erosion, which is already a manifest threat to Andean communities.

A. Rosales Hernández et al., *Agriculture Under Climate Change*, The Latin American Studies Book Series, https://doi.org/10.1007/978-3-032-06368-7_4

Projections from various regional studies in Latin America and Venezuela indicate that the country's agriculture is sensitive to climate change. This sensitivity has triggered soil erosion, glacier retreat, and loss of vegetation cover. It has also increased the intensity of rainfall and disrupted crop production. These impacts affect both highland crops (such as potatoes, coffee, and vegetables) and lowland crops (including corn, rice, sugarcane, cassava, legumes, fruit trees, vegetables, and pastures). In particular, the increase in local temperature, the potential for evapotranspiration, and the scarcity of water would cause the loss of important crops such as vegetables. Climate change will affect future access to drinking water, as well as the supply for irrigated agriculture in mountainous areas (Andean and Coastal) and lowland areas (Vuille et al. 2008; Peralvo et al. 2012; Postigo et al. 2012). In the Venezuelan Andes, a region of major agricultural importance (especially for coffee, vegetables, and fruit trees), projections suggest a rise in temperatures. This warming could push the upper limit of cultivable land to higher altitudes, opening new areas for agriculture. However, it would also increase the demand for irrigation water. At the same time, it would threaten fragile ecosystems such as moorlands (Páramos), cloud forests, and water sources in the upper watersheds (García 2022).

It is also assessed, as an effect of the predicted higher temperatures and deficits of water, changes in planting and harvesting dates, alterations in crop cycles, and incidence of pests and diseases (Paredes 2014; Paredes and Chacón-Moreno 2021). A considerable expansion of areas with a negative annual water balance is projected, which would lead to an increase in arid and semi-arid areas and the loss of land agricultural suitability for a diversity of crops. The projections for the RCP 4.5 and RCP 8.5 scenarios allowed us to verify the reduction of the area occupied by humid climates and a spatial increase occupied by dry climates, attributed to changes in the rainfall. Water deficits will alter the quality of natural pastures and increase the vulnerability of livestock farming. Without a doubt, climate change will transform the geography and suitability of Venezuela's agricultural lands (Medina et al. 2016).

The essential intention of this chapter is to reveal the impacts of climate change on Venezuelan agriculture through a comprehensive picture, based on the review of the documents produced through research carried out in the country and the Latin American region. The chapter will expose the main projected impacts on the Venezuelan agricultural sector and their repercussions on the different agricultural production systems existing in Venezuela and on food security. For its geographical understanding, a general approach of regionalization of the country has been adopted at two levels or scales, understood as two visions and degrees of detail of the territory.

In this chapter, a first level is developed, on a very general scale: (1) mountainous areas and mountain ranges, particularly the Andes and the Coastal Range, stand out for their unique agricultural suitability. These regions offer optimal conditions for specific crops or crop groups such as coffee, fruit trees, and vegetables. Their high agroecological potential makes them especially valuable for diversified and sustainable agricultural practices, and (2) the lowlands are characterized by predominantly flat or gently undulating topography. These areas include large alluvial plains (Llanos), high plateaus, such as the Plio-Pleistocene Mesa Formation and the eastern high plains, as well as tectonic depressions like the Unare Depression. They also encompass valleys and broad peneplains south of the Orinoco River, especially in the northern part of Bolívar State, within the Guayana or Amazonian region. These landscapes offer important agricultural lands, often found in slightly hilly terrains. In addition, the impacts on forestry, livestock, and fisheries production are well thought out.

Table 4.1 and Fig. 4.1 illustrate the main geostructural regions comprising Venezuela, where three large geomorphological domains can be visualized:

Table 4.1 Geostructural arrangement and great regions of Venezuela, according to Zinck et al. (2016)

Geostructure	Code	Geomorphological landscapes	Region
Mountain ranges/ Cordilleras	1	Mountains, intramontane valleys, piedmonts, and high plains	Andean and Coastal
Geosynclinal (Sedimentary basin)	2	Alluvial plains, valleys, and high plains or plateaus	Western plains
			Central plains
			Eastern plains
			Orinoco Delta
Guayana Shield	3	Peneplains, high plains	Guayana

Fig. 4.1 Map of geostructural arrangement and great regions of Venezuela: 1 = Mountain ranges/ Cordilleras; 2 = Geosynclinal (Sedimentary basin); and 3 = Guayana Shield, according to Zinck et al. (2016)

Cordilleras (1), Geosynclines or depositional environments (landscapes of alluvial plains, high plains) (2), and the Guayana Shield (3), where occur the extensive peneplains located north of the state of Bolívar. These lands represent the territories where groups of crops such as coffee, vegetables, cereals, oilseeds, grains, sugarcane, cocoa, coconut, pastures (natural and artificial), and fruit trees are cultivated, mainly under rainfed agriculture, where cattle ranching has also been developed, especially in the vast region of the plains (Central, Western, and Eastern alluvial plains). Areas dedicated to forest production (plantations and primary forests in Forest Reserves) are also included.

The spatial arrangement for regionalizing impacts makes sense because these vary across the national geography. They differ according to the various physiographic regions. Consequently, the effects on the responses of cultivated crops also vary by region. Therefore, mitigation and adaptation measures must be developed and implemented specifically for the different regional agricultural production systems.

4.2 Climate Change and Current Accessible Specialized and Scientific Information in Venezuela

Most of the national information presented in this chapter comes from research conducted in the regions described in Sect. 4.1. These areas represent the country's traditional agricultural zones. They are home to several universities, agricultural research centers, institutions for animal husbandry and veterinary medicine, and regional offices of the National Institute of Agricultural Research. However, it is important to note that both research activity and the production of scientific publications have declined significantly in recent years in Venezuela.

The research papers reported in this chapter are published in specialized journals and documents produced by official institutions and international organizations. Some articles show the results of regional studies and research carried out by international institutions such as the Inter-American Development Bank (IDB), Comisión Económica para América Latina y el Caribe (ECLAC), FAO, Consorcio para el Desarrollo Sostenible de la Ecorregión Andina (CONDESAN), among others. Added articles are derived from studies carried out in Venezuela by different universities and national research centers.

The period of publication and the number of publications reviewed and incorporated into each section of this chapter are displayed in Table 4.2. It is important to note that some of the articles reviewed may be mentioned in different sections, and for this reason, the total number of publications reviewed is not totaled in Table 4.2. According to the references cited, the total number of publications reviewed in this chapter is 104, with publications dated from 2004 to 2025. As shown, there is a significant imbalance in the number of articles collected for each section of this chapter. This reflects the existing information gaps in the country. It also highlights

Table 4.2 Summary of publications reviewed in sections of this chapter

Section	No. publications
4.1 Introduction to the topic	7
4.3 Impacts of climate change on the suitability of agricultural lands and specific crops	6
4.3.1 Impacts on mountainous regions and high-altitude agriculture	20
4.3.2 Impacts on lowland agriculture areas	27
4.3.3 Expansion of arid and semi-arid areas in Venezuela as a result of climate change	5
4.4 Impacts on forest production	13
4.5 Impacts on livestock production	12
4.6 Impacts on fisheries	22
4.7 Conservation of watersheds and their impact on agriculture in mountainous and lowlanda areas	5
4.8 Incidence of pests and diseases driven by climate change	1
4.8.1 In mountainous agricultural areas	2
4.8.2 In lowland agricultural areas	8

the urgent need for more research and data to reduce the uncertainty Venezuela faces regarding climate change impacts on agriculture and related issues. Additionally, it points to the need for better information to support the design and implementation of effective mitigation and adaptation measures. These efforts must be integrated into public policies for both the agricultural and environmental sectors.

As can be seen, there is a significant disproportion in the number of articles by sections that could be collected for the preparation of this chapter, which is indicating the information gaps that exist in the country and the demands for research and information to decipher and reduce the uncertainty that the country faces to address climate change in terms of its impacts on agriculture and associated issues, as well as for the design and implementation of mitigation and adaptation measures, which must be reflected in public policies for the agricultural and the environmental sector.

4.3 Impacts of Climate Change on the Suitability of Agricultural Lands and Specific Crops

An overview of Venezuela's current climate was developed in Chap. 3 (Sect. 3.2). This information allows us to compare with the projected changes in the temperature and precipitation variables, whose variations will have a direct impact on Venezuelan agriculture regardless of the geomorphological context (geostructural domains) considered in this chapter (Fig. 4.1).

Table 4.3 summarizes the current conditions of the temperature and precipitation variables, taking as a spatial reference the geostructures and regions included in each of them, as illustrated in table and Fig. 4.1. The highest temperature values can

Table 4.3 Current average temperature and precipitation in the geostructures of Venezuela according to the Climate Change Knowledge Portal (2025)

Geostructure	Region	Current climate	
		Mean temperature (°C)	Precipitation (mm)
Mountain ranges	Andean and Coastal	19.1–25.5	1.133–1.539
Geosynclinal (Sedimentary basin)	Western plains	26.9–28.1	1.539–1.753
	Central plains	24.6–27.7	900–1.400
	Eastern plains	27.2–27.4	1.171–1.425
	Orinoco Delta	27.1	1.700
Guayana Shield	Guayana	24.4–24.5	2.400–3.048

(a) (b)

Fig. 4.2 (**a**) Andean mountain landscape; (**b**) croplands on mountain terrace. (Pictures by Pedro García) (Mérida State, Venezuela, 2014)

be observed in the western and eastern plains, while the highest precipitation values occur in the Orinoco Delta and the Guayana shield, in the south of the country. Geomorphological landscapes, topographic conditions, land use, and vegetation cover have a significant impact on climatic and hydrological conditions within the different regions.

It is expected that changes in temperature and precipitation variables will affect agricultural lands and the resources of the Andean and Coastal range ecosystems and the lowlands of Venezuela. The country has a high to moderate dependence on family farming, predominantly located in the Andean Mountain range (Fig. 4.2a, b). Crops such as potatoes, coffee, and vegetables are mainly grown in this landscape, which is potentially vulnerable to climate change due to the particularity of its geological, relief, edaphic, ecological, and socioeconomic conditions (García 2022).

According to Vuille et al. (2008), model projections of future climate change in the tropical Andes indicate prolonged warming throughout the twenty-first century, with temperature increase being highlighted in higher elevations. An increase in rainfall during the wet season and a decrease during the dry season are projected, which would effectively reinforce the seasonal hydrological cycle in the tropical Andes. Venezuela is a country with a Caribbean facade, but also with a significant

Andean landscape throughout a good part of the country. The narrative of impacts of climate change in the Venezuelan Andean region in this chapter is a result of studies mainly carried out in other countries of the region, particularly about the agricultural sector. Unlike precipitation, in which case the uncertainty is so wide that the evidence on rising temperature and evapotranspiration suggests a negative effect on water availability. This effect is distributed throughout the year and affects the rainy and dry seasons equally (Postigo et al. 2012).

Zhang and Cai (2013) evaluated the change in crop water deficit of 26 crops (including rainfed and irrigated crops) in the current climate (1961–1990) and in the projected climate (2070–2099). They found that, despite the widespread increase in average temperature, crop water deficit is likely to decrease slightly globally, although changes vary across regions. While increased rainfall and the change in its intra-annual distribution in many areas may lead to more effective rainfall for crop growth, the decrease in daytime thermal amplitude will play a key role in offsetting the effect of warming on a global scale. These authors consider that, in South America, Venezuela could have greater irrigation needs to maintain agricultural production. These authors believe that it is necessary to address the uncertainty in the effects of climate change through downscaling processes and/or the improvement of Global Climate Models (GCMs). In the perspective of Altieri et al. (2015), farmers living in harsh environments in Latin America have developed and/or inherited complex agricultural systems that are ingeniously managed to continuously deal with extreme weather events and climate variability over the centuries. These systems have enabled small farming families to meet their subsistence needs despite environmental variability, without count on modern agricultural technologies. The continued existence of millions of hectares dedicated to traditional agriculture is living proof of the success of the indigenous agricultural strategy, which is a testimony to the "creativity" of smallholder farmers throughout the developing world. One manifestation of this creativity is the thousands of hectares of raised bed cropping systems, seasonally flooded lands, savannahs, and upland basins, as is the case in Venezuela.

Figure 4.3a illustrates the projected anomaly in annual precipitation across Venezuela for the period 2080–2099, compared to the reference baseline of 1995–2014, under the SSP5-8.5 high-emissions scenario. A generalized decline in precipitation is evident throughout much of the national territory, with particularly severe reductions in the western and southern regions, where annual anomalies exceed −600 mm in some areas. This projection points toward a drier future, with significant implications for water availability, especially in climate-sensitive agricultural zones such as the central Llanos, the Andean piedmont, and the Guayana Shield.

The projected intensification of water deficits could exacerbate soil degradation, reduce the yields of rainfed crops, and compromise water supply for both irrigation and human consumption. The most affected areas overlap with biodiversity hotspots and indigenous territories in the southern region, as well as with major agricultural zones in the west (Zulia, Trujillo, and Lara). These projections underscore the urgency of implementing adaptation strategies centered on integrated water resource

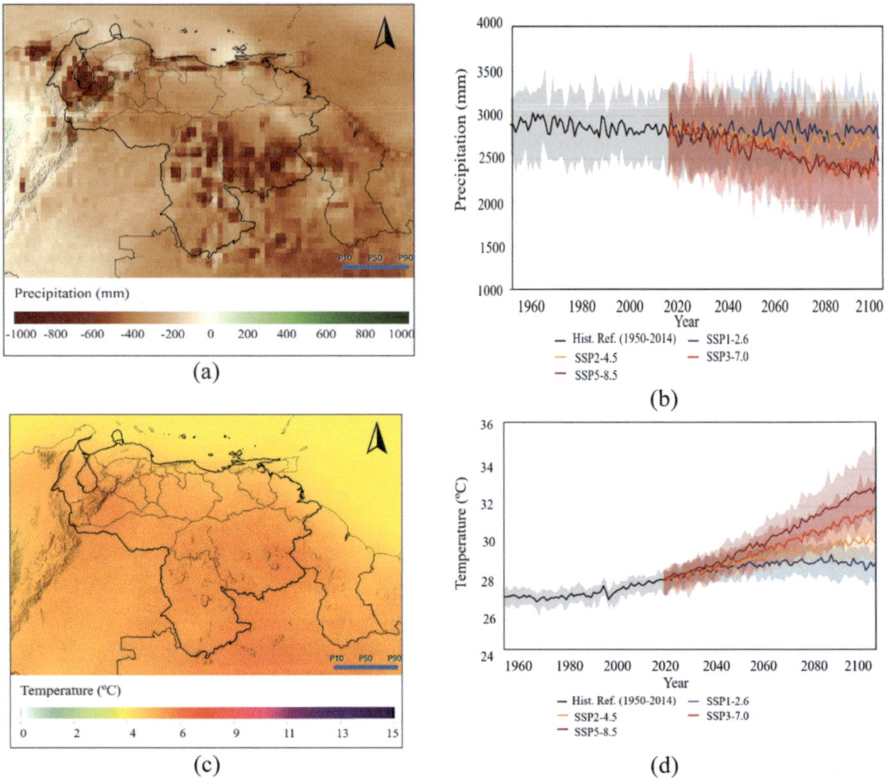

Fig. 4.3 (**a**) Projected annual precipitation anomaly for Venezuela (2080–2099), relative to the reference period 1995–2014, under the high-emissions scenario SSP5-8.5 (multi-model ensemble); (**b**) projected annual precipitation for Mérida, Venezuela (1950–2100), using multi-model ensemble simulations under four SSP scenarios, relative to the historical reference period (1995–2014); (**c**) projected average annual surface air temperature anomaly for Venezuela (2080–2099), relative to the reference period 1995–2014, under the SSP5-8.5 scenario (multi-model ensemble); (**d**) projected average annual surface air temperature in the lowland Cojedes state, Venezuela (1950–2100), based on multi-model ensemble simulations under four SSP scenarios, relative to the 1995–2014 reference period, according to the data source from Eyring et al. (2016) and available in World Bank, Climate Change Knowledge Portal (2025)

management and enhancing local capacities to address the regionally differentiated impacts of climate change on Venezuela's agriculture and ecosystems (García 2022).

Figure 4.3b presents the projected evolution of annual precipitation in Mérida state, located in the Venezuelan Andes, from 1950 to 2100, based on a multi-model ensemble under four Shared Socioeconomic Pathways (SSP1-2.6, SSP2-4.5, SSP3-7.0, and SSP5-8.5). The historical reference period (1950–2014) shows relatively stable precipitation values around 2800–3000 mm/year. However, from mid-century onward, all emission scenarios project a declining trend in annual precipitation, with the sharpest reductions under the SSP5-8.5 and SSP3-7.0 pathways. These high-emission scenarios suggest a potential drop below 2000 mm/year by the end of the century, signaling significant hydrological stress for this highland region.

The uncertainty ranges represented by the shaded envelopes highlight growing variability over time, especially under scenarios SSP5-8.5 and SSP3-7.0. Under low-emission conditions (SSP1-2.6), the annual precipitation stabilizes closer to the historical baseline, indicating the potential for mitigation strategies to preserve climatic stability in this key agricultural zone. Given Mérida's reliance on rainfed systems and its role in national food production, the projected declines in rainfall pose serious risks to water availability, crop yields, and the sustainability of rural livelihoods.

Figure 4.3c illustrates the projected average annual surface air temperature anomaly across Venezuela for the period 2080–2099, under the high-emission scenario SSP5-8.5, relative to the 1995–2014 baseline. The country is expected to experience widespread warming, with anomalies ranging between +4 °C and +6 °C throughout most regions. The spatial distribution indicates relatively homogeneous temperature increases, including in both lowland and highland areas, posing challenges for diverse agroecological zones.

Such elevated warming levels are likely to significantly affect agricultural productivity, especially in rainfed systems and regions already experiencing climatic stress. Increased temperatures will lead to higher evapotranspiration rates, reduced water availability, and shifts in crop development cycles, exacerbating the vulnerability of essential crops such as maize, beans, cassava, and vegetables.

Figure 4.3d displays the projected trajectory of average annual surface air temperature in Cojedes state (Central plains). The historical baseline indicates relatively stable temperatures around 27–28 °C. However, all future scenarios project a marked increase in temperature from mid-century onwards, particularly under high-emission pathways. Under SSP5-8.5, average temperatures could exceed 34 °C by the end of the century, reflecting an increase of more than 6 °C above historical levels.

Such warming poses critical challenges for agriculture in Cojedes state, a lowland region with extensive cropping systems dependent on optimal thermal ranges for yield permanency. Rising temperatures are expected to increase evapotranspiration rates, exacerbate soil moisture deficits, and reduce the viability of heat-sensitive crops. Furthermore, the extreme heat projected under SSP3-7.0 and SSP5-8.5 could compromise livestock productivity, the physiology of animal species, and increase the incidence of crop pests and diseases.

4.3.1 Impacts on Mountainous Regions and High-Altitude Agriculture

The Venezuelan Andes represent the geographic space where intensive agriculture has been developed in small farms with a predominance, in many cases, of family labor. These cultivation areas have been distributed occupying a diversity of geoforms, mountain slopes (Fig. 4.4a), mountain summits (Fig. 4.4b), and intramontane valleys (Fig. 4.4c, d), among others. By the end of the twenty-first century, according to the SRES A2 emissions scenario, the tropical Andes could experience massive warming in the order of 4.5–5.0 °C. An increase in rainfall during the wet season and a decrease during the dry season is projected, which would effectively affect the seasonal hydrological cycle (Vuille et al. 2008). In a study conducted by Pérez et al. (2010), it is concluded that climate change represents one of the current threats to sustainable agriculture in the Andes. Tendency and forecast suggest that climate-related pressures will increase due to climate change, requiring changes in land use, production systems, local knowledge systems, managing mechanisms, and livelihood strategies.

In the last 50 years, an increase in the altitudinal floor has been observed at the upper limit of high mountain agriculture (Fig. 4.5). Several studies predict that this trend will continue throughout the century, with a general increase of at least 500 m. The risk for Andean communities, in the short term, is not so much the reduction of water availability in general, rather, a change in the seasonal distribution and regularity of water supply together with an increase in torrential rainfall during the rainy season and a decrease in minimum flows during the dry season. All of these affect the quality and availability of water for domestic consumption and agriculture.

(a) (b) (c) (d)

Fig. 4.4 (**a**) Cropland area on step mountain landscape, (**b**) croplands on mountain summits, (**c**, **d**) agricultural areas on terraces and alluvial plain in intramontane Andean valley. (Pictures by Pedro García) (Mérida state, Venezuela, 2014)

(c)

Fig. 4.5 (**a**) Altitudinal expansion of agricultural lands in the Andean mountains; (**b**) vegetable cultivation in the Páramo ecosystem (>3000 m asl); (**c**) croplands with high vulnerability to water erosion. (Pictures by Pedro García) (Mérida state, Venezuela, 2014)

Climate change in the Andean region is projected to intensify soil erosion due to more severe rainfall in the wet season, agricultural expansion into higher elevations, and intensified land use, increasing the vulnerability of already fragile production systems. Peralvo et al. (2012), identify the Tropical Andes as a highly sensitive region, particularly in high-altitude zones where climate models predict stronger warming and greater environmental exposure. These climate changes are expected to interact with land-use changes, threatening agricultural productivity, social systems, and ecosystem services. However, Buytaert and Ramírez-Villegas (2012) note high uncertainty in regional climate models, especially regarding precipitation, due to complex topography and mountain climate processes.

In the Venezuelan Andes, projections suggest a temperature increase of 1–3 °C and a possible 10% rise in precipitation, though overall trends may still lead to more drought and warming. Melting of the remaining glacial cirques, reduced by over 95% since 1952, has already disrupted the hydrological cycle, with future water shortages expected during the dry season, affecting irrigation and domestic supply. According to Vuille (2013), páramo ecosystems, crucial for regulating water flow to lowland areas, will be severely impacted by rising temperatures. These changes will reduce soil moisture and water retention, alter decomposition dynamics, and limit the availability of water for high-altitude crops such as potatoes and carrots, particularly during critical dry periods.

Figure 4.6a presents the projected monthly anomalies in average surface air temperature for the state of Trujillo, located in the Venezuelan Andes, under four shared socioeconomic pathways (SSP1-2.6, SSP2-4.5, SSP3-7.0, and SSP5-8.5). Under the high-emission SSP5-8.5 scenario, the median anomaly exceeds +5 °C throughout the year, reaching peaks above +6 °C during September and October. In contrast, under the low-emission SSP1-2.6 scenario, anomalies remain below +1.5 °C, with relatively stable values across months. The scenarios SSP2-4.5 and SSP3-7.0 depict intermediate warming trajectories, with median values of approximately +2.5 °C and +4 °C, respectively.

The marked increase in temperatures, especially under the SSP5-8.5 and SSP3-7.0 scenarios, could have profound impacts on high-altitude agriculture in Trujillo state. Warmer conditions may alter phenological cycles, reduce the thermal suitability for crops like potatoes and coffee, and increase heat stress and pest incidence. The seasonal pattern of warming, with pronounced peaks in the late dry and early rainy seasons, may also disrupt planting and harvesting schedules.

Figure 4.6b shows a heatmap of projected monthly temperature anomalies for Trujillo state, Venezuela, from 1951 through 2100, using the SSP5-8.5 high-emissions scenario. The color scale reflects the progressive warming over time, with early decades (1951–2000) remaining close to or below the baseline reference period. However, from the mid-twenty-first century onward, anomalies increase steadily, surpassing +2 °C by 2040 and reaching over +5 °C in the final decades (2080–2100). This warming is consistently projected across all months, indicating a year-round intensification of thermal stress.

The spatial uniformity of the warming trend throughout the annual cycle underscores the systemic nature of climate change in the Andean region of Trujillo. Months such as September and October exhibit the highest projected anomalies by the end of the century, which may coincide with critical periods for crop growth and harvest. Such shifts threaten the viability of traditional high-altitude agriculture, particularly for temperature-sensitive crops like potatoes and coffee. This long-term projection reinforces the urgency of transitioning to climate-resilient agricultural systems and integrating temperature risk into land-use and food security planning for mountain communities.

Figure 4.6c displays the projected monthly precipitation anomalies for Táchira state for the period 2080–2099 under four emissions scenarios. Under SSP5-8.5 and SSP3-7.0, the state is projected to experience consistent negative anomalies

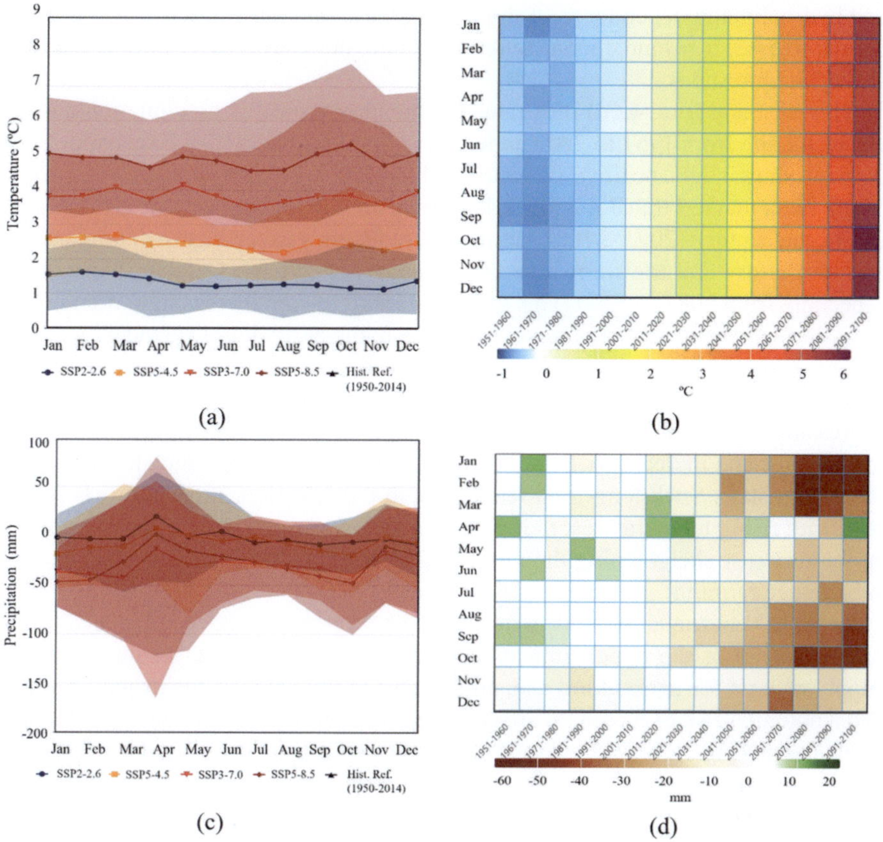

Fig. 4.6 (**a**) Projected average monthly surface air temperature anomaly in Trujillo state, Venezuela for the period 2080–2099, relative to the 1950–2014 baseline, under four SSP scenarios (multi-model ensemble); (**b**) projected monthly average surface air temperature anomaly in Trujillo, Venezuela, from 1951 to 2100, relative to the 1995–2014 reference period, under the SSP5-8.5 scenario (multi-model ensemble); (**c**) projected average monthly precipitation anomaly for Táchira state, Venezuela (2080–2099), relative to the 1950–2014 reference period, under four SSP scenarios (multi-model ensemble); (**d**) projected monthly precipitation anomaly in Táchira state, Venezuela (1951–2100), relative to the 1995–2014 reference period, under the SSP5-8.5 scenario (multi-model ensemble), according to the data source from Eyring et al. (2016) and available in World Bank, Climate Change Knowledge Portal (2025)

throughout the year, with median deficits ranging from −20 mm to −60 mm, especially during the dry season (January–March) and transitional months like October. The SSP1-2.6 and SSP2-4.5 scenarios indicate more stable or near-zero anomalies, reflecting the potential benefits of emission mitigation pathways. The negative precipitation anomalies in the high-emission scenarios suggest heightened drought risk and reduced hydrological input to mountain watersheds. These effects may compromise water security and agricultural productivity, particularly for crops dependent on seasonal rainfall patterns such as maize, beans, and plantains. Given the terrain and hydroclimatic dependence of the region, these findings reinforce the necessity of adaptive water resource management and climate-resilient agricultural practices in the Táchira Andes.

This heatmap (Fig. 4.6d) illustrates the monthly projected precipitation anomalies in Táchira from 1951 to 2100 under the high-emission SSP5-8.5 scenario. Early decades show slight positive or neutral anomalies in several months (e.g., April, May), but a marked transition toward negative anomalies begins after 2030. By the end of the century (2081–2100), significant reductions exceeding −40 mm are evident in the dry and transitional seasons (January, February, October, November), especially from 2081 to 2090. This indicates a shift toward increasingly drier conditions during months that are already sensitive in the agricultural calendar.

The observed pattern suggests an intensification of the dry season and reduced rainfall during key agricultural periods, potentially shortening growing seasons and decreasing soil moisture reserves. These changes may severely affect rainfed farming systems and increase irrigation demand in highland valleys and sloped terrains of Táchira state. Such long-term reductions in precipitation, especially when coupled with rising temperatures, may reduce water availability for both domestic and agricultural use, heightening the vulnerability of rural communities dependent on subsistence farming.

In some Venezuelan Andean zones, warmer temperatures have led to an altitudinal expansion of the upper limit of potentially cultivated land, leading to a new supply of land for agricultural use, an increase in the demand for irrigation water, and a potential risk of increased erosion processes on step land under cultivation on steep slopes (Fig. 4.5c). As a result of the projected higher temperatures, planting and harvest dates could be earlier and crop cycles shorter, but overall production would be lower, particularly for certain important crops such as potatoes (*Solanum tuberosum* L.).

Postigo et al. (2012) appraised in their study that the areas with the greatest loss of climatic suitability for 25 selected crops are in the piedmont areas of the Andean countries. The areas with the greatest gain in crop suitability are located in the highland zones of the Andes, which suggests an altitudinal shift of some crops, the loss of aptitude for highland crops, and an expansion of lowland crops. According to the results, the potato (*Solanum tuberosum* L.) would experience a significant loss of suitability only in Venezuela (about 5%).

According to Ortiz (2012), a study carried out by the Inter-American Development Bank (IADB) concluded that the increase in projected temperatures (especially when increases exceed 3 °C) will considerably affect agricultural productivity, producer incomes, and food security, increasing vulnerability in the geography of Latin America and the Caribbean. Average food availability per capita could decrease by at least 300 calories (a 12% reduction) by 2050. Projections of changes in climatic variables suggest that in the tropical Andes region, agricultural productivity could fall by between 12% and 50%. Maize yields in the Venezuelan piedmont could be reduced to levels close to zero. The potato will also be vulnerable to heat that affects the development of the plant and tuber, which will cause a drop in yields.

Cuesta et al. (2012) admit that there are clear trends in climate variations. For precipitation, a change of less than 10% is expected over the Andes. The strongest impact is in the Caribbean region, under the influence of the trade winds coming from the Atlantic. That region expects a dramatic decrease in precipitation to less than 40%, which will affect the Andes of Venezuela around the city of Mérida. It is not possible to observe clear relationships with topographic indices such as height, slope, or orientation. The modeled impacts on precipitation (in total and temporal terms) are spatially differentiated, while the impact of temperature, in spatial terms, is relatively homogeneous and causes the increase in evapotranspiration to be greater in hot areas than in cold areas.

De Bièvre et al. (2012), contrary to other authors, estimate that in most of the high tropical Andean zone, very little change in water availability will be perceived, because two phenomena are counteracted, (1) the increase in precipitation of up to 10% and (2) a similar increase in current evapotranspiration due to the increase in temperature. Therefore, the water requirements for crop production will increase, in turn increasing the demand for water for irrigation.

Postigo et al. (2012) established a loss of suitability for high-altitude crops represented by the potato (*Solanum tuberosum* L.); in Venezuela alone, this loss represents approximately 5%. In the case of Venezuela, 27.0% of the population below the poverty line lives in areas that lose climatic suitability for crops such as creole celery (*Arracacia xanthorriza* Bancr.) and sweet potato (*Ipomoea batatas* L.). These patterns of future change in the climatic suitability of high-altitude crops are important for the food security of rural populations in the Andes, whose livelihoods depend heavily on these crops.

The results of Vuille (2013) showed that future changes in the amount of precipitation and its seasonality could increase the variability of flows and their possibilities of agricultural use for other high-floor crops (potatoes, carrots, etc.). In some Andean areas, warmer temperatures have led to an altitudinal expansion of the upper limit of potentially cultivated land, leading to a new supply of land for agricultural use, an increase in the demand for irrigation water, and a potential risk of increased erosion processes on land under crops. As a result of the projected higher temperatures, planting and harvesting dates could be earlier and growing cycles shorter, but overall production would be lower, particularly for certain important crops such as potatoes.

Santaella (2013) analyzed vegetation dynamics under climate change scenarios in the cloud forest–páramo ecotone of the Sierra Nevada de Mérida, concluding that the páramo is projected to shrink significantly or even face extinction, gradually being replaced by preparamo shrubland and later by preparamo forest and cloud forest. The models indicate an upward shift of the timberline, with colonization and succession processes altering plant structure over time, even under scenarios assuming constant greenhouse gas emissions from the year 2000.

Paredes (2014) analyzes climate variability in Venezuela, noting an increase in minimum temperatures of 0.37 °C per decade and a decline in precipitation between 1940 and 2002, with significant implications for highland agriculture in the Andean region. Rising temperatures and altered rainfall patterns are expected to shift the distribution of key crops, such as shaded coffee, which will likely move to higher altitudes, potentially conflicting with protected areas. Climate models suggest that coffee will remain viable in mid-altitude zones like the Mocotíes Valley and Pueblos del Sur, while potatoes and carrots may shift to elevations above 2000 and even 3500 m, respectively. A temperature increase of 1.5 °C could reduce suitable areas for potato cultivation, though a slight rainfall decrease might expand potential zones. In contrast, carrot cultivation could face near-total loss of suitable areas by 2100 if greenhouse gas emissions continue, with projected warming of 6.3 °C and a 28.5% drop in annual precipitation severely impacting viability.

According to the IPCC (2018), global warming is likely to reach 1.5 °C between 2030 and 2050 if it continues to increase at the current rate, and climate-related risks to natural and human systems will be greater with global warming of 1.5 °C than currently exist. Climate models predict significant differences in regional climate characteristics between now and when global warming of 1.5 °C occurs, and between global warming of 1.5 °C and 2 °C. The number of warm days is projected to increase in most terrestrial regions, and the largest increases will occur in the tropics.

Paredes and Chacón-Moreno (2021) consider that present-day spatial models have a high climate potential to project the distribution of potato and shade coffee crops in climate change scenarios. Coffee would decrease the suitable area, while potato cultivation would lose 69% of the potential area by 2100, in scenarios of continuous emissions. In the case of coffee, it is observed that the current suitable areas change indicating that any alteration in precipitation will bring with it a reduction in the areas with current climatic aptitude, but these will probably ascend in the altitudinal gradient, which is why an increase in new potential areas is observed in any climate change scenario. But any change in the lands suitable for both coffee and potatoes would expand the degree of intervention and fragmentation of natural ecosystems. In banana cultivation, any change in precipitation due to climate change would affect the distribution of potential areas, until the disappearance of current areas, which is an indication of being a crop sensitive to any climate change.

Romero and Millano (2021) found that temperatures in the coastal range bioregion show a tendency to increase, both in maximum and minimum temperatures, but not in a decisive way and in different degrees of magnitude. However, locally there are series of times with decreasing trends as in the case of Colonia Tovar area, where maximum temperatures show a slight decrease, which could be associated with local factors that would influence temperatures to a greater degree. This finding is relevant because of the effects that these temperature increases could have on the crops that are planted on the slopes and intramontane valleys of the Coastal Range, such as coffee, some fruit trees, and vegetables.

Ortiz (2012) argues that the increase in estimated temperatures (>3.0 °C) will considerably affect agricultural productivity, producer incomes, and food security, increasing vulnerability in the geography of Latin America and the Caribbean. It is estimated that, in South America, the values of establishments specialized in agricultural production and those dedicated to both agricultural and livestock production (mixed system) will fall by 19.0% and 34.0% respectively. Although this decline will be smaller in the case of the mixed system (15%) by 2060 (considering a temperature increase of 3.0 °C and a decrease in rainfall of 10%), average food availability per capita could decrease by at least 300 calories (12% reduction) by 2050, as a result of climate change. Estimates of changes in climatic variables suggest that in the tropical Andes region, agricultural productivity could fall by between 12% and 50%. While maize yields may remain unchanged in Colombia, predictions indicate that they will fall to levels close to zero in the Venezuelan Piemont.

According to Hagen et al. (2022), in their review of articles on climate change produced since 2012, climate change in South America during the twenty-first century may increase the severity for the risks of (1) food insecurity; (2) floods and landslides; (3) water scarcity; (4) epidemics of vector-borne diseases; (5) change of biome of the Amazon rainforest, among others. Warming is expected to reach 2–3 °C under RCP 4.5 and 3–5 °C under RCP 8.5 by 2100, with stronger warming for the Amazon basin and central Andes. The food security of indigenous and rural communities dependent on the resources of the Amazon has been strongly affected due to the high dependence on fishing and small-scale agriculture, two sectors highly vulnerable to extreme droughts.

Llambí and Garcés (2021), highlight significant gaps in the understanding of the spatio-temporal dynamics of climate and its ecological and social impacts in the Andean region, including Venezuela. They emphasize the high vulnerability of mountain soils to erosion and organic matter loss due to warming and extreme events such as droughts, floods, and landslides. As water demand for agriculture, domestic use, and energy grows, especially in areas with projected rainfall decline, conflicts over water use are expected to intensify. While Andean populations have historically developed adaptive strategies, such as managing diverse ecological zones and water systems, there is a lack of detailed knowledge about high-altitude hydrological processes and the role of elevation in temperature increases. Moreover, the authors emphasize a deficiency in the communication of climate change impacts on ecosystems and argue that adaptation research in the Andes remains fragmented and outdated, lacking a comprehensive analysis of current knowledge needs. The authors conclude that there are gaps in the basic knowledge of the spatio-temporal variability of temperature and precipitation and of hydrological processes in high mountain environments, including key aspects such as the effect of elevation on the rate of temperature increase.

Núñez et al. (2018) emphasize that climate change impact assessments in Andean agriculture, particularly in the Department Norte de Santander, Colombia, bordering Venezuela, have largely focused on large-scale, technician crops, overlooking the fact that a significant portion of national food production stems from small-scale, polyculture systems on hillside soils, conditions similar to those in Venezuela. The authors highlight that family farming dominates Andean agriculture, yet remains underrepresented in climate assessments, despite its high vulnerability to changes in temperature and precipitation. These climatic shifts are expected to increase pressure on Andean ecosystems, with subsistence farmers, especially the poorest, being the most exposed to climate risks and least equipped to adapt. In the Venezuelan context, such projected impacts could exceed the adaptive capacity of local production systems, while uncertainty remains high regarding the magnitude and spatial distribution of future climate effects.

Parra et al. (2018), analyzing the history of data series of the monthly Standardized Precipitation Index (SPI), conclude that the drought events observed in the mid and late 1990s–2000s were the most severe in duration and intensity, and geographic extension that have been recorded within the Venezuelan Andes during the period 1980–2014.

In the opinion of Paredes and Chacón-Moreno (2021), a series of possible climate change scenarios for the Venezuelan Andes, based on integrated climate models, show that for a moderately optimistic scenario (RCP 4.5) we could have by the end of the century, temperatures with 2–3 °C above the average of the reference period (1986–2005), and a decrease in rainfall close to 0.1 mm day^{-1}. The state of Mérida has a high climatic potential for the spatial distribution of potato and shade coffee crops, although there are specific ideal areas for cocoa and banana crops. For banana and cocoa crops, its climatic suitability area is in the lowland agricultural zone known as Sur de Lago, while in the case of coffee cultivation the potential areas, depending on the ideal climatic conditions, are at intermediate altitudes, located in agricultural areas of the Mocotíes Valley and the Pueblos del Sur of the state. Also, for potato and carrot crops, the potential areas according to climatic conditions are located above 2000 m asl in the agricultural areas of the páramo.

Lozano-Povis et al. (2021) report that rising temperatures in the Andes are accelerating glacier retreat, disrupting the hydrological cycle and water availability for agriculture. This climate-driven decline in productivity has contributed to rural migration, reducing labor availability and threatening future harvests. The complex mosaic of agroclimatic zones, diverse production systems, and socioeconomic dependencies increases the vulnerability of Andean populations, especially in countries like Venezuela, to water scarcity and the loss of key crops such as rice. Furthermore, limited meteorological data in mountainous regions hampers the verification and refinement of climate projections.

Paredes and Chacón-Moreno (2021) found that, in Mérida state, projected climate change scenarios indicate a significant reduction in suitable areas for highland crops. By 2100, potato cultivation could lose up to 69% of its potential area, while shaded coffee would also see declines, with its distribution shifting toward higher elevations. Their spatial models suggest that while some areas may remain viable—such as the Mocotíes Valley and the Pueblos del Sur—any shift in climatic suitability would increase pressure on fragile ecosystems (intervention and fragmentation). These findings emphasize the dual challenge of agricultural adaptation and environmental conservation in Venezuela's high Andes.

4.3.2 Impacts on Lowland Agriculture Areas

This section describes the impacts of climate change on agriculture in lowlands, such as floodplains, tectonic depressions, valleys, the plateau or highplains of the Plio-Pleistocene Mesa Formation in the east of the country (Fig. 4.7), landscapes on

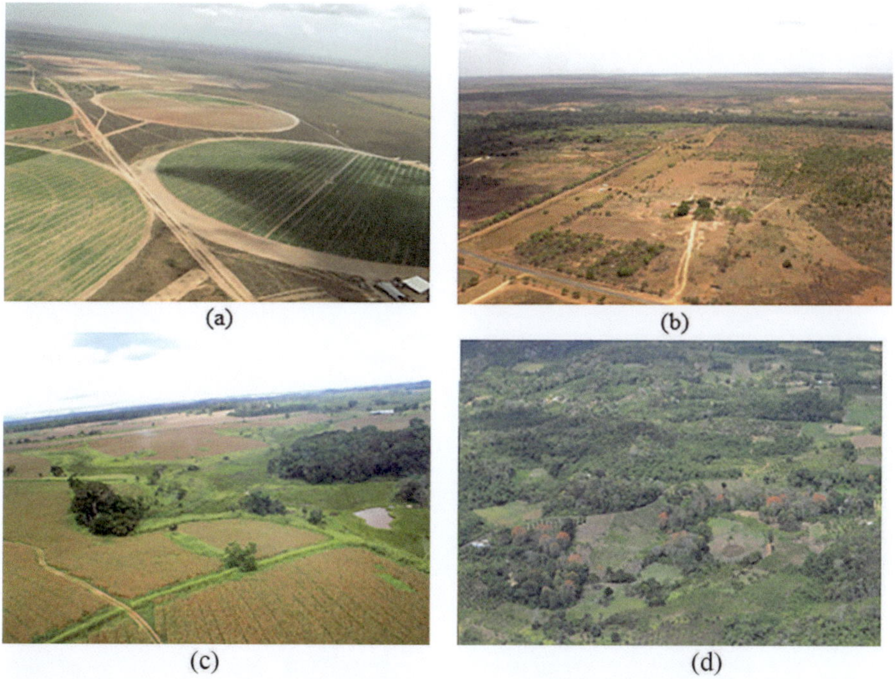

(a) (b) (c) (d)

Fig. 4.7 (**a**, **b**) Plateau or high plains, south of the Monagas and Anzoátegui states (Eastern plains); (**c**, **d**) Crop areas on peneplain landscape, north of Bolívar state (Guayana Shield). (Pictures by Pedro García) (Monagas, Anzoátegui and Bolívar states, Venezuela, 2013–2015)

tertiary geological formations and hilly environments of the Guayana Shield, among others. These agricultural landscapes, physiographically recognized as lowlands, correspond to the vast region of the Venezuelan Central, Western and Eastern plains (Venezuelan llanos), the peneplains north of the Bolivar state, and the depression of Lake Maracaibo.

According to the projections resulting from the application of the different models of global climate change, water will be one of the resources affected in several physiographic regions of the country and, consequently, negatively impacting agricultural activity in the lowlands (Viloria et al 2023). By the year 2055, in the opinion of López (2009), it is expected that many small producers in South America will suffer a drop in corn production, estimated at between 10% and 20%, on average. It is even suggested that there would be areas such as the piedmont in Venezuela, where the yield of corn would be reduced by up to 100%, forcing production to be transferred to other areas. It is estimated that, for rice, the reduction will be broad-based. Buytaert et al. (2011) estimated that the expected changes in rainfall are much more variable.

Figure 4.8a exemplifies projected monthly temperature anomalies for Portuguesa state during the period 2080–2099 under four emission scenarios. Under SSP5-8.5, the median anomaly exceeds +5 °C for much of the year, with October reaching a peak near +7 °C. The SSP3-7.0 scenario also shows a consistent warming trend,

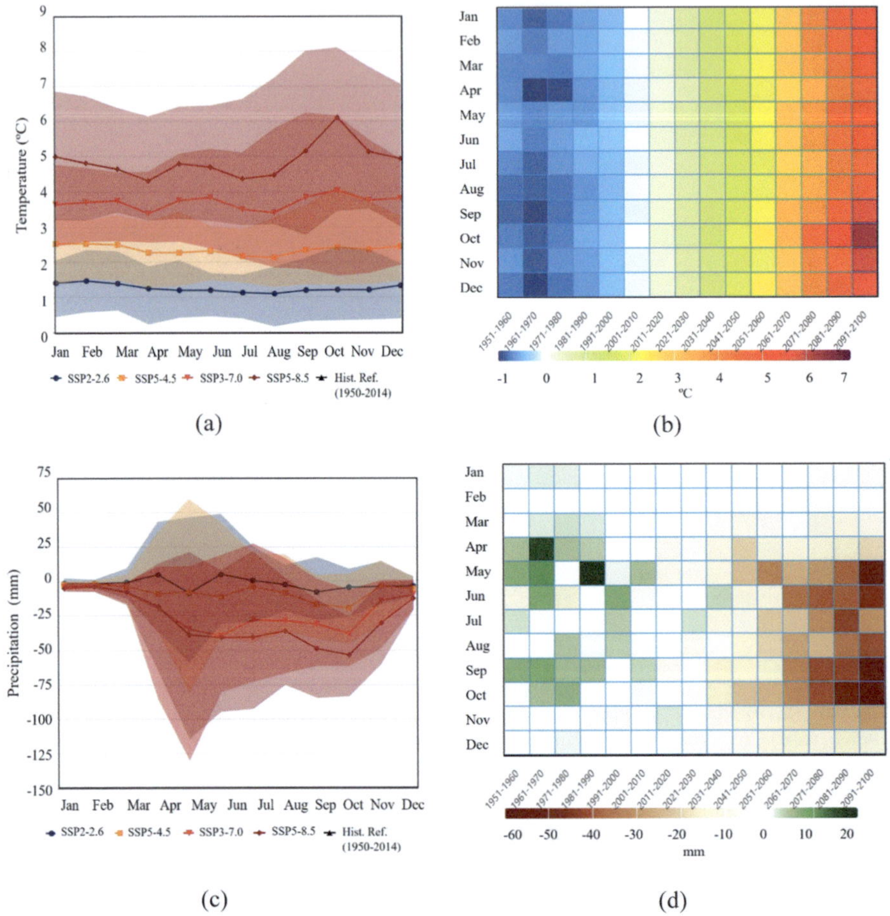

Fig. 4.8 (**a**) Projected average monthly surface air temperature anomaly for Portuguesa, Venezuela (2080–2099), relative to the 1950–2014 baseline, under four SSP scenarios (multi-model ensemble); (**b**) projected monthly surface air temperature anomaly in Portuguesa, Venezuela (1951–2100), relative to the 1995–2014 reference period, under the SSP5-8.5 scenario (multi-model ensemble); (**c**) projected average monthly precipitation anomaly for Guárico, Venezuela (2080–2099), relative to the 1950–2014 reference period, under four SSP scenarios (multi-model ensemble); (**d**) projected monthly precipitation anomaly in Guárico, Venezuela (1951–2100), relative to the 1995–2014 reference period, under the SSP5-8.5 scenario (multi-model ensemble), according to the data source from Eyring et al. (2016) and available in World Bank, Climate Change Knowledge Portal (2025)

with monthly anomalies between +3.5 °C and +5 °C. In contrast, the low-emission scenario SSP1-2.6 projects a more moderate increase of approximately +1.3 °C annually, demonstrating the potential benefits of mitigation.

The implications for agriculture in Portuguesa state are substantial, given that it is one of Venezuela's main cereal-producing regions. Higher temperatures could accelerate crop maturation, reduce yields, and increase water stress. The sharp rise in temperatures during critical periods (e.g., October and November) may also affect flowering and grain-filling stages for crops like maize and rice. These findings highlight the urgency of adopting heat-tolerant crop varieties, optimizing planting schedules, and reinforcing irrigation systems to sustain agricultural productivity in the face of future climate extremes.

This heatmap (Fig. 4.8b) shows the evolution of monthly temperature anomalies in Portuguesa over 15 decades, highlighting the steady progression of warming under the high-emissions SSP5-8.5 scenario. While early decades remain close to the reference average, from 2030 onwards, the warming trend intensifies across all months. By the end of the century (2081–2100), anomalies exceed +6 °C in several months, especially between September and December, indicating a pronounced late-year thermal stress.

This sustained rise in surface temperatures across the calendar year suggests long-term shifts in the agroclimatic regime of the region. Crops traditionally adapted to the tropical savanna climate may face reduced suitability, and increased evapotranspiration rates may exacerbate soil moisture deficits. Additionally, the overlap between peak warming months and critical crop development stages poses a compound risk to food security and farmer livelihoods. Adaptation planning should consider these spatial-temporal patterns when designing resilient agricultural and water management strategies.

Figure 4.8c depicts the projected monthly precipitation anomalies for the state of Guárico during the period 2080–2099 under four emission scenarios. Under the SSP5-8.5 and SSP3-7.0 pathways, significant precipitation deficits are projected between April and October, with monthly anomalies reaching up to (−75 mm) during the core of the rainy season (May–September). In contrast, the SSP1-2.6 scenario maintains near-zero anomalies, while SSP2-4.5 projects moderate but persistent reductions. The high variability bands in SSP5-8.5 accentuate growing uncertainty and the potential for extreme drought conditions. The timing of these projected rainfall reductions is especially concerning for Guárico, a central lowland state that plays a key role in Venezuela's rainfed cereal and legume production. Reduced rainfall during the planting and early growth stages may result in crop failures, reduced yields, and food insecurity.

This heatmap (Fig. 4.8d) shows the evolution of monthly precipitation anomalies for Guárico from 1951 to 2100, relative to the 1995–2014 baseline, under the high-emissions SSP5-8.5 scenario. During the historical period (1951–2000), the region exhibited periods of positive anomalies, particularly in April through September. However, beginning in the second half of the twenty-first century, anomalies shift increasingly negative. By the 2081–2100 period, most months, especially the rainy season, experience sharp declines of over more than 40 mm, with May and October showing deficits beyond 60 mm.

The spatial and temporal progression from surplus to deficit indicates a transformation in Guárico's hydrological regime. Rainy seasons are projected to become shorter and drier, challenging the sustainability of traditional agricultural systems. Water stress, compounded by high temperatures, may lead to increased competition for irrigation, deterioration of soil fertility, and overall reduction in agricultural resilience. These projections reinforce the necessity of regional climate adaptation policies, including water governance reforms and climate-smart agriculture.

For the north of Venezuela, where the trade winds from the Caribbean Sea predominate, a decrease in rainfall and its consequent impacts on the climatic suitability for some crops are appraised. However, the discrepancies between the different IPCC models are very high, often exceeding 50%. In the opinion of Samaniego et al. (2017), the estimated potential losses for the agricultural sector in Latin America and the Caribbean would be greater during the second half of the century, affecting poverty since climate change reduces productivity and agricultural output and, in this sense, reduces the stride at which poverty can be reduced. In a simulated scenario with a projected loss of 5.6% of Latin America's agricultural output by 2025, it is assessed that 6.9% of the total population would remain in poverty in relation to the baseline scenario. These authors propose that with or without CO_2 fertilization, Venezuela would have 10.5 million people below the poverty line by 2025.

According to the IPCC (2018, 2020), if warming is limited to 1.5 °C instead of 2 °C, net reductions in yields of maize, rice, and wheat, and possibly other cereals, especially in Central and South America, and in the food quality of rice and wheat, are estimated to be smaller.

Bárcena et al. (2020) warn that climate change in Latin America and the Caribbean will lead to lower agricultural production, reduced food quality, declining incomes, and rising prices due to increased temperatures and extreme weather events. For Venezuela, a projected temperature rise of 2.5 °C by 2050 could reduce GDP per capita by 10%. These impacts would particularly affect rural households, intensify poverty, and reduce food availability, especially under a 2 °C warming scenario, which would be more damaging than a 1.5 °C increase.

Paredes (2014) emphasizes that small-scale, subsistence producers are among the most vulnerable to climate change, especially in regions like the lowlands of Mérida. Cocoa and banana crops, which require warm lowland conditions, are at risk due to rising temperatures. Climate models project that with a 2 °C increase and reduced rainfall, suitable areas for these crops—particularly the internationally valued Porcelana cocoa and bananas—could decline sharply or even disappear, especially in the Sur del Lago region below 1000 m asl, where thermal conditions are becoming increasingly unfavorable.

Medina and Laliberte (2017), in a review of the research carried out on the effects of drought stress and temperature and the increase of CO_2 on cocoa (*Theobroma cacao* L.), reported the results of an irrigated field trial at an experimental station of the National Institute of Agricultural Research (INIA) in Mérida, Venezuela, where they evaluated the drought tolerance of four-year-old trees (no specific genotypes were mentioned). For the treatment of water deficit, the trees were left 3, 12, and 5 days without irrigation. With an average rainfall of only 570 mm, this Venezuelan region receives far less rainfall than is considered suitable for cocoa cultivation. At the end of the study, tolerant trees had lower rates of leaf transpiration and considerably lower stomatal conductance than control plants, and also showed osmotic adjustment responses. The results show that the capacity of osmotic adjustment to provide tolerance to water deficit to the evaluated cocoa genotypes could still be limited.

In another study reported by Medina and Laliberte (2017) conducted by the Universidad Central de Venezuela and Universidad de Los Andes indicate drought tolerance of criollo-type cacao trees (Porcelana, Guasare, Porcelana, and Criollo merideño) that had recently been established under an agroforestry system. The study concludes that stomatal conductance decreased in all genotypes by 60%, resulting in a 73% decrease in photosynthetic activity, with no significant differences in water potential. However, the variation observed in the level of osmotic adjustment achieved between the genotypes evaluated was sufficient to categorize some as more tolerant than others. The use of physiological parameters such as stomatal conductance, photosynthetic response, and osmotic fit appears to provide valuable insights into the level of adaptability of different genotypes, and they remain favorites in terms of candidate traits for evaluation in ongoing trials. Studies indicate a positive response of cocoa to increased shade toward greater climate resilience.

Agreeable to Paredes and Chacón-Moreno (2021), working with cocoa, in a scenario with a strict emissions trajectory, there is evidence of high loss of potential areas, but maintaining a small area by the year 2100; however, in a scenario of continuous emissions by 2100, cocoa cultivation would disappear for the state of Mérida. For banana cultivation, any change in precipitation due to climate change would affect the distribution of potential areas, until the disappearance of current areas, which is an indication of being a crop sensitive to any climate change.

Barrios et al. (2011) consider that the unimodal distribution of precipitation generates periods of water deficit and periods of excess moisture that are reflected in the production of oil palm (*Elaeis guineensis* Jacq) in the Palmera Vuelta Larga-La Hormiga area, in the state of Monagas. This area had a water deficit of 450 mm per year, concentrated between January and May. The deficit particularly negatively affected the phenological phases of fruit filling, central arrow state and sexual differentiation of inflorescences, and consequently the production of oil palm. The authors conclude that the application of irrigation represents an alternative to mitigate the adverse effects of water deficit that would be a consequence of climate change.

When analyzing the patterns of suitability loss and gain by country resulting from the study by Postigo et al. (2012) for the Andean region, important differences are observed in terms of potential opportunities or threats resulting from climate change. A gradient of potential impacts that decreases from north to south is present in the Andean axis. Venezuela is losing suitability for 17 crops, Colombia and Bolivia for 10 crops, and Ecuador and Peru for 8 crops. In Venezuela, only 6 crops would make a profit. An important process will be the potential altitudinal displacement of the climatic suitability for the crops analyzed. The lowland crops that will experience an increase in suitability in Andean countries are rice (*Oryza sativa* L.), bananas (*Musa* sp. L.), plantain (*Musa balbisiana* colla), sorghum (*Sorghum bicolor* var. sweet (L.) Moench.), and cassava (*Manihot esculenta* Crantz). The sweet potato (*Ipomoea batatas* L.) has modest loss values (about 2.5%) in the Venezuelan Andes. In the case of lowlands, these authors estimate that less than 3% of people live in areas that are becoming suitable for crops such as rice (*Oryza sativa* L.), bananas (*Musa* sp. L.), sorghum (*Sorghum bicolor*), and cassava (*Manihot esculenta* Crantz).

In the context of Latin America and, of course, valid for Venezuela, López-Feldman (2015) reports that the agricultural sector will suffer the greatest economic effects from climate change and that it will affect food security by impacting the availability and access to food, as well as the stability of food reserves and price volatility. This author estimates that at first, moderate global warming will harm semi-arid and tropical regions and that developing countries are more vulnerable to the effects of climate change than developed countries, since small farmers will be the most affected given their low access to technologies, inputs, information, and monetary resources to take adaptation measures. He believes that the impact of climate change on agriculture may be greater, as the potential effects of extreme weather events are included.

Gourdji et al. (2015) assessed climate change impacts on crop yields using the Decision Support System for Technology Transfer (DSSAT) and EcoCrop models. Their findings suggest that rising temperatures and declining rainfall will heighten agricultural drought risk, especially for rainfed crops, with Venezuela among the most severely affected countries. Rainfed beans and maize are particularly vulnerable, with notable yield declines, while rainfed rice may experience gains. Crop cycles for corn, rice, beans, and soybeans are projected to shorten by around 10%, though beans and soybeans may lengthen slightly in some areas, influencing yield variability.

Water stress is expected to increase across nearly all regions and crops, especially for rainfed maize in Venezuela, where yield reductions are closely tied to moisture deficits. The cited study highlights Venezuela as a critical hotspot for climate-related yield losses, particularly for rainfed maize and dry beans, which face extensive areas of risk under future conditions. Authors also suggest that, for Venezuela, changes in yield are more influenced by changes in water stress. Unlike other crops, rainfed rice shows yield increases in the vast majority of growing areas. In the case of corn, both rainfed and irrigated, Venezuela is considered a hotspot for the production of this item, additionally, dry rainfed beans represent the cropping system with the second largest number of critical areas.

Rodríguez-De Luque et al. (2016), in their analysis of the socioeconomic impacts of climate change in Latin America and the Caribbean and using the "Impact" model, for 39 crops, estimated that the growth of corn and bean productions would fall significantly. To some extent, corn and bean affect food security in Venezuela. They observe that the increase of their yields of corn and beans would be negatively affected, while for rice, the growth of yields would be positive, while the cultivated area would decrease. In the case of increment in demand for beans and soybeans, there would be no major changes with respect to the scenario without climate change, while in the case of the demand for corn and rice, it would be negatively affected. Venezuela, according to the Impact model, is among the countries with the highest increases in the percentage of the population at risk of hunger (food insecurity indicator) and the highest rate in the region (17.8%).

In a study conducted by López-Hernández and Infante (2016) in a sugarcane (*Saccharum officinarum*) plantation, they determined that burning, together with crop removal, leads to considerable losses of N in this agroecosystem. Thus, approximately 98% of the total nitrogen contained in the aboveground biomass leaves the system through these pathways. A very significant loss of N is associated with the pre-harvest burning operation, an operation that could be reduced or eliminated and thus avoiding or reducing greenhouse gas (GHG) emissions and other pollutants associated with this practice. This study concludes that a change in the management of sugarcane fields, without pre-harvest burning, would significantly reduce greenhouse gas emissions and increase soil organic matter and associated organic N contents.

Jarvis et al. (2019) advise that climate change will negatively impact food security in Latin America and the Caribbean by reducing yields of staple crops such as corn, rice, beans, cassava, and sugarcane as well as increasing pests, diseases, and food prices—especially in import dependent countries- like Venezuela. A 2 °C rise in temperature, expected around 2040–2070, could result in over 5% yield losses in Venezuela and heightened food insecurity. Early exceedance of this threshold is projected in areas such as the Amazon, Brazil, and Venezuela, likely worsening droughts and rainfall variability.

According to Venezuela's Second National Communication on Climate Change (MINEA 2017a), under an intermediate scenario (RCP 4.5), yield reductions for rice, beans, and corn in Portuguesa and Barinas are minimal, but in the pessimistic scenario (RCP 8.5) for 2076–2099, losses become severe, up to 70.3% for sugarcane, 55.3% for beans, 49.1% for corn, and 31.1% for rice. Similar patterns are observed in fruit trees, oilseeds, pastures, and cotton, with irrigation increasingly required under all scenarios. These projections highlight the growing vulnerability of Venezuelan agriculture to climate change, especially under high-emission pathways.

FAO-CIAT (2018) declared that maize production, a fundamental crop in the diet of the Latin American population, will be affected by a reduction in its yield of at least 10% by the middle of this century. The modeling results in the projection of change in future yields show, in the case of corn and bean cultivation, a reduction in yields, especially in rainfed conditions, with the most negative effect occurring especially in northern Venezuela. In commercial terms, sugarcane and coffee are crops of great importance in the northern region, and with climate change, a significant decrease in their production. The areas that become unsuitable are located especially in the north of Venezuela.

Garruña et al. (2021) opine that one of the problems in the horticultural sector is the vulnerability of many species to the increase in temperature due to climate change. Of all the climatic factors, temperature has the greatest influence on the growth and development of vegetables. Likely, the benefits that plants will obtain from increasing atmospheric CO_2 will not be enough to counteract the damaging effects that the increase in temperature will cause on agricultural production. Vegetables grown in tropical regions can tolerate temperatures close to 40 °C, under photosynthetically active radiation above 2000 μmol m^{-2} s^{-1}. It is difficult to understand that some vegetables, such as tomatoes (*Solanum lycopersicum*), chili peppers (*Capsicum* spp.), pumpkins (*Cucurbita* spp.), watermelon (*Citrullus lanatus*), melons (*Cucumis melo*), and cucumbers (*Cucumis sativus*) can perform their normal physiological functions under the characteristic environmental conditions of the tropics (temperature close to 40 °C and relative humidity greater than 75%). With sufficient water availability, vegetables can continue their normal functions, even if temperatures are close to 40 °C. Tropical vegetables are considered "super" plants and are an excellent model for studying the effects of climate change.

For Gouveia et al. (2023), food security continues to depend largely on the practices of smallholder farmers, who are often the most vulnerable to climate extremes. They consider that the use of dual-cropping systems, although very successful in recent years, may also be at risk within a few decades, mainly due to the expected decrease in rainfall, the time pause in the start of the rainy season and the increase in temperature. For the 60 years, 1961–2021, a generalized pattern of significant and positive trends has been observed for most countries. However, for sorghum and sugarcane crops in Venezuela, the trends are in the opposite direction, but not in the case of rice.

Marquina et al. (2015) consider that despite the rapid expansion of tropical agriculture in Latin America and the Caribbean, little is known about its impact on atmospheric emissions of trace gases, such as nitrogen oxides (NOx), nitrous oxide (N_2O), and carbon dioxide (CO_2), which are produced in soils by microbial processes which are accelerated in tropical climates. In their research in two of Venezuela's largest cereal-producing regions during the growing season, they measured soil NO, N_2O, and CO_2 emissions along with soil variables from low-till corn fields and no-till agriculture. Mean NO and CO_2 emissions were higher in the under-till than no-till sites, while N_2O emissions showed the opposite. CO_2 emissions under tillage were 1.6 times higher than those of no-tillage, while N_2O emissions were 0.5 times higher than those of no-tillage. The authors conclude that the greater mitigating effect of no-till practices in this monoculture system is mainly due to the reduction in CO_2 emissions. They suggest that agricultural mitigation actions for tropical monoculture systems should aim at improving no-till practice management practices along with reducing the N fertilization rate to compensate for the higher N_2O emissions.

Olivares et al. (2019) analyzed future climate impacts on banana production in Venezuela using two General Circulation Models (UKTR and CCC-EQ) and MAGICC/SCENGEN software, supported by IPCC data. Their projections indicate increased average temperatures, decreased total rainfall, and fewer wet months in key banana-producing areas like Aragua and Carabobo states, leading to heightened risks of extreme events, soil erosion, and pest and disease outbreaks. These changes are expected to directly reduce yields and alter agricultural calendars and practices. Specifically, the banana-growing area in Aragua (4481 ha) is projected to face a drop in rainfall to 800–1200 mm/year and a substantial rise in water deficit from 1 to 4.7 million hectares in the central region. These climatic shifts threaten to alter phenological phases, lower fruit quality, and disrupt sowing, fertilization, and harvesting schedules, pointing to a drier and more challenging future for banana cultivation in Venezuela. In general, there is evidence of a drier future for the banana-growing areas of the country, in addition, it is expected alteration of phenological phases, reduction of yields, deterioration of the quality of products, impact on agronomic practices (time and method for soil preparation; planting season; fertilization levels), impact on harvest time, and other threats of pests and diseases.

In the view of Aggarwala et al. (2019), many assessments of the impacts of climate change on crop yields globally foresee declines in Latin America, including Colombia, Venezuela, and Brazil, as early as the 2020s, and that losses will increase over time, up to 50% in the 2080s. The results are based on the review and comparison of published projections of climate change impacts from 34 studies and 4500 data points for the 2020s for maize, rice and wheat at the country level. They found that the observed changes in yield are considerably higher than the expected changes in yield from climate change. Technological improvements appear to have a large performance improvement impact compared to the negative effects of climate change, at least in the short term. The authors argue that assessments of the effects of climate change on crop yields show low-latitude and low- and middle-income countries as highly vulnerable, but, on the other hand, these countries have shown the highest growth in yields observed during the same reference period. For rice, the differences between observed and projected changes in yields due to climate change were smaller compared to maize.

Ray et al. (2019) estimate that crop yields will decrease under future climatic conditions, and, certainly, recent research suggests that yield has already been affected. However, the current effects on various crops at the subnational scale and their implications for food security remain unclear. They assessed the potential impact of observed climate change on the yields of several major crops globally, including barley, cassava, maize, rapeseed, oil palm, rice, sorghum, soybeans, sugarcane, and wheat in approximately 20,000 localities. They point out in their study that, while in North America and some South American countries, benefits are observed for crop yields derived from climate change, especially in crops such as corn, oil palm, soybeans, and sugarcane. However, in nearly half of the food-insecure countries, estimated caloric availability decreased. In fact, in the case of Venezuela, recent climate change has produced significant losses in consumable dietary calories.

Correa et al. (2020) highlight microalgae-based biofuel systems as a highly sustainable alternative due to their high productivity per unit area and capacity to grow on non-arable land. Evaluating four Neotropical countries, including Venezuela, they estimate that to meet 30% of transport energy demand by 2050, microalgal systems would require significantly less land than sugarcane or oil palm. For Venezuela, a potential production of 9.7 million tons of oil equivalent was projected using microalgae, sugarcane, and oil palm.

In Venezuela, the most suitable areas for microalgae cultivation, using freshwater, brackish, or seawater, are the dry lowlands along the Caribbean coast, especially in Falcón and Zulia. These zones present minimal conflict with high-value agricultural land or biodiversity. In contrast, oil palm would be best located in the lowland wetlands of Zulia, Táchira, Monagas, and Delta Amacuro. The authors conclude that microalgae systems could reduce land use by 52% compared to oil palm and 67% compared to sugarcane, making them a strategic option for biofuel development in low-productivity or ecologically sensitive regions.

4.3.3 Expansion of Arid and Semi-arid Areas in Venezuela as a Result of Climate Change

Medina et al. (2016) argue that one of the potentially most detrimental consequences of climate change is the expansion of the world's dry areas, with projections for northern South America of a substantial expansion of areas where seasonal drought will be greatest. These authors, applying the Bailey Index, estimated for Venezuela a considerable expansion of the areas with a negative annual water balance, concluding that the semi-arid categories will increase by about 106,000 km^2 and the dry sub-humid by 107,000 km^2, located especially in the northwestern region (Zulia and Falcón states) and the states of Guárico, Anzoátegui, and Monagas. The largest absolute increase in the area of the arid domain is observed in the states of Falcón, Zulia, and Lara. The states of Anzoátegui, Cojedes, and Monagas will have an increase in semi-arid areas greater than 10,000 km^2 derived from the reduction of dry sub-humid areas, while Guárico will experience greater increases in semi-arid and dry sub-humid areas.

Figure 4.9a summarizes projected values for the maximum annual number of consecutive dry days in Falcón by the year 2040. Median values range from 29.99 days under SSP1-2.6 to 35.72 days under SSP5-8.5, with upper uncertainty limits reaching up to 56.48 days. The consistent increase across all scenarios reflects a robust signal of drying trends, with more severe outcomes expected under high-emissions trajectories.

Even under moderate scenarios (SSP2-4.5 and SSP3-7.0), the projected extension of dry spells by several days could substantially affect seasonal crop performance and natural vegetation dynamics. These findings underscore the critical importance of reducing emissions to minimize future aridification and ensuring that Falcón's rural sectors are equipped to cope with changing climate extremes.

This heatmap (Fig. 4.9b) represents the projected monthly change in the maximum number of consecutive dry days in Lara state from 1951 to 2100 under the SSP5-8.5 scenario. While earlier decades show neutral or negative anomalies (fewer dry days), future projections indicate a marked increase, particularly from the 2060s forward. By 2081–2100, anomalies exceed +7 to +8 days in months like February and March, which already form part of the region's dry season.

The accumulation of consecutive dry days may intensify drought conditions and increase evapotranspiration stress on crops, particularly for rainfed agriculture. These extended dry spells may also trigger shifts in planting calendars, delay

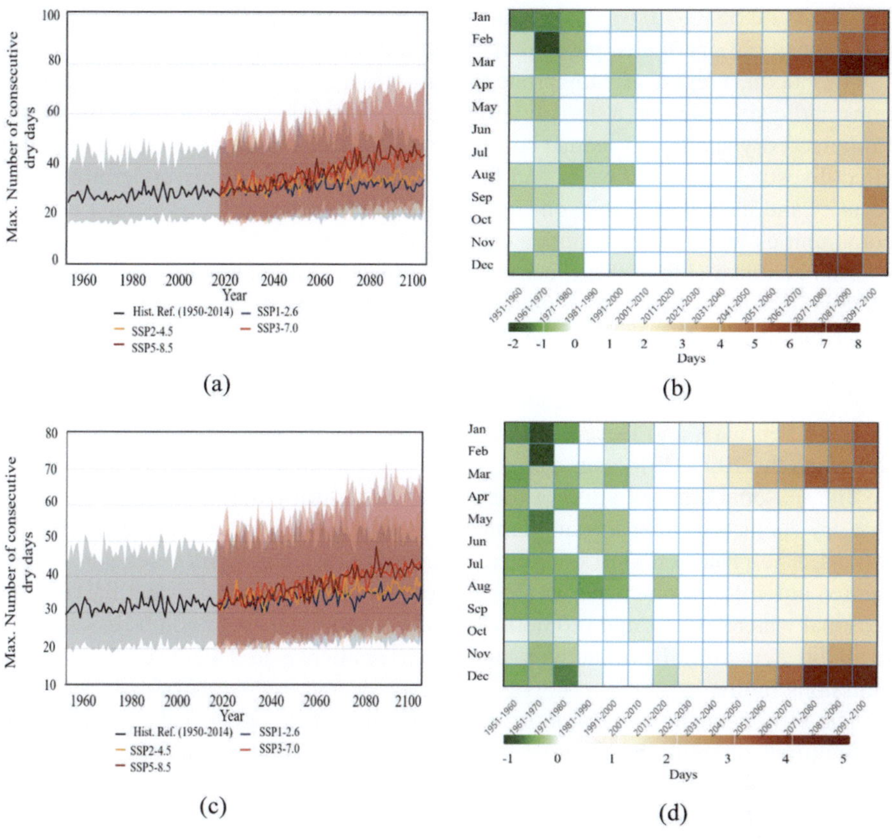

Fig. 4.9 (**a**) Projected maximum number of consecutive dry days anomaly in Lara, Venezuela (1951–2100), relative to the 1995–2014 reference period, under the SSP5-8.5 scenario; (**b**) projected monthly maximum number of consecutive dry days anomaly in Lara, Venezuela (1951–2100), relative to the 1995–2014 reference period, under the SSP5-8.5 scenario; (**c**) projected maximum number of consecutive dry days in Falcón, Venezuela (1950–2100), under four SSP scenarios (multi-model ensemble); (**d**) projected monthly maximum number of consecutive dry days anomaly in Falcón, Venezuela (1951–2100), relative to the 1995–2014 reference period, under the SSP5-8.5 scenario, according to the data source from Eyring et al. (2016) and available in World Bank, Climate Change Knowledge Portal (2025).

sowing periods, and threaten pasture regeneration for livestock. For Lara—a state with significant vulnerability to water scarcity—such changes imply a heightened need for adaptation measures targeting seasonal water storage, supplemental irrigation, and drought-tolerant crop varieties.

Figure 4.9c shows the long-term projections of the maximum number of consecutive dry days (CDD) per year in the state of Falcón from 1950 to 2100, under SSP1-2.6, SSP2-4.5, SSP3-7.0, and SSP5-8.5 scenarios. By the year 2040, the median values range from 32.01 days (SSP1-2.6) to 38.05 days (SSP5-8.5), with uncertainty ranges reaching up to 54.18 days in the highest emission pathway. The

SSP5-8.5 and SSP3-7.0 trajectories display a marked upward trend in the frequency and intensity of dry spells, particularly from mid-century onward.

For Falcón, a region already prone to semi-arid and arid conditions, these projections indicate growing risks of agricultural droughts and hydrological stress. Prolonged dry spells can reduce soil moisture availability, delay crop sowing, and affect livestock access to water and forage. These conditions may undermine food security and increase reliance on costly irrigation infrastructure, especially in rural communities with limited adaptive capacity.

This heatmap (Fig. 4.9d) illustrates the projected anomalies in maximum consecutive dry days across each month and decade from 1951 to 2100 for Falcón, under the SSP5-8.5 scenario. The shift from negative anomalies (fewer dry days) during earlier decades toward strong positive anomalies is particularly evident from 2050 onward. By the final decades of the century (2081–2100), months such as January, February, and December are projected to have 4–5 more consecutive dry days than the 1995–2014 average.

The seasonal concentration of these anomalies during the dry months suggests that dry spells will intensify and lengthen in times when water scarcity is already critical. The projected shift could exacerbate desertification processes and hinder rainfed agricultural practices. Strategic interventions such as rainwater harvesting, seasonal water storage, and promotion of drought-tolerant crops will be essential to maintain agricultural productivity and water resilience in Falcón under a high-emissions future.

The effects of the expansion of arid and semi-arid areas reveal differences in agricultural suitability for the four critical crop areas of Venezuela: corn, sugarcane, rice, and cassava. It is estimated that in 2050, in the state of Portuguesa around 10,260 km^2 (68% of its surface) will become drier areas, affecting the rainfed crops produced in this entity. Except the states of Barinas and Lara, the other states are among those that will show the most marked increases (>50%) in the area of dry areas. In the case of cassava cultivation in Monagas, the main producer of this item, the increase in dry areas expected by 2050 would reach 75% of its extension, negatively impacting the production of this important crop. Sugarcane and rice crops have the highest values of crop coefficients (Kc), with a higher water requirement, while cassava has the lowest demands associated with its greater efficiency in water use, giving it an advantage over the rest as a crop to be developed as a source of carbohydrates in these scenarios of increases in arid and semi-arid zones. The results of this study allow to presume that the disappearance of the humid tropical climate (Af) and the increase in the Bsh climate (warm semi-arid) would indicate a reduction in rainfall in the wettest areas, as well as an increase in the number of dry months for most of the territory, and in the spatial variability of rainfall that would increase and decrease in a very concentrated way in some areas of the Plains. The projections for the RCP 4.5 and RCP 8.5 scenarios allowed us to verify the reduction of the area occupied by humid climates and an increase in the space occupied by dry climates (Medina et al. 2016).

Silva and Mendoza (2019, 2021) evaluated spatial climate changes in the Venezuelan Plains according to the Köppen classification, using the mean temperature and the 75th percentile of precipitation from historical records and future series under climate change. Future series were generated from the results of the CCAFS climate change program, for the NCAR_CCSM4 model, RCP4.5 and RCP 8.5 scenarios, and periods 2050s and 2080s. The results revealed a reduction in the space occupied by humid climates and an increase in dry climates in the medium and long term for both future scenarios compared to the reference period, attributed to probable changes in the behavior of rainfall. The proportion occupied by dry climates would be lower for the RCP 4.5 scenario than for the RCP 8.5 scenario. These results allowed us to deduce detrimental effects on the natural ecosystems of the region for both scenarios.

Fernández (2020) applied three climate classification systems Köppen-Geiger, Thornthwaite, and Holdridge Life Zones, in the Aroa, Tocuyo, and Yaracuy River basins. He used both observed climate data from 1971 to 2000 and simulated data for the same historical period and two future periods (2016–2045 and 2046–2075). The simulations were based on two emission scenarios: RCP 4.5 and RCP 8.5. He concluded that the usefulness of climate classifications for spatial and temporal analyses depends heavily on the quality of the climate data. In Venezuela, this is a major limitation due to the sparse network of weather stations. Despite this, his research revealed important trends in the study area. These include the expansion of arid climates (BW) according to the Köppen-Geiger classification, a decline in the Thornthwaite humidity index, and the disappearance of Humid Life Zones. He also concludes that it is a priority to propose adaptation measures for agricultural and livestock activities envisioned for the populations located in the upper basin of the Aroa and Yaracuy rivers.

By way of summary and taking into consideration the spatial arrangement of the geostructures considered as a reference frame in Table 4.1 and Fig. 4.1, Table 4.4 shows the ranges of the projected values of the variations of the temperature (°C) and precipitation (%) variables in the different regions for the period 2080–2099, under the SSP5-8.5 scenario.

Significant temperature increases can be observed, as well as deficits in terms of the amount of rainfall. The smallest ranges of variation are observed in the projected values for temperature; however, in precipitation, wider ranges are observed in the projected

Table 4.4 Projected temperature and precipitation changes in the geostructures of Venezuela. (Ref. Period 1995–2014) using SSP5-8.5 of Period 2080–2099, according to Climate Change Knowledge Portal (2025)

Geostructure	Region	Code in Fig. 4.1	Scenario SSP5-8.5	
			Temperature (°C)	Precipitation (%)
Mountain ranges	Andean and Coastal ranges	1	+4.81/+5.06	−14/−23
Geosynclinal (Sedimentary basin)	Western plains	2	+4.78/+5.01	−21/−23
	Central plains		+3.89/+4.78	−19/−31
	Eastern plains		+4.34/+4.46	−37/−40
	Orinoco Delta		+4.19	−30
Guayana Shield	Peneplains	3	+4.94/+4.99	−15/−33

values, with the Eastern plains, Central plains, the Orinoco Delta, and the Guayana Shield being the regions where the greatest decreases in the projected values are presented.

4.4 Impacts on Forest Production

Some geographical areas of Venezuela's lowlands have a vast forest cover, and parts of it have been declared Forest Reserves with diverse objectives, protection, management, and use (Fig. 4.10). This important supply of forest resources, both primary and plantations, are severely threatened by the effects of anthropogenic activeness (fire, deforestation, land-use change), climatic variability, and climate change), making these forest lands endangered with the resultant environmental, social and economic impacts. The Amazon region of Venezuela, accomplished by the states of Bolivar, Amazonas, and Delta Amacuro, is home to diverse ethnic communities whose livelihoods depend significantly on the use of primary forests. In this sense, alterations to the forest communities derived from climate change will negatively impact the survival of these communities as a result of potential modifications to wildlife habitats. At the regional level, many authors are pointing out the effects of climate change on the forests of Latin America and consequently, on the forestlands of Venezuela.

Rojas et al. (2010) found that during the severe El Niño years of 1997–1998, pine plantations of *Pinus caribaea* in Uverito (Monagas state) were significantly affected due to decreased rainfall and extended dry seasons, leading to increased tree mortality. While a general correlation between El Niño and dry years was not evident across the entire plantation area, a mesoscale relationship was observed in the western expansion zones, where extreme events aligned with increased vulnerability.

Similarly, Rebolledo and Albarrán (2014), using the UNKO-HAD-CM3 model, identified a relationship between forest cover changes and climate variability in

(a) (b)

Fig. 4.10 (**a**, **b**) Forestry lands in the Guayana Shield. (Pictures by Pedro García) (Bolivar state, Venezuela, 2015)

Venezuela during the last two decades of the twentieth century. Their analysis showed that climate impacts varied by forest type, with anomalies occurring most frequently in January, May, and June, especially in the central-western plains and river confluences. Although focused on variability rather than long-term climate change, the findings offer a basis for projecting future forest dynamics under climate change scenarios.

Nobre et al. (2016) suggest that a number of driving forces of large-scale environmental change, such as land-use change and climate change due to global warming and deforestation, can induce a higher frequency of extreme weather events and vegetation fires, which adds to greater exposure and vulnerability of tropical forests. The rainfall pattern in the Amazon shows strong interannual and interdecadal variability. Severe droughts are associated, in general, with the occurrence of strong El Niño events that mainly affect the central and eastern portions of the Amazon. Extreme drought events have the potential to increase forest mortality, and the synergistic association between severe droughts, deforestation, and fires can be very damaging to the Amazon rainforest.

Zanetti et al. (2017) highlight that climate change will alter precipitation and runoff patterns, reducing the effectiveness of forests in upper watersheds to control erosion and sustain water quality. Latin American forests face increasing vulnerability due to more frequent and intense extreme events, floods, droughts, rising temperatures, higher atmospheric CO_2 levels, pests, and wildfires. Although CO_2 fertilization could increase tropical forest productivity by up to 35% at 550 ppm, there remains limited knowledge on how climate change will affect these ecosystems. The authors also stress that adaptation and mitigation measures in forestry often overlap, particularly in the case of forest conservation, which supports multiple environmental services. Authors conclude that there is little knowledge about the likely impacts of climate change on Latin American forests.

Pacheco-Angulo et al. (2017) reveal that Venezuela's Caparo Forest Reserve lost one-third of its forest area between 1990 and 2000 due to deforestation and degradation, leading to annual CO_2 emissions of 2.21 Mt during that period. Despite this, 30,749 ha remained in 2015, with a potential carbon stock of 27 Mt CO_2 if degradation were halted. A complete ending of deforestation and forest degradation in Caparo would potentially maintain an estimated total of 27 Mt of CO_2 stored in these forests. Meanwhile, Marrero and Rodríguez-Olarte (2017), warn that sea level rise, projected to reach 0.5 m by 2090, will endanger mangroves and coastal forests, particularly along the Atlantic and Caribbean coasts. This would increase saline intrusion, intensify flooding, erode coastlines, and disrupt wetlands and estuarine ecosystems, ultimately affecting biodiversity and fisheries.

Longo et al. (2018) project that increasing drought recurrence in the Amazon, particularly in the Venezuelan states of Bolívar and Delta Amacuro, will raise evapotranspiration and impose water stress on tall trees, reducing growth and increasing mortality, thus threatening carbon stocks and biodiversity. Biomass loss becomes significant when droughts recur every 2–7 years. Similarly, Pabón-Caicedo et al. (2018) emphasize that floods and droughts are the most impactful hydroclimatic extremes in the Amazon, with more frequent events in the south and longer cycles (5–6 years or more) in the northern Amazon, including Venezuela. Although precipitation trends remain uncertain, global warming may intensify extreme events, posing risks to agriculture and water availability.

Silva and Mendoza (2019) warn of unfavorable future conditions for native vegetation such as gallery and deciduous forests in the central-eastern plains and piedmont regions of Venezuela, with an expansion of semi-arid climates threatening biodiversity and increasing forest fire risks, particularly in Uverito's pine plantations. The increase in the extension of the semi-arid climate areas toward the eastern highplain (mesas) and plains would create limiting environmental conditions for the current highly heterogeneous vegetation, made up of dense "morichales" (palm-dominated communities with poor drainage) and deciduous forests.

Meanwhile, Esquivel-Muelbert et al. (2019), based on 30 years of forest inventory data, observed an increasing dominance of drought-tolerant tree genera and a concurrent decline in moisture-dependent species. These trends indicate a gradual shift toward a drier Amazon. However, changes in species composition lag behind shifts in climate, largely due to the long generation times of trees. This temporal delay may weaken the Amazon's capacity to maintain key ecosystem functions, such as carbon storage and biodiversity conservation. The authors argue that although the floristic composition is beginning to change in response to climate factors, the overall community structure remains relatively stable. If this compositional lag persists, intact Amazon forests could increasingly be dominated by species not optimally adapted to their climatic niches. This misalignment could compromise the ecosystem's ability to deliver critical services, including biodiversity protection and carbon sequestration.

Chacón-Moreno et al. (2021) found that Venezuelan mountain species might respond to climate change through upward altitudinal migration, with forests shifting vertically by approximately 13 meters per decade to counteract temperature increases of around 3 °C. This migration is linked to changes in woody vegetation density and structure, potentially altering the distribution of the tree line and threatening the existence of high mountain ecosystems such as the páramo forest. The study warns that Venezuela is highly vulnerable and currently unprepared to cope with climate variability, with projected warming of 2–3 °C and a decrease in rainfall (~0.1 mm/day) under RCP 4.5 by century's end.

Complementing these findings, Hagen et al. (2022) project a 15% reduction in tropical forest extent by 2050 under RCP 8.5, increasing to 60% when deforestation and fire are factored in, placing the Amazon at risk of shifting from a carbon sink to a carbon source. Pörtner et al. (2022) emphasize that terrestrial ecosystems face growing limitations in carbon sequestration due to warming and drought, potentially reversing their climate-mitigating role. Widespread tree mortality, wildfires, and ecosystem feedbacks could transform key biomes like the Amazon into drier, lower-biomass systems, threatening biodiversity and accelerating climate change impacts.

4.5 Impacts on Livestock Production

Climate change significantly influences agricultural and livestock systems in South America, affecting not only productivity but also the distribution and emergence of animal diseases. The IPCC (2007) and Pinto et al. (2008) highlight that global warming, coupled with more frequent El Niño events, will alter the geographic and temporal spread of vector-borne diseases such as bluetongue, West Nile fever, and vesicular stomatitis. Factors such as habitat fragmentation, altered host density, and livestock displacement due to droughts may increase contact rates between wild and domestic animals, facilitating pathogen transmission. Weak disease surveillance systems across the region further compound this risk. Climate-driven ecological disturbances, like deforestation and land-use changes, could expose animals to new disease vectors, expanding the range of shared pathogens like rabies and brucellosis. The expansion of the tropical belt is expected to broaden habitats for vector-borne and parasitic diseases across northern and southern South America.

The effects of climate change on livestock vary by species and scenario. Seo et al. (2010) project that under a hot, dry 2060 scenario, Venezuela and other Andean countries may see declines in cattle and pig populations but an increase in sheep, which are more adaptable to such conditions. In contrast, under wetter scenarios, cattle numbers may drop due to excessive moisture. Salazar-Lizán (2014) emphasizes that grazing-based systems will be particularly vulnerable, with tropical grasslands more sensitive than livestock themselves to changes in precipitation and temperature. Reduced water and soil nutrients, lower pasture biomass, and heat stress are key limiting factors. While both *Bos taurus* and *Bos indicus* are physiologically affected by prolonged heat stress, *Bos indicus* and its hybrids display greater resilience, making them more suitable for warmer conditions projected in the region.

López-Feldman (2015) emphasizes that livestock production is highly sensitive to climatic conditions, noting that variables such as air temperature, humidity, and wind speed directly affect animal appetite, growth, reproduction, and productivity (e.g., milk and wool). Additionally, temperature and precipitation influence the quality and availability of pastures as well as disease and parasite dynamics. He points out that, under warmer and drier conditions, cattle herds are expected to decline while sheep production may increase, due to their greater adaptability. Despite this, most climate impact studies have focused on crops, leaving livestock comparatively understudied.

Christopher et al. (2015) and Delgado (2017) add that heat stress reduces feed intake and milk yield, negatively affects reproductive performance, and raises mortality risks. As global temperatures rise by 1–2 °C, the range of viable livestock species in Venezuela and the region is expected to decrease, with sheep adoption potentially increasing by up to 20%. Natural grasslands, crucial for extensive cattle systems, are threatened by woody plant invasion, a process accelerated by climate change and land-use pressures, which diminishes forage quality. The IPCC (2018) also warns that rising temperatures could undermine livestock systems through altered forage composition, increased disease transmission, and reduced water availability, particularly in humid tropical grasslands prone to rapid ecological shifts.

Silva and Mendoza (2019) identify the reduction of rainfall in the alluvial plains of Apure and Barinas as a key example of climate change impacts in Venezuela, threatening the hygrophilous savannas that sustain livestock production. Such changes could disrupt traditional agricultural systems based on pasture and adapted livestock breeds. Jarvis et al. (2019) also highlight that climate change, particularly heat stress, will negatively affect livestock and poultry productivity, with projections showing declines in beef cattle and sheep in Venezuela under hot and dry scenarios, contrary to the more optimistic view of Seo et al. (2010). Rising temperatures will further influence pest and disease distribution, reducing both the quality and quantity of animal production.

Additionally, Lee et al. (2012) note that Venezuela is among the countries with favorable future conditions for the common vampire bat (*Desmodus rotundus*), which harms livestock through blood loss, reduced milk and weight gain, and increased disease risk, including rabies. Rodríguez-Morales (2013) emphasizes that climate change may facilitate the spread of zoonotic diseases like brucellosis through habitat alterations and greater contact between domestic and wild species. However, limited research exists on the epidemiological dynamics of such diseases in Venezuela, pointing to a need for improved data and surveillance systems to address the growing health risks in livestock systems under changing climatic conditions.

4.6 Impacts on Fisheries

Deforestation and climate change pose significant threats to tropical freshwater ecosystems and river fisheries. Wright and Flecker (2004), studying the Las Marías River in Venezuela's Andean foothills, found that deforestation drastically reduces the input of woody debris, which is crucial for maintaining aquatic biodiversity. Key commercial species like *Prochilodus mariae* (coporo) were more abundant in habitats with large woody remains, suggesting that habitat loss directly impacts fish populations and, consequently, fisheries' ecological and economic value.

On a broader scale, Allison et al. (2009) ranked Venezuela among the 33 "highly vulnerable" countries in terms of climate change impacts on capture fisheries. Vulnerability stems from a combination of high exposure to climatic shifts, moderate adaptive capacity, and reliance on fisheries for food and income. Anderson and Maldonado-Ocampo (2010) further highlight that tropical Andean fish species are particularly sensitive to environmental change, with warming, glacier retreat, and hydrological disruptions expected to affect species distributions, river flows, and high-altitude wetlands. These transformations could result in significant social and economic consequences for communities reliant on these freshwater resources.

Coastal and marine ecosystems in Venezuela, particularly those supporting fisheries and mangroves, are highly vulnerable to climate change. The presence of 332

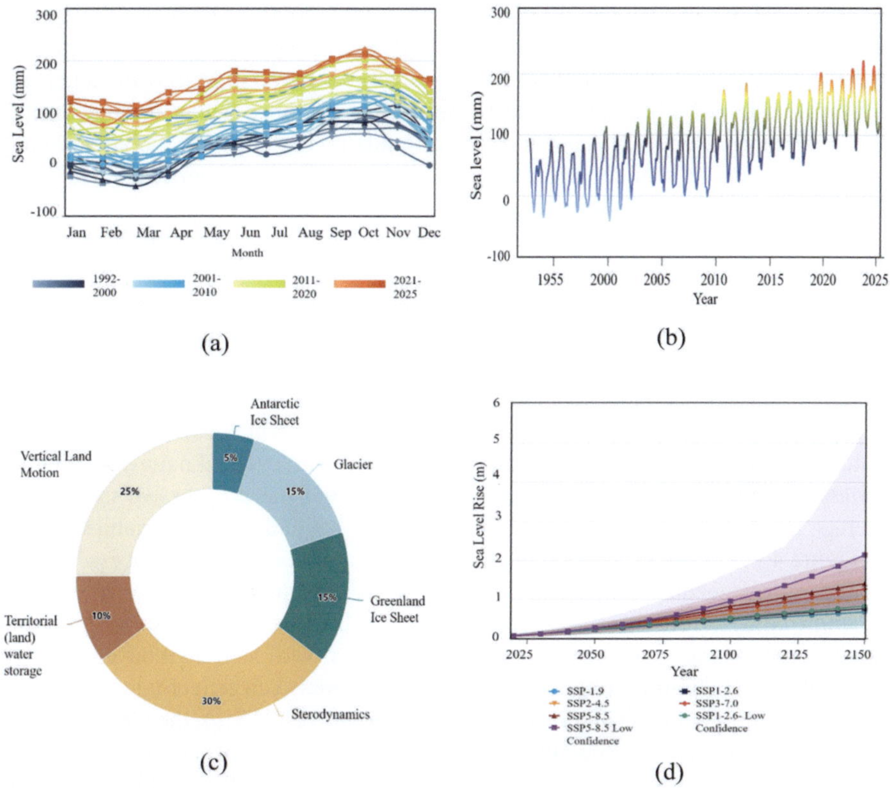

Fig. 4.11 (**a**) Monthly total sea level change in the Venezuelan Exclusive Economic Zone (1993–Present); (**b**) monthly sea level change (1993–2025); (**c**) contributions to sea level rise (1993–2024); (**d**) projected sea level rise (2025–2150), according to the data source from Dangendorf et al. (2024) and available in World Bank, Climate Change Knowledge Portal (2025)

fishing settlements along the Venezuelan coast underscores the socioeconomic importance of artisanal fisheries, which are at risk due to rising sea temperatures that drive migratory species into deeper, cooler waters, resulting in increased fishing effort and reduced access for local communities. Changes in capture, production, and marketing are expected, alongside heightened food insecurity and risk exposure for communities reliant on fishing and aquaculture (Ministerio del Poder Popular para el Ambiente 2011). Rising sea levels also threaten mangrove ecosystems (e.g., *Rhizophora*, *Avicennia*, *Laguncularia*), which provide critical ecological functions and livelihoods for coastal and indigenous populations.

Figure 4.11a presents the temporal evolution of sea level changes across months in the Venezuelan Exclusive Economic Zone from 1993 to 2025, categorized by time periods. The color scale represents different decades: dark blue (1992–2000), light blue (2001–2010), yellow (2011–2020), and orange (2021–2025).

The data clearly indicate a progressive rise in sea level over time, with the most recent years (2021–2025) consistently showing the highest sea levels across all months, peaking in October. Earlier periods (1992–2000) exhibit the lowest sea levels, with values ranging close to or below zero in some months. The seasonal cycle is evident, with minimum sea level typically observed in March and a general upward trend until a maximum around September–October. These findings reflect both interannual variability and long-term trends consistent with global sea level rise due to climate change, suggesting increasing risks for Venezuela's low-lying coastal and marine ecosystems.

Figure 4.11b illustrates the monthly variability in sea level within Venezuela's Exclusive Economic Zone from 1993 to 2025, categorized by periods using a color gradient. A clear upward trend in sea level is evident, particularly pronounced from 2011 onward, with the most recent years (2021–2025) showing values exceeding 200 mm in several months. While seasonal fluctuations persist, the multi-decadal increase underscores the growing influence of global climate change. This upward trend is consistent with rising global temperatures, thermal expansion of ocean waters, and the increased melting of land-based ice. The cyclical nature of monthly variability reflects ocean-atmosphere interactions, but the progressive increase in average levels indicates persistent climatic forcing beyond natural variability.

Figure 4.11c presents a decomposition of the major contributing processes to sea level rise in Venezuela's marine territory between 1993 and 2024. The largest contribution comes from sterodynamic processes (30%), which encompass thermal expansion and large-scale ocean circulation changes. Vertical land motion accounts for another 25%, highlighting local geophysical factors such as tectonic subsidence or uplift. Additional contributions stem from glacier melt (15%), the Greenland Ice Sheet (15%), and the Antarctic Ice Sheet (5%), reflecting global cryospheric dynamics. Terrestrial water storage (10%), including changes in soil moisture, groundwater extraction, and reservoir storage, also plays a notable role. This breakdown highlights the importance of both global drivers and local/regional processes in determining relative sea level rise, with implications for coastal vulnerability assessments and long-term planning.

Figure 4.11d displays projected sea level rise in the Venezuelan Exclusive Economic Zone from 2025 to 2150 relative to a baseline of 1995–2015, under a range of Shared Socioeconomic Pathways (SSPs). Under the high-emissions scenario SSP5-8.5, sea level is projected to exceed 2 m by 2150, with the potential for more extreme outcomes (up to 5 m) when considering the low-confidence range. In contrast, the most optimistic pathways (SSP1-1.9 and SSP1-2.6) limit sea level rise to under 1 m. These scenarios suggest that the extent of future sea level rise is strongly contingent upon mitigation strategies and global emissions trajectories. For Venezuela, which has extensive low-lying coastal zones, including deltaic and estuarine environments, the projected rise presents major risks to infrastructure, human settlements, and coastal ecosystems. These findings stress the urgency of climate adaptation and coastal resilience planning in national policy.

Vegas-Vilarrubia et al. (2015) project erosion, flooding, and saline intrusion in the Orinoco Delta with a 1-m sea level rise by 2100, jeopardizing mangroves, wetlands, and the ancestral territories of the Warao people. Although sediment accretion and vegetation growth may raise deltaic surfaces in some areas, this is unlikely to fully offset the impacts of sea level rise. Ward et al. (2016) further highlight that mangroves are especially sensitive to changes in flood frequency and salinity; prolonged flooding and salt stress can reduce species diversity and productivity, diminishing ecosystem services. Variability in local elevation (e.g., Delta Amacuro's rise of 1.8 mm/year) and altered rainfall patterns may influence sediment dynamics, but during droughts, reduced freshwater input will likely impair mangrove resilience, leading to shifts toward less productive riparian vegetation. These transformations threaten not only ecological integrity but also the cultural and economic stability of coastal populations.

In Venezuela, mangroves and other wetlands in its marine-coastal environments will be negatively affected, both ecologically and in terms of their use. Khan et al. (2017) have reported variations in sea level rise (RSL) in the marine-coastal environments (wetlands) of Venezuela. For the western part of the country, Venezuela's record begins in the mid-Holocene, placing the rise (RSL) at −0.3 ± 0.4 m at 4.4 ka. Thereafter, the RSL slowly increased to 0.2 ± 0.3 m/ka until it reached −0.1 ± 0.4 m at 0.9 ka. In the case of the Orinoco Delta, the RSL record of the Orinoco Delta, on the eastern coast of Venezuela, shows good temporal coverage during the last ~8 ka, with a rapid increase in the RSL from 2.7 ± 0.6 m/ka in the early to mid-Holocene. Sea level rise went from −6.8 ± 3.0 m at 7.8 ka to −1.0 ± 0.9 m at 5.9 ka. The position of RSL remained relatively constant during the middle and late Holocene, slowly increasing at a rate of 0.2 ± 0.2 m/ka to −0.3 ± 1.1 m at 0.5 ka.

Marrero and Rodríguez-Olarte (2017) consider that environments such as coastal lagoons and deltas are vulnerable to climate change and variation in sea levels; the rise in sea levels would increase saline intrusion into estuaries and deltas, modifying the properties of the waters, affecting plant and animal communities. The worsening, or increase of vulnerability, would not only affect these biological systems in their functioning and biodiversity, but would consequently affect the ecosystem services, the economy, and the quality of life of the inhabitants of a large sector of the Venezuelan population that has settled in these regions. It would impact the provision of biological resources, including fishing, as well as recreational and tourism services, and atmospheric carbon sequestration, among others.

Villamizar et al. (2017) remark that the South American region is particularly vulnerable to climate change and weather events such as El Niño (ENSO), which threaten its economy and sustainable development. Mangrove losses are encouraged by anthropogenic factors, for example, deforestation and land conversion, agriculture, and shrimp ponds, which added to climate change can lead to total loss over the next 100 years. If the Coastal Vulnerability Index (UNEP-CVI) is recalculated, Venezuela would be classified as one of the countries with the least exposed coasts (very little vulnerable). At least 6% of South America's population lives in low-lying coastal areas exposed to sea level rise, ENSO variability and storms, so climate adaptation is a regional priority.

According to MINEA (2017b), climate change scenarios project divergent trends for key fishery species in Venezuela. Simulations using RCP 4.5 and RCP 8.5 scenarios with temperature data from MIROC5 and HADGEM2 models show a consistent decline in Catch per Unit Effort (CPUE) for sardines (*Sardinella aurita*), a species that thrives in cooler waters and benefits from strong coastal upwelling. Reduced upwelling and warmer sea surface temperatures could lower plankton availability, negatively affecting sardine abundance. Conversely, CPUE for tuna (*Thunnus albacares*) is projected to increase, particularly under RCP 8.5 and HADGEM2 simulations, as warmer waters around 28 °C favor its distribution. These contrasting trends are expected to affect artisanal and industrial fisheries, especially in communities dependent on sardine fisheries, such as in Nueva Esparta, which has already seen steep declines in sardine catches following years of elevated sea surface temperatures. Any change in the catch rates of one of these two species can have an impact on various sectors of Venezuelan society, possibly on the target species of artisanal and industrial fisheries, along with the communities that benefit from both sectors.

Isaac and Ferrari (2017) further warn that the Northern Shelf of Brazil, which includes Venezuela's coastal waters, is highly sensitive to warming due to riverine inputs from the Orinoco and Amazon. Marine productivity has declined from reduced light and increasing temperatures, potentially reducing maximum fish catch by 16–50% by mid-century. Villamizar and Cervigón (2017) emphasize that Venezuela is the leading fishing nation in the Atlantic Caribbean, yet its eastern coastal upwelling system, vital for sardine productivity, is under increasing stress from climate change. Historical warming events (e.g., 1995, 1998, 2005) have coincided with major drops in *Sardinella aurita* landings, severely impacting fishers, processors, and consumers. The collapse of sardine populations exemplifies the broader vulnerability of Venezuela's coastal fisheries to climate-induced changes in oceanographic processes.

Ding et al. (2017) analyzed the vulnerability of 109 countries in terms of the implications of climate shocks on marine fisheries for food security. For countries most vulnerable to climate-induced impacts on marine fisheries, more than two-thirds of them relied on domestic marine fisheries as their main source of fish supply. These authors conclude that, in the case of Venezuela, there is a high level of exposure, a low level of sensitivity, and a moderate capacity for adaptation, which classifies the country as moderately vulnerable to climate change.

According to the IPCC (2018), model projections of global mean sea level rise (compared to 1986–2005) suggest an indicative range of 0.26–0.77 m by 2100 if global warming is 1.5 °C, exposing small islands, low-lying coastal areas and deltas to the risks associated with sea level rise, in particular saltwater intrusions and flooding. If global warming is limited to 1.5 °C instead of 2 °C, it is estimated that impacts on terrestrial, coastal, and freshwater ecosystems will be lower and ecosystem services will be conserved. An increase of 1.5 °C will reduce the raise in temperature in the oceans, the associated increase in their acidity, and the decrement in their oxygen level. The risks are expected to be lower for biodiversity, fisheries and marine ecosystems, as well as the ecosystem functions and services they provide. However, they are also expected to cause the loss of coastal resources and reduce the productivity of fisheries and aquaculture, especially at low latitudes (IPCC 2018).

Jarvis et al. (2019) identify Venezuela as one of the most vulnerable countries in Latin America to climate change impacts on marine fisheries, posing risks to food security. Nagy et al. (2019) emphasize that mangroves, which provide essential ecosystem services, face increased vulnerability due to both climatic (e.g., sea level rise) and non-climatic (e.g., land-use change) stressors. These pressures compromise coastal ecosystem services and exacerbate social inequalities. Cortés et al. (2020) advise that rising sea levels, temperature increases, acidification, and extreme events threaten the survival of Caribbean mangroves, while Venezuela's low marine protection rate and poor Ocean Health Index ranking underscore its environmental fragility. The degradation of coastal ecosystems will have direct socioeconomic impacts, particularly in tourism, fisheries, and food security.

Villamizar (2020) reports evidence of sea level rise and mangrove loss in low-elevation coastal areas such as the Hueque-Sauca Wetland, where a 3 mm/year rise and a 375-ha loss of mangrove cover have been recorded. Venezuela's extensive low-lying coastline (80% of 6068 km) increases its exposure to coastal hazards. This degradation compromises the resilience and productivity of mangrove ecosystems, essential for rural livelihoods. Rull (2022) documents accelerating sea level rise in the Caribbean, with regional variability, and projects warming of 1–3.5 °C and a potential 20–30% decline in rainfall, coupled with more frequent extremes.

Finally, FAO (2022) and Pörtner et al. (2022) warn that ocean warming will significantly reduce aquatic food supplies, with global marine animal biomass projected to decline by up to 15.5% by 2080–2099 under high-emission scenarios. These changes will alter fish catch composition, impact income and food security, and particularly affect coastal populations unable to adapt or diversify. Villamizar et al. (2017) stress the urgent need for adaptation policies in Venezuela's coastal zones to address mounting climate-related risks.

4.7 Conservation of Watersheds and Their Impact on Agriculture in Mountainous Areas and Lowlands

The management and conservation of watersheds in the mountainous areas of Venezuela will be decisive of projections of an increase in water-deficient lands and irrigation needs in low-lying agricultural areas and even in mountain landscapes. Climate change will affect future access to drinking water, as well as the supply for irrigated agriculture in the tropical Andes and in other watersheds of the country, where multiple-use reservoirs have been built (close to 100) in Venezuela. The five glacial cirques that persist (of the 10 that existed in 1952) registered a rapid retreat during the twentieth century, losing more than 95% of their surface since 1850, with a direct impact on the hydrological cycle and replacement of water for irrigation and other downstream uses. The most important changes in flow are expected to occur during the dry season, affecting the supply of water for numerous uses, including agricultural activities (Vuille et al. 2008; Vuille 2013; Braun and Bezada 2013; Larsen 2017).

According to Paredes-Trejo et al. (2018), precipitation's spatial and temporal distribution over the large Venezuelan hydrographic regions has changed in recent decades. A global warming of 1.5 °C could lead to a relative decrease of 5% to 10% in average rainfall across several regions, including the Andes, the Coriano and Deltaic systems, the Central and Eastern Mountain Ranges, and the Lake Maracaibo basin. As a result, river hydrological regimes are expected to experience reduced flows due to diminished rainfall inputs. Specifically, projections indicate a decline in rainfall contributions to the hydrographic basins of the Motatán, Misoa, Machango, Torondoy, Chirurí, and Tucaní rivers in the Lake Maracaibo region, as well as in the upper Cuyuní basin in Bolívar State. This reduction could negatively affect aquifer recharge and limit water availability for existing reservoirs, which are crucial for irrigation and agricultural use within these basins.

4.8 Incidence of Pests and Diseases Driven by Climate Change

A foreseen consequence of climate change is the occurrence of new agricultural pests or diseases or the intensification of these where they are already present. With the advent of new diseases or pests that could potentially intensify with climate change and that could affect crops in the country, it was not possible to assemble many articles that would allow to have a generalized and updated idea of the incidence of diseases and pests in crops in Venezuela. Projections indicate that the wide high thermal range could increase pest risks and that changes in precipitation will probably create favorable environmental conditions for many pathogens (fungi and bacteria) or reduce the defense capacity of plants against pathogen attacks (Paredes 2014).

4.8.1 In Mountainous Agricultural Areas

Lozada et al. (2008) determined the dates of low climatic risk of spread for “late potato blight” (*Phytophthora infestans* Mont de Bary) in the Andean region of Venezuela, using an agrometeorological model of the disease and geographic information system (GIS) tools. The disease model used in this study was developed by Hyre in 1954, which requires daily precipitation and temperature data, which were obtained from 106 meteorological stations, located in the states of Mérida, Táchira and Trujillo, for 31 years. The potato (*Solanum tuberosum* L.) is one of the main crops in the Andean region of Venezuela. The current climatic conditions are very favorable for “late potato blight,” which is the main limitation for local potato production, causing losses of up to 100% in some years. It would be expected, that in the face of a projected increase in temperature and precipitation in the Andean potato production areas, favorable conditions for the spread and degree of infestation of this disease will intensify. The potato will also be vulnerable to heat that affects the development of the plant and tuber, which will cause a drop in yields.

The increasing temperature and humidity will favor the development of potato "late blight," one of the pests that can cause the most damage to this tuber and that could spread to areas located above 3000 m asl, where it is currently non-existent.

Kroschel et al. (2013) estimated that climate change will increase the distribution and abundance of *Phthorimaea operculella* (potato tuber moth or tomato leaf miner). Currently, this pest already has suitable temperature conditions in most potato-growing regions of Venezuela and Colombia. By 2050, its distribution range could expand from 94.3% to 97.3% of the potato production area in Venezuela and from 87.7% to 96.0% in Colombia. In the Andean countries, the area with >4 generations per year is projected to grow from 29% to 46.6%, affecting an additional 38,032 ha with 4–6 generations and 72,148 ha with severe infestations of 7–15 generations annually. The authors conclude that the damage potential of *P. operculella* will progressively increase, especially in warmer tropical and subtropical regions.

4.8.2 *In Lowland Agricultural Areas*

Pörtner et al. (2022) project that climate change will intensify the emergence and spread of pests, diseases, and invasive species, increasing biotic stress on crops, forests, and livestock, with negative impacts on food security and rising production costs due to greater use of control measures. Ortega Andrade et al. (2017) warn that global warming could expand the distribution of *Moniliophthora roreri*, the fungus responsible for cocoa rot, especially in southern Ecuador and parts of Venezuela (Bolívar, Amazonas, Delta Amacuro), with projections showing persistence through 2080 under optimal conditions of precipitation, temperature, and humidity. The main conditions for determining the growth of *M. roreri* are related to precipitation, minimum temperature in the coldest month, maximum temperature in the warmest month, and relative humidity. Their combination allows optimal microenvironmental conditions for the fungus, *Theobroma cacao* itself and its distribution in Latin America.

Olivares et al. (2021) used ecological niche modeling (MaxEnt) to assess the risk of *Fusarium oxysporum* f. sp. *cubense* TR4 (Foc TR4) in Venezuela, identifying moderate to high suitability in key banana-producing states like Zulia, Trujillo, Miranda, and Sucre. While regions like the Andes and parts of Bolívar show low risk, most northern zones show moderate to high potential for Foc TR4 establishment. Olivares (2022) highlights that human activity is a major factor in the spread of Foc and stresses the importance of soil management and sustainable agricultural practices to mitigate its impacts.

Rodríguez-Román et al. (2023) report that during the pepper production seasons of 2008, 2014, and 2022, virus occurrence was observed in the states of Lara (central west) and Miranda (center) of Venezuela. In their study, it was discovered for the first time that the pepper is a natural host of AMV (alfalfa mosaic virus) and TRV (tobacco rattlevirus) in Venezuela. They also detected another divergent isolate of Potyvirus in 23.75% of the pepper plants in Lara state. Based on the taxonomic criteria of Potyvirus, these results suggest that the isolate represents a distinct species of virus, for which it was proposed the name "severe pepper mottling virus" (PepSMoV). The authors indicate that occurrences of the virus could be attributed to agricultural and environmental factors, such as climate change, wastewater use, use of uncertified seeds, misuse of agricultural chemicals, transmission with food trade networks, and the development of new viral strains due to mutations and recombination, and the spread of pathogens.

The potential incidence of new pests and diseases or the intensification and expansion of existing ones in Venezuela is a threat to crops in the face of imminent climate change, a situation that is aggravated by the existing crisis in the institutions responsible for phytosanitary and zoosanitary research.

Marys and Rosales (2021) argue that plant health services in Venezuela have been severely weakened due to the country's economic collapse and the deterioration of public institutions. As a result, many plant pathogens remain unidentified, and surveillance and crop protection strategies are lacking. The emergence and spread of both new and known pathogens, such as rice blight, potato moth, coffee rust, banana wilt, and cocoa pod rot, have disrupted food production and threatened rural livelihoods. Notably, the citrus industry has been devastated by huanglongbing (HLB), and the arrival of *Fusarium* TR4 poses a severe risk to banana and plantain crops.

The authors warn that the spread of invasive plant diseases is a growing national and regional crisis with long-lasting effects on food security. Limited infrastructure, equipment, and trained personnel hinder effective phytosanitary responses. Broader factors, such as globalization, climate change, evolving pathogens, and political instability, further exacerbate the risks. Additionally, Venezuela has lost 16% of its scientific workforce to emigration, weakening its research capacity. The authors suggest that building partnerships with the Venezuelan scientific diaspora could help develop international research programs to support national phytopathology efforts.

In the opinion of Maytín et al. (2019, 2020), it is a relevant challenge to integrate the knowledge of those potentially affected by climate change, such as Venezuelan farmers, into research and scientific knowledge, under participatory approaches. In other words, the science of national climate change must be developed as "science with the people" and not only as "science for the people," which is highly recommended to achieve adequate levels of design and implementation of measures for adaptation to climate change.

4.9 Conclusions

Climate projections for Venezuela envisage an increase in temperature and changes in precipitation patterns that will have negative consequences for Venezuelan agriculture, including reduced suitability of cropland for large numbers of crops, intensification of pests and diseases, and decreased crop yields, which could lead to greater food insecurity and rural poverty.

The tropical Andes will be one of the regions identified as highly sensitive and exposed to the effects of global warming, with greater changes in the highest parts, where exposure to impacts would increase. The potential displacement of agricultural aptitude to higher altitudes in the Venezuelan Andes is recognized, and the incidence of pests and diseases exacerbated by climate change is revealed.

Climate change will increase local temperatures and evapotranspiration in mountainous areas, leading to water scarcity, glacier retreat, and hydrological changes that will undermine irrigation and water security, reduction (and even losses) of important crops such as high-altitude vegetables, high-altitude fruit trees, and coffee in several mountain agricultural regions. Potatoes and coffee could be forced to move to higher altitudes in search of suitable climatic conditions, which could incite the loss of areas suitable for cultivation and an increase in the intervention and fragmentation of natural ecosystems, many of which are protected under the figure of Areas Under Special Administration Regime (ABRAE).

The impacts of climate change in the form of more severe rainfall, the expansion of the agricultural frontier toward higher levels, and the intensification of agriculture would generate production systems more sensitive to soil erosion, which is already a manifest menace to Andean communities. Venezuela would represent the most serious case of the tropical Andean countries, since 19 of the 25 most common crops in the region would present a loss of climate suitability in the future in one or both emission scenarios. Research is deficient in these areas of the country, reflected in the scarce publication of updated works on the specific impacts of climate change on Venezuelan mountain agriculture.

The projections for the RCP 4.5 and RCP 8.5 scenarios made it possible to verify, for the country, the reduction of the area occupied by humid climates and an increase in the space occupied by dry climates (arid and semi-arid). In the context of South America, Venezuela could have greater irrigation needs to maintain agricultural production.

In the lowlands, climate change will negatively impact agricultural activity, with projections indicating reductions in yields of crops such as corn, beans, beans, rice, low-floor vegetables, and sugarcane. Climate change represents a serious threat to Venezuelan agriculture, with differentiated regional impacts that require adaptation and conservation measures to mitigate the negative effects. Expected variations in temperature and precipitation will increase pressure on agricultural land and natural resources.

There are important gaps in our basic knowledge of the spatio-temporal patterns of climate and the ecological and social impacts of climate change in the Venezuelan Andes. There is no updated analysis of the knowledge needs around adaptation in the Andes, which would allow the identification of priority areas of work to face the most urgent challenges imposed by global environmental change.

For several climate change scenarios, coffee would decrease the suitable area, while potato cultivation would lose 69% of the potential area, by 2100. Current coffee-friendly areas change indicating that any alteration in precipitation will bring with it a reduction in the areas with current climatic aptitude, but these will probably climb in the altitudinal gradient. From an environmental point of view, any change in the lands suitable for both coffee and potatoes would expand the degree of intervention and fragmentation of the natural ecosystems that are at present not intervened. Shade coffee will probably be forced to move up the altitudinal gradient competing with some protected areas, due to a decrease in climate-friendly land for planting. In either scenario, coffee cultivation will likely tend to lose its current suitable areas for possible development.

Projections of changes in climatic variables suggest that in the tropical Andes region, agricultural productivity could fall by between 12% and 50%. A decrease in food availability and an average per capita reduction of at least 300 calories (a 12% reduction) are expected by 2050.

Various studies project for Venezuela a considerable expansion of the areas with a negative annual water balance. The categories of semi-arid will increase by about 106,000 km^2 and that of dry sub-humid by 107,000 km^2, located especially in the northwestern region (Zulia and Falcón) and the states of Guárico, Anzoátegui, and Monagas. These changes in semi-arid and dry sub-humid reveal differences in agricultural suitability for the four critical crops in Venezuela: corn, sugarcane, rice, and cassava. The water deficit will increase from 1,000,000 to 4,700,000 ha in the central region of the country, whose drought events will have repercussions on the yields of bananas and other crops native to the region. In general, a drier future is evident for banana areas, which could lead to a reduction in the suitability of rainfed lands for this crop. In a scenario of continuous emissions by 2100, cocoa cultivation would disappear in the state of Mérida. In relation to banana cultivation, any change in precipitation due to climate change would affect the distribution of potential areas, until the disappearance of the current areas.

Climate change is likely to alter pest and disease dynamics and the emergence of new pathogens, potentially increasing risks. Climate scenarios suggest heightened risks of pests adapting, expanding their ranges, and causing economic losses. Warmer temperatures favor the spread and lifecycle acceleration of pests like late potato blight, tomato moths, and cocoa rot among others. Diseases such as *Moniliophthora roreri* (cocoa) and Fusarium wilt threaten key crops, with models predicting expansion of vulnerable areas. The potential for *Fusarium oxysporum* f. infestation in the northern states of the country is estimated to be moderate. However, most of the banana-growing areas of the states of Zulia, Trujillo, Miranda, and Sucre have climatic conditions that make them highly suitable for the establishment of Foc TR4. Few specific studies describe the impact of environmental change on livestock and wildlife diseases or the emergence of pathogens, especially in the South American context. The disease situation in some South American countries remains uncertain because animal disease surveillance and vector surveillance systems are weak in most countries.

Phytosanitary services and thus the control of plant pests and diseases have been considerably weakened in recent years in the country, because of the collapse of the economy and the substantial deterioration of government services. Most of the pathogens associated with symptom-causing diseases remain unidentified or uncharacterized, and surveillance and crop protection strategies have not been implemented.

Floods and droughts are the most important extreme hydroclimatic phenomena in the Amazon forest lands. A slow shift toward a drier Amazon is taking place, with changes in the dynamics of floristic composition, consistent with climate change factors. The Amazon rainforest is at risk of becoming a different biome and moving from a carbon sink to a source of carbon emissions. Most of the forests north of the Orinoco are at risk of danger due to the effects of climate change, while in the south of the country the situation is less worrying.

Coastal ecosystems, such as mangroves, are vulnerable to sea level rise and erosion, impacting fisheries and local livelihoods. The use of forests that occupy marine-coastal environments and wetlands (mangroves and flooded forests) could be impacted considering a sea level rise trend scenario by 2090 of approximately 0.5 m, exacerbating flooding, saline intrusion and deteriorating wetlands and estuarine systems of the Caribbean and Atlantic facades, and the Maracaibo estuary fisheries.

In Latin America, grasslands in tropical areas will be more vulnerable to climate change. The low availability of water and the consequent reduction of nutrients in the soil seem to be the most important limiting factors for the quality of the grasslands in this region. Climate change would cause a reduction in rainfall in part of the Llanos de Apure and Barinas, with potential negative effects on the hygrophilous savannas of these plains, which represent the necessary food environment for livestock activity, reducing their production in the region. Venezuela is among the 33 "highly vulnerable" countries due to its high exposure to climate change by 2050, and moderate sensitivity of fishermen and adaptive capacity for the IPCC's B2 scenario, when compared to other countries.

Communities dependent on fishing and aquaculture activity could become more vulnerable in terms of food stability, decreased availability and quality of fish, and risks, for example, fishing in adverse weather conditions and away from usual spaces. The Catch per Species Unit (CPUE) projections under two hypothetical climate change scenarios (RCP 4.5 and RCP8.5) generally show a trend toward a decrease in the annual CPUE of the sardine (*Sardinella aurita*). Regarding the species *Thunnus albacares* (Tuna), there is a tendency toward increasing catch rates with higher CPUE values when using temperature data. In Latin America and the Caribbean, Venezuela is one of the country's most vulnerable to climate change in marine fisheries, which has an impact on its food security.

Notably, there are significant gaps in region-specific research, emphasizing the need for localized data, technological advances, and participatory adaptation strategies. Overall, climate change poses a severe, multifaceted threat to Venezuelan agriculture and ecosystems. Addressing these challenges demands integrated, regionally tailored responses, increased research, and sustainable practices to safeguard food security and environmental resilience.

The projected scenarios indicate that, if decisive measures are not taken, Venezuela will face an intensification of the effects of climate change, including loss of biodiversity, decrease in agricultural productivity, increase in rural poverty, and deterioration of vital ecosystems such as moors and mangroves. However, with coordinated actions, investment in research, and active participation of all social and productive actors, it is possible to build a more resilient future. Adaptation based on scientific knowledge, sustainable practices and integrated policies will be key to mitigate the impacts and take advantage of the opportunities that arise in a context of accelerated global change.

4.10 Future Outlook

Some recommendations and viewpoints to face climate change in Venezuela include:

(a) Strengthening regionalized research: It is essential to increase and update specific studies on the effects of climate change in the different Venezuelan regions, especially in the Andean mountains and lowlands, to design effective and contextualized adaptation strategiesAdaptation strategies.
(b) Implementation of sustainable agricultural practices: Promote resilient agricultural systems, such as agroecology, regenerative and conservation agriculture, irrigationIrrigation, integrated pest and disease management, and soil conservation techniques, which reduce vulnerability to extreme weather events.
(c) Efficient management of water resources: Develop and implement comprehensive water management plans, including the conservation of glaciers, the protection of watersheds, and the optimization of the use of water for irrigationIrrigation and human consumption, to guarantee water security in a scenario of growing scarcity.
(d) Protection and conservation of key ecosystems: Prioritize the conservation of moorlands, forests, mangroves, and wetlands, which act as carbon sinksCO2 sinks, regulators of the hydrological cycle and biodiversity habitats, strengthening ecological protection and restoration policies.
(e) Adaptive infrastructure development: Invest in infrastructure resilient to extreme weather events, such as drainage systems, levees, and reservoirs, to reduce risks of flooding, erosion, and crop loss.
(f) Training and community participation: Promote environmental education and the active participation of rural and peasant communities in resource management and decision-making, promoting traditional knowledge and appropriate technologies.
(g) Integrated public policies: Design and strengthen national policies that integrate climate change adaptationClimate change adaptation and mitigationMitigation at all levels of agricultural, forestry, and natural resource management planning, with a focus on sustainability and social equity.
(h) Expand and modernize the hydrometeorological station network.
(i) Encourage technological innovation: Promote the use of satellite monitoring technologies, climate prediction systems, and precision agriculture to anticipate adverse events and optimize responses.

(j) Productive diversification: Incentivize the diversification of crops and economic activities to reduce dependence on a few vulnerable products and increase the economic resilience of communities.
(k) Regional and international cooperation: Leverage partnerships with international agencies, neighboring countries, and regional organizations to share knowledge, resources, and strategies for adaptation to the effects of climate change.

References

Aggarwala P, Vyasa S, Thorntona P, Campbella BM, Kropffe M (2019) Importance of considering technology growth in impact assessments of climate change on agriculture. Glob Food Secur 23:41–48. https://www.researchgate.net/publication/332494819

Allison EH, Perry AL, Badjeck M, Adger WN, Brown K, Declan CD, Halls AS, Pilling GM, Reynolds JD, Andrew NL, Dulvy NK (2009) Vulnerability of national economies to the impacts of climate change on fisheries. Aust Fish Fisher 10:173–196. https://doi.org/10.1111/j.1467-2979.2008.00310.x

Altieri MA, Nicholls CI, Henao AH, Lana MA (2015) Agroecology and the design of climate change-resilient farming systems. Agron Sustain Dev 35:869–890. https://doi.org/10.1007/s13593-015-0285-2

Anderson EP, Maldonado-Ocampo JA (2010) Regional perspective on the diversity and conservation of tropical Andean fishes. Conserv Biol 25(1):30–39. https://www.academia.edu/15349173/A

Bárcena A, Samaniego JL, Peres W, Alatorre JE (2020) La emergencia del cambio climático en América Latina y el Caribe: ≃seguimos esperando la catástrofe o pasamos a la acción? Libros de la CEPAL, No. 160 (LC/PUB.2019/23-P). Comisión Económica para América Latina y el Caribe (CEPAL), Santiago. https://n9.cl/yaf7x

Barrios MR, Mark Del V, Rivas E, Fariñas J, Salazar J, Rodríguez G (2011) Efecto del déficit hídrico sobre el ciclo productivo de la palma aceitera en el estado Monagas, Venezuela. Agron Trop 61(3–4):267–274. http://ve.scielo.org/pdf/at/v61n3-4/art09.pdf

Braun C, Bezada M (2013) The history and disappearance of glaciers in Venezuela. J Lat Am Geogr 12(2):85–124. https://www.researchgate.net/publication/265202332

Buytaert W, Ramírez-Villegas J (2012) Generación de escenarios desagregados del cambio climático para los Andes Tropicales. In: Cuesta F, Bustamante M, Becerra MT, Postigo J, Peralvo J (eds) Panorama andino de cambio climático: vulnerabilidad y adaptación en los Andes Tropicales. CONDESAN, SGCAN, Lima, pp 37–58. https://www.researchgate.net/publication/261216340

Buytaert W, Cuesta-Camacho F, Tobón C (2011) Potential impacts of climate change on the environmental services of humid tropical alpine regions. Glob Ecol Biogeogr 20:19–33. https://doi.org/10.1111/j.1466-8238.2010.00585.x

Chacón-Moreno E, Rodríguez-Morales M, Paredes D, Suárez del Moral P, Albarrán A (2021) Impacts of global change on the spatial dynamics of treeline in Venezuelan Andes. Front Ecol Evol 9:74. https://doi.org/10.3389/fevo.2021.615223

Christopher POR, Adams S, Albrecht T, Baarsch F, Boi A, Canales NT, Cartsburg M, Coumou D, Eden A et al (2015) Climate change impacts in Latin America and the Caribbean and their implications for development. Reg Environ Change. https://doi.org/10.1007/s10113-015-0854-6

Cortés J, Villamizar A, Nagy GJ, Girot PO, Miglioranza KSB, Villasante, S (2020) Ecosistemas marino-costeros. In: Adaptación frente a los riesgos del cambio climático en los países iberoamericanos—Informe RIOCCADAPT [Moreno, J.M., C. Laguna-Defior, V Barros,

E. Calvo Buendía, J.A. Marengo and U. Oswald Spring (eds.)]. McGraw-Hill, Madrid, España (pp. 131–160., ISBN: 9788448621643. https://rioccadapt.com/wp-content/uploads/2020/07/04_Cap_4_CambioClimatico.pdf)

Correa DF, Beyer HL, Possingham HP, García-Ulloa J, Ghazoul J, Schenk PM (2020) Freeing land from biofuel production through microalgal cultivation in the Neotropical region. Envirom Res Lett 15 (2020) 094094. https://doi.org/10.1088/1748-9326/ab8d7f

Cuesta CF, Bustamante M, Becerra MT, Maldonado G (2012) Introducción: cambio climático y los Andes Tropicales. In: Cuesta F, Bustamante M, Becerra MT, Postigo J, Peralvo J (eds) Panorama andino de cambio climático: vulnerabilidad y adaptación en los Andes Tropicales. CONDESAN, SGCAN, Lima, pp 12–17. https://condesan.org/recursos/panorama-andino-cambio-climatico/

Dangendorf S, Sun Q, Wahl T, Thompson P, Mitrovica JX, Hamlington B (2024) Probabilistic reconstruction of sea-level changes and their causes since 1900. Earth Syst Sci Data 16:3471–3494. https://doi.org/10.5194/essd-16-3471-2024

De Bièvre B, Bustamante M, Buytaert W, Murtinho F, Armijos MT (2012) Síntesis de los impactos de los efectos del cambio climático en los recursos hídricos en los Andes Tropicales y las estrategias de adaptación desarrolladas por los pobladores. P5. In: Cuesta F, Bustamante M, Becerra MT, Postigo J (eds) Panorama Andino de cambio climático: vulnerabilidad y adaptación en los Andes Tropicales. Condesa SGCAN, Lima, pp 59–101. https://www.researchgate.net/publication/261216340

Delgado SA (2017) Assessment of long-term effects of woody plants expansion in humid, tropical grassland of Venezuela. Theses and dissertations, Graduate College of the University of Texas, Rio Grande Valley, 241 p. https://scholarworks.utrgv.edu/etd/241

Ding Q, Chen X, Hilborn R, Chen Y (2017) Vulnerability to impacts of climate change on marine fisheries and food security. Yellow Sea Fisheries Research Institute, Chinese Academy of Fisheries Sciences. Reporter: Qi Ding. https://doi.org/10.1016/j.marpol.2017.05.011

Esquivel-Muelbert A, Bake TR, Dexter KG, Lewis SL, Brienen RJW, Feldpausch TR et al (2019) Compositional response of Amazon forests to climate change. Glob Change Biol 25:39–56. https://doi.org/10.1111/gcb.14413

Eyring V, Bony S, Meehl GA, Senior CA, Stevens B, Stouffer RJ, Taylor KE (2016) Overview of the Coupled Model Intercomparison Project Phase 6 (CMIP6) experimental design and organization. Geosci Model Dev 9:1937–1958. https://doi.org/10.5194/gmd-9-1937-2016

FAO (2022) Versión resumida de El estado mundial de la pesca y la acuicultura 2022. Hacia la transformación azul. Food and Agriculture Organization of the United Nations FAO, Roma. https://doi.org/10.4060/cc0463es

FAO-CIAT (2018) Retos del Cambio Climático para la Agricultura en América Latina y el Caribe. Informe para responsables de políticas (Borrador). Organización de las Naciones Unidas para la Agricultura y la Alimentación. 71 p. https://n9.cl/9ddje

Fernández GGA (2020) Determinación de áreas prioritarias para la adaptación al cambio climático mediante clasificacion esclimáticas. Tesis de Maestría, Universidad Simón Bolívar, Caracas, Venezuela. http://www.usb.ve/home/node/6222

García MP (2022) El cambio climático: posibles impactos en la agricultura en el contexto de América Latina y Venezuela. Agroalimentaria 28(55) https://n9.cl/gscq6

Garruña R, Pereyda-González J, Oliva-Ruíz M, Rodríguez-Castellanos A, Castillo-Collí M, Ríos-Bolívar FM, Cetina-Escalante RR (2021) Hortalizas tropicales: súperplantas ante el cambio climático. Bioagro Ciencias 14(1):46–55. https://n9.cl/qvnhho

Gourdji S, Mesa J, Moreno P, Navarro C, Obando D, Fisher MJ, Ramírez-Villegas J (2015) Modeling of present-day and future agricultural yields for regionally important crops. International Center for Tropical Agriculture (CIAT). https://n9.cl/q6wk6

Gouveia CM, Justino F, Gurjao C, Zita L, Alonso C (2023) Revisiting climate-related agricultural losses across South America and their future perspectives. Atmosphere 14:1303. https://doi.org/10.3390/atmo.14081303

Hagen CH, Ramajo L, Chacón N, Ometto JP, Postigo JC, Castellanos EJ (2022) Climate change-related risks and adaptation potential in Central and South America during the 21st century. Environ Res Lett 17:033002. https://doi.org/10.1088/1748-9326/ac5271

IPCC (2007) Cambio climático: impacto, adaptación y vulnerabilidad. Resumen para responsables de Políticas Informe del Grupo de Trabajo II del Grupo Intergubernamental de Expertos sobre Cambio Climático. https://www.ipcc.ch/site/assets/uploads/2020/02/ar4-wg2-sum-vol-sp.pdf

IPCC (2018) Resumen para responsables de políticas. In Masson-Delmotte V, Zhai P, Pörtner H-O, Roberts D, Skea J, Shukla PR, Pirani A, Moufouma-Okia W, Péan C, Pidcock R, Connors S, Matthews JBR, Chen Y, Zhou X, Gomis MI, Lonnoy E, Maycock T, Tignor M, Waterfield T (eds) Calentamiento global de 1.5 °C, Informe especial del IPCC sobre los impactos del calentamiento global de 1.5 °C con respecto a los niveles preindustriales y las trayectorias correspondientes que deberían seguir las emisiones mundiales de gases de efecto invernadero, en el contexto del reforzamiento de la respuesta mundial a la amenaza del cambio climático, el desarrollo sostenible y los esfuerzos por erradicar la pobreza. https://www.ipcc.ch/site/assets/uploads/sites/2/2019/09/IPCC-Special-Report-1.5-SPM_es.pdf

IPCC (2020) El cambio climático y la tierra. Informe especial del IPCC sobre el cambio climático, la desertificación, la degradación de las tierras, la gestión sostenible de las tierras, la seguridad alimentaria y los flujos de gases de efecto invernadero en los ecosistemas terrestres. Resumen para responsables de políticas. Grupo Intergubernamental de Expertos sobre el Cambio Climático. https://www.ipcc.ch/site/assets/uploads/sites/4/2020/06/SRCCL_SPM_es.pdf

Isaac VJ, Ferrari S (2017) Assessment and management of the North Brazil shelf large marine ecosystem. Environ Dev 22:97–110. https://www.sciencedirect.com/science/article/pii/S2211464516302317

Jarvis A, Loboguerrero A, Martinez-Baron DPS, Ramírez-Villegas J, Eitzinger A, Born L, González C, Tarapues J (2019) Situación rural de América Latina y el Caribe con 2 grados de calentamiento. In: 2030—Alimentación, agricultura y desarrollo rural en América Latina y el Caribe, No. 10. FAO, Santiago de Chile. 41 p. https://openknowledge.fao.org/server/api/core/bitstreams/ab6158c6-5725-474e-9ed2-23e19528aa90/content

Khan NS, Ashe E, Horton BP, Dutton A, Kopp RE, Brocard G, Engelhart SE, Hill DF, Peltier WR, Christopher H, Vane CH, Scatena FN (2017) Drivers of Holocene sea-level change in the Caribbean. Quat Sci Rev 155:13–36. https://doi.org/10.1016/j.quascirev.2016.08.032

Kroschel J, Sporleder M, Tonnang HEZ, Juarez H, Carhuapoma P, Gonzales JC, Simon R (2013) Predicting climate-change-caused changes in global temperature on potato tuber moth *Phthorimaea operculella* (Zeller) distribution and abundance using phenology modeling and GIS mapping. Agric For Meteorol 170:228–241. https://cgspace.cgiar.org/handle/10568/41979

Larsen MC (2017) Forested watersheds, climate change, ecosystem services, and natural hazards. In: Raynal Villaseñor JA (ed) Facing the threat: climate change, proceedings of the second international conference on hydrometeorological risks and climate change 2015, Coleción Memorias. Universidad de las Americas Puebla, Cholula, Mexico, pp 13–43. https://n9.cl/iycyr9

Lee DN, Papes M, Van Den Bussche RA (2012) Present and potential future distribution of common vampire bats in the Americas and the associated risk to cattle. PLoS One 7(8):e42466. https://doi.org/10.1371/journal.pone.0042466

Llambí LD, Garcés A (2021) Adaptación al cambio climático en Los Andes: Vacíos y prioridades para la gestión del conocimiento. CONDESAN, Quito-Ecuador. https://n9.cl/oz8g2

Longo M, Knox RG, Levine NM, Alves LF, Bonal D, Camargo PB, Fitzjarrald DR, Hayek MN, Restrepo-Coupe N, Saleska SR, da Silva R, Stark SC, Tapajos RP, Wiedemann KT, Zhang K, Wofsy SC, Moorcrof PR (2018) Ecosystem heterogeneity and diversity mitigate Amazon forest resilience to frequent extreme droughts. New Phytol 219:914–931. https://doi.org/10.1111/nph.15185

López SR (2009) Acerca de los impactos del cambio climático en Sudamérica. Instituto para el Desarrollo Rural de Sudamérica (IPDRS). 57 p. https://ipdrs.org/images/exploraciones/archivos/exploraciones_3.pdf

López-Feldman AJ (2015) Cambio climático y actividades agropecuarias en América Latina. Unidad de Cambio Climático de la División de Desarrollo Sostenible y Asentamientos Humanos de la Comisión Económica para América Latina y el Caribe (CEPAL). https://www.cepal.org/sites/default/files/publication/files/39824/S1501286_es.pdf

López-Hernández D, Infante C (2016) N cycle in a Venezuelan sugarcane plantation. How biogeochemical processes contribute to supply N needs. STJ Agri Sci 1(1):1003. https://www.researchgate.net/publication/313639789
Lozada GBI, Sentelhas PC, Tapia LR, Sparovek G (2008) Climatic risk for potato late blight in the Andes region of Venezuela. Sci Agric (Piracicaba, Braz) 65(Special Issue):32–39. https://doi.org/10.1590/S0103-90162008000700007
Lozano-Povis A, Álvarez-Montalván CE, Moggiano N (2021) El cambio climático en los Andes y su impacto en la agricultura: una revisión sistemática. Scientia Agropecuaria 12(1):101–108. http://www.scielo.org.pe/pdf/agro/v12n1/2306-6741-agro-12-01-101.pdf
Marquina S, Pérez T, Donoso L, Giuliante AR, Rasse R, Herrera F (2015) NO, N_2O and CO_2 soil emissions from Venezuelan corn fields under tillage and no-tillage agriculture. Nutr Cycl Agroecosyst 101(1):123–137. https://doi.org/10.1007/s10705-014-9659-0
Marrero C, Rodríguez-Olarte D (2017) Los humedales costeros venezolanos en los escenarios de cambios climáticos: vulnerabilidad, perspectivas y tendencias. In: Botello AV, Villanueva S, Gutiérrez J, Rojas Galaviz JL (eds) Vulnerabilidad de las zonas costeras de Latinoamérica al cambio climático. UJAT, UNAM, UAC, pp 461–476. 476 p. https://www.researchgate.net/publication/320947213
Marys E, Rosales LC (2021) Plant disease diagnostic capabilities in Venezuela: implications for food security. Front Sustain Food Syst 5:715463. https://doi.org/10.3389/fsufs.2021.715463/full
Maytín C, Carrasco R, Rodríguez RJ (2019) Contexto internacional y necesidades de investigación sobre cambio climático y agricultura en Venezuela. Terra Nueva Etapa XXXV(58) http://www.redalyc.org/articulo.oa?id=72164777001
Maytín C, Aquino RC, Rodríguez RJ, López J (2020) Una aproximación a la historia y situación actual de las investigaciones científicas sobre cambio climático en Venezuela. Algunos avances y retos temáticos. Terra Nueva Etapa 36(60) http://saber.ucv.ve/ojs/index.php/rev_terr/article/view/22991
Medina V, Laliberte B (2017) A review of research on the effects of drought and temperature stress and increased CO_2 on Theobroma cacao L., and the role of genetic diversity to address climate change. Biodiversity International, Costa Rica. https://core.ac.uk/download/pdf/132697857.pdf
Medina E, Velásquez G, Hernández I (2016) Impacto del calentamiento global y enriquecimiento atmosférico de CO_2 sobre cultivos tropicales: la perspectiva para Venezuela. Rev Fac Agron (UCV) 42(1):25–37. https://www.researchgate.net/publication/306013946
Ministerio del Poder Popular de Ecosocialismo y Aguas-MINEA (2017a) Segunda Comunicación Nacional ante la Convención Marco de las Naciones Unidas sobre Cambio Climático. Capítulo 3. In: Efectos del cambio climático sobre el sector agrícola, pp 237–245. https://n9.cl/z2255
Ministerio del Poder Popular de Ecosocialismo y Aguas-MINEA (2017b) Segunda Comunicación Nacional ante la Convención Marco de las Naciones Unidas sobre Cambio Climático. Capítulo 3. In: Efecto potencial del cambio climático sobre los principales recursos pesqueros ícticos de Venezuela, pp 247–257. https://n9.cl/kc2mip
Ministerio del Poder Popular para el Ambiente (2011) Implicaciones del Cambio Climático en las Zonas Costeras y el Espacio Acuático de Venezuela. In: Despacho del Viceministro de Ordenación y Administración Ambiental. Dirección General de Planificación y Ordenación Ambiental. Dirección Técnica de las Zonas Costeras, Caracas, Venezuela. 62 pp. https://www.researchgate.net/publication/331889626
Nagy GJ, Gutiérrez O, Brugnoli E, Verocai JE, Gómez-Erache M, Villamizar A, Olivares I, Azeiteiro UM, Lea-Filho FW, Amaro N (2019) Climate vulnerability, impacts and adaptation in Central and South American coastal areas. Reg Stud Mar Sci 29:100683. https://doi.org/10.1016/j.rsma.2019.100683
Nobre CA, Sampaio G, Borma LS, Castilla-Rubio JC, Silva JS, Cardoso M (2016) Land-use and climate change risks in the Amazon and the need for a novel sustainable development paradigm. Proc Natl Acad Sci 113(39):10759–10768. https://www.researchgate.net/publication/308214927
Núñez J, Carvajal JC, Mendoza O, Carrero D (2018) Indicadores del impacto del cambio climático en la agricultura familiar andina colombiana. Revista Iberoamericana de Bioeconomía y

Cambio Climático 4(7):824–831. http://portal.amelica.org/ameli/jatsRepo/394/3941754002/index.html

Olivares BO (2022) Determination of the potential influence of soil in the differentiation of productivity and the classification of susceptible areas to banana wilt in Venezuela. PhD Thesis, University of Cordoba. Doctoral Program in Agricultural, Food, Forestry, and Sustainable Rural Development Engineering. https://helvia.uco.es/handle/10396/22355

Olivares BO, Rey JC, Lobo D, Gómez JA, Landa BB (2019) El cambio climático en zonas bananeras de la Región Central de Venezuela: El futuro de los bananos con un escenario hídrico incierto. In: III Simposio Venezolano de Cambio Climático, Agricultura y Seguridad Alimentaria. 08 y 09 de octubre 2019. Universidad Metropolitana Caracas, Venezuela. https://n9.cl/tuyvc

Olivares BO, Rey JC, Lobo D, Navas-Cortés JA, Gómez JA, Landa BB (2021) Fusarium wilt of bananas: a review of agro-environmental factors in the Venezuelan production system affecting its development. Agronomy 11:986. https://doi.org/10.3390/agronomy11050986

Ortega Andrade S, Páez GT, Feria TP, Muñoz J (2017) Cambio climático y riesgo de expansión del hongo *Moniliophthora roreri* en *Theobroma cacao* en América Latina. Neotrop Biodivers 3(1):30–40. https://doi.org/10.1080/23766808.2016.1266072

Ortiz R (2012) El cambio climático y la producción agrícola. Banco Interamericano de Desarrollo. Unidad de Salvaguardias Ambientales (VPS/ESG). NotasTécnicas # ESG-TN-383, 36 p. https://n9.cl/5iq02

Pabón-Caicedo JD, Ycaza R, Friend F, Espinoza D, Fenzl N, Apostolova M (2018) Vulnerabilidad de la cuenca amazónica ante fenómenos hidroclimáticos extremos. Cuadernos de Geografía: Revista Colombiana de Geografía 27(1):27–49. https://doi.org/10.15446/rcdg.v27n1.56027. https://revistas.unal.edu.co/index.php/rcg/article/view/56027/html

Pacheco-Angulo C, Vilanova E, Aguado I, Monjardin S, Martínez S (2017) Carbon emissions from deforestation and degradation in a forest reserve in Venezuela between 1990 and 2015. Forests 8(8):291. https://www.researchgate.net/publication/319072164

Paredes MY (2014) Distribución potencial de los principales cultivos agrícolas en escenarios de cambio climático en el estado Mérida, Venezuela. Trabajo Especial de Grado para optar al Título de Licenciado en Biología. Universidad de Los Andes, Facultad de Ciencias, Departamento de Biología, Mérida. 60 p. http://www.saber.ula.ve/bitstream/handle/123456789/39966/Paredes%2CY.2014.pdf?sequence=1&isAllowed=y

Paredes MY, Chacón-Moreno E (2021) Distribución potencial de cultivos agrícolas en el estado Mérida (Venezuela), en escenarios de cambio climático. Rev Geogr Venezolana 62(2):428–444. https://www.researchgate.net/publication/356970049

Paredes-Trejo F, Barbosa-Alves H, Moreno-Pizani MA, Farías-Ramírez A (2018) Influencia de las sequías sobre el régimen hídrico de los ríos en Venezuela. Ríos En Riesgo de Venezuela 2:171–184. https://n9.cl/3xs3d

Parra RM, Olivares BO, Cortez A, Lobo D, Rey JC, Rodríguez MF (2018) Características de la sequía meteorológica (1980–2014) en dos localidades agrícolas de los Andes Venezolanos. Rev Invest 42(95):38–55. https://www.redalyc.org/journal/3761/376160247003/html/

Peralvo M, Bustamante M, Cuesta F, Becerra MT (2012) Adaptación al cambio climático en los Andes Tropicales-Discusión y conclusiones. In: Cuesta F, Bustamante M, Becerra MT, Postigo J, Peralvo J (eds) Panorama andino de cambio climático: vulnerabilidad y adaptación en los Andes Tropicales. CONDESAN, SGCAN, Lima, pp 263–281. https://n9.cl/m1z74

Pérez C, Nicklin C, Dangles O, Vanek S, Sherwood S, Halloy S, Garrett K, Forbes G (2010) Climate change in the high Andes: implications and adaptation strategies for small-scale farmers. Int J Environ Cult Econ Soc Sustain 6(5). https://doi.org/10.18848/1832-2077/CGP/v06i05/54835

Pinto J, Bonacic C, Hamilton-West C, Romero J, Lubroth J (2008) Climate change and animal diseases in South America. Rev Sci Tech Off Int Epiz 27(2):599–613

Pörtner HO, Roberts DC, Adams H, Adelekan I, Adler C, Adrian R, Aldunce P (2022) Climate change 2022: impacts, adaptation and vulnerability. In: Pörtner H-O, Roberts DC, Tignor M, Poloczanska ES, Mintenbeck K, Alegría A, Craig M, Langsdorf S, Löschke S, Möller V, Okem A, Rama B (eds) Contribution of Working Group II to the sixth assessment report of the

Intergovernmental Panel on Climate Change. Cambridge University Press, Cambridge, UK, pp 37–118

Postigo J, Peralvo M, López S, Zapata-Caldas E, Jarvis A, Ramírez J, Lau C (2012) Adaptación y vulnerabilidad de los sistemas productivos andinos. In: Cuesta F, Bustamante M, Becerra MT, Postigo J, Peralvo J (eds) Panorama andino de cambio climático: vulnerabilidad y adaptación en los Andes Tropicales. CONDESAN, SGCAN, Lima, pp 141–171. https://n9.cl/3dkn5

Ray DK, West PC, Clark M, Gerber JS, Prishchepov AV, Chatterjee S (2019) Climate change has likely already affected global food production. PLoS One 14(5). https://doi.org/10.1371/journal.pone.0217148

Rebolledo WR, Albarrán TA (2014) Cambios de la cobertura boscosa en Venezuela y su relación con cambios en el clima. Acta Biol Venez 33(1–2):125–144. http://saber.ucv.ve/ojs/index.php/revista_abv/article/view/7676

Rodríguez-De Luque JJ, González C, Gourdji S, Mason-D'Croz D, Obando BD, Mesa-Diez J, Prager SD (2016) Impactos socio económicos del cambio climático en América Latina y el Caribe: 2020–2045. Cuadernos de Desarrollo Rural 13(78):11–34. https://revistas.javeriana.edu.co/index.php/desarrolloRural/article/view/16482

Rodríguez-Morales AJ (2013) Climate change, climate variability and brucellosis. Recent Pat Antiinfect Drug Discov 8(1):4–12. https://core.ac.uk/reader/71397382?utm_source=linkout

Rodríguez-Román E, León Y, Pérez Y, Amaya P, Mejías A, Montilla JO, Ortega R, Zambrano K, Olivares BO, Marys E (2023) Peppers under siege: revealing the prevalence of viruses and discovery of a novel Potyvirus species in Venezuela. Sustainability 15:14825. https://www.mdpi.com/2071-1050/15/20/14825

Rojas M, Saavedra H, Márquez J, Torres A (2010) Análisis e incidencias de la variabilidad climática en las áreas de plantación de pino caribe al sur de los Estados Monagas y Anzoátegui, Venezuela. Rev Geogr Venezolana 51(2):249–268. https://www.redalyc.org/pdf/3477/347730385009.pdf

Romero AJ, Millano JL (2021) Comportamiento de las temperaturas en la Bioregión de la Cordillera de la Costa, Venezuela. Ciencia y Tecnología Agropecuaria 6(1):9–16. https://ojs.unipamplona.edu.co/index.php/rcyta/article/view/1076/5935

Rull V (2022) Responses of Caribbean mangroves to quaternary climatic. Eustatic and anthropogenic drivers of ecological change: a review. Plants 11:3502. https://www.preprints.org/manuscript/202211.0091/v1

Salazar-Lizán S (2014) Efecto del cambio climático en la ganadería doble propósito. In: Buenas prácticas en Ganadería Doble Propósito. Cuadernos Científicos Girarz, pp 305–312. https://www.researchgate.net/profile/Yenen_Villasmil/publication/342927055

Samaniego J, Galindo LM, Mostacedo Marasovic SJ, Carbonell JF, Alatorre JE, Reyes O (2017) El cambio climático, la agricultura y la pobreza en América Latina. División de Desarrollo Sostenible y Asentamientos Humanos de la Comisión Económica para América Latina y el Caribe (CEPAL). https://n9.cl/4p4pv

Santaella W (2013) Modelos espaciales de la dinámica de la vegetación en escenarios de cambio climático en el "Timberline" selva nublada—paramo de la vertiente norte de la Sierra Nevada de Mérida (Venezuela); escenario B1 y situación constante. Revista de Teledetección 39:63–82. https://dialnet.unirioja.es/servlet/articulo?codigo=7392075

Seo SN, McCarl BA, Mendelsohn R (2010) From beef cattle to sheep under global warming. An analysis of adaptation by livestock species choice in South America. Ecol Econ 69(12):2486–2494. https://doi.org/10.1016/j.ecolecon.2010.07.025

Silva M, Mendoza N (2019) Uso de la clasificación climática de Köppen en la evaluación de los efectos del cambio climático en los llanos venezolanos. Geominas 47(80):153–160. https://www.academia.edu/50792233/

Silva MI, Mendoza NM (2021) Evaluación de la sequía meteorológica en locaciones agrícolas de Venezuela. Rev Fac Agron Univ Zulia 38(3):548–572. https://doi.org/10.47280/RevFacAgron(LUZ).v38.n3.05

Vegas-Vilarrubia T, Hernández E, Rull V (2015) The Orinoco megadelta as a conservation target in the face of the ongoing and future sea level rise. Sci Total Environ 515–516:129–142. https://doi.org/10.1016/j.scitotenv.2015.01.056

Villamizar GAV (2020) Medidas de adaptación al aumento del nivel del mar considerando las Trayectorias de Concentración Representativa. Boletín de la Academia de Ciencias Físicas, Matemáticas y Naturales LXXX(2):41–77. https://n9.cl/e8zqj

Villamizar GEY, Cervigón F (2017) Variability and sustainability of the Southern Subarea of the Caribbean Sea large marine ecosystem. Environ Dev. https://doi.org/10.1016/j.envdev.2017.02.005

Villamizar A, Gutiérrez ME, Nagy GJ, Caffera RM, Leal FW (2017) Climate adaptation in South America with emphasis on coastal areas: the state-of-the-art and case studies from Venezuela and Uruguay. Clim Dev 9(4):364–382. https://www.researchgate.net/publication/299386639

Viloria JA, Olivares BO, García P, Paredes-Trejo F, Rosales A (2023) Mapping projected variations of temperature and precipitation due to climate change in Venezuela. Hydrology 10:96. https://doi.org/10.3390/hydrology10040096

Vuille M (2013) El cambio climático y los recursos hídricos en los Andes tropicales. Banco Interamericano de Desarrollo. Technical Note 517. https://ponce.sdsu.edu/Glaciar_CCRRR.pdf

Vuille M, Francou B, Wagnon P, Juen I, Kaser G, Mark BG, Bradley RS (2008) Climate change and tropical Andean glaciers: past, present and future. Earth Sci Rev 89(2008):79–96. https://www.researchgate.net/publication/223835402

Ward RD, Friess DA, Day RH, Mackenzie RA (2016) Impacts of climate change on mangrove ecosystems: a region-by-region overview. Ecosyst Health Sustain 2(4):1–25. https://www.researchgate.net/publication/301732207

World Bank, Climate Change Knowledge Portal (2025). https://climateknowledgeportal.worldbank.org/. Accessed 5 Apr 2025

Wright JP, Flecker AS (2004) Deforesting the riverscape: the effects of wood on fish diversity in a Venezuelan piedmont stream. Biol Conserv 120:443–451. https://documentoskoha.s3.amazonaws.com/7178.pdf

Zanetti EA, Gómez García JJ, Mostacedo J, Reyes O (2017) Cambio climático y políticas públicas forestales en América Latina. Una visión preliminar. Unidad de Cambio Climático, de la División de Desarrollo Sostenible y Asentamientos Humanos de la Comisión Económica para América Latina y el Caribe (CEPAL). https://n9.cl/ie5ki

Zhang X, Cai X (2013) Climate change impacts on global agricultural water deficit. Geophys Res Lett 40:1111–1117. https://doi.org/10.1002/grl.50279

Zinck JA, Metternihct G, Bocco G, Del Valle HF (eds) (2016) Geopedology. An integration of geomorphology and pedology for soil landscape studies. Springer, Switzerland. 556 p

Chapter 5
Regional Variability of Climate Impacts in Venezuela

5.1 Introduction

The increase in air temperature and the alteration of the hydrological cycle due to climate change can significantly affect agriculture and water resources. However, the impacts of climate change will vary across different countries and even within regions of the same country influenced by the pre-existing local climate conditions (Medina et al. 2016; Cabré et al. 2016; Zhang et al. 2022). Understanding the spatial distribution of temperature and precipitation changes within a country is essential for identifying specific vulnerabilities, directing research efforts, and formulating effective adaptation strategies (Rodríguez et al. 2013; Lescher Soto et al. 2024).

This chapter presents an analysis of the regional variability in projected changes in annual temperature and precipitation for Venezuela, comparing historical data from 1970–2000 and 2041–2060. The analysis is based on the research conducted by Viloria et al. (2023), which utilizes the MPI-ESM 1.2-LR model under the SSP3–7.0 scenario. Given the potential uncertainties inherent in general circulation model (GCM) projections, the chapter also contrasts these findings with projections from two additional GCMs. Furthermore, it explores how these climatic changes could influence the suitability of land for maize cultivation in Venezuela's western plains, a key agricultural region where maize is one of the predominant crops.

A. Rosales Hernández et al., *Agriculture Under Climate Change*, The Latin American Studies Book Series, https://doi.org/10.1007/978-3-032-06368-7_5

5.2 Climate Change Effects and Regional Physiographic Characteristics

5.2.1 Conceptual and Methodological Framework

The appraisal of spatial variation in projected changes in annual temperature and precipitation in Venezuela between the early and mid-twenty-first century presents significant challenges due to the scarcity of reliable climate data. The existing meteorological networks in the region suffer from low spatial coverage, substantial gaps in data, and limited historical records (Guenni et al. 2008; Vila et al. 2009; Olivares et al. 2017). To mitigate these issues, Viloria et al. (2023) utilized WorldClim data, which offers a 1 km resolution grid covering the period from 1970 to 2000 as a historical climate data source. This data has been instrumental in projecting climate change impacts on species of agricultural and forestry interest (Dutta 2014; Bunn et al. 2015; Läderach et al. 2017; Babalik et al. 2021; Sondermann et al. 2022). Nonetheless, concerns persist regarding the accuracy of WorldClim data as it is derived from interpolation between distantly located weather stations and subsequently downscaled to 1 km resolution. Such interpolation may overlook significant spatial climate features that become prominent at finer scales, particularly in regions with complex topography or significant coastal influences, making accurate mapping challenging. Furthermore, areas with sparse weather station density may experience high levels of uncertainty due to interpolation shortcomings (Karger et al.,,2021; Zheng et al. 2022). Consequently, validation of the WorldClim model is essential before its application. Fernández and Milla, (2018) statistically compared WorldClim estimates with data from 185 weather stations and found that monthly precipitation estimates did not accurately reflect Venezuelan conditions. Conversely, annual precipitation estimates exhibited a low error rate and can be reliably used as independent variables in environmental studies.

5.2.2 Principal Results

The MPI-ESM 1–2-LR model under the SSP3–7.0 scenario predicts an average increase of 1 °C (almost 4%) in mean annual temperature from 1970–2000 to 2041–2060 in Venezuela. However, the magnitude of this increase varies from one physiographic region to another (Table 5.1). It shows a discernible gradient in both north–south and east–west directions from nearly 1% in the northern and eastern peripheries of the country to almost 9% in the southernmost region. Additionally, the projected rise in mean annual temperature is expected to be more pronounced at higher altitudes of the Andes mountains (physiographic region 7) compared to the lowlands (Fig. 5.1). A Kruskal and Wallis (1952) test shows that southern Venezuela and the western and central plains will have a significantly higher increase in mean annual temperature ($p = 0.05$) compared to other regions of the country.

Table 5.1 Variation in mean annual temperature between 1970–2000 and 2041–2060 by physiographic region projected by the MPI-ESM 1–2-LR model under the SSP3–7.0 scenario

Physiographic region	1970–2000 (°C)	2041–2060 (°C)	Variation	
			°C	%
1 Lake Maracaibo depression	28.0	28.5	0.5	1.8
2 Western plains	27.2	28.9	1.7	6.3
3 Central Plains	27.0	28.2	1.2	4.4
4 Unare depression	26.4	27.2	0.8	3.0
5 Eastern plains	26.7	27.2	0.5	1.9
6 Orinoco-deltaic	26.4	26.7	0.3	1.1
7 Andes and Perijá	20.3	21.0	0.7	3.4
8 Falcón-Lara mountains	25.2	25.4	0.2	0.8
9 Central-coastal range	24.0	24.5	0.5	2.1
10 Eastern-coastal range	23.7	24.1	0.4	1.7
11 Intrusive Amazonian shield	25.7	27.7	2.0	7.8
12 Guayana shield	25.8	26.8	1.0	3.9
13 Ancient Roraima basin	23.4	24.8	1.4	6.0
14 Casiquiare shield	26.2	28.5	2.3	8.8
Mean	25.4	26.4	1.0	3.9

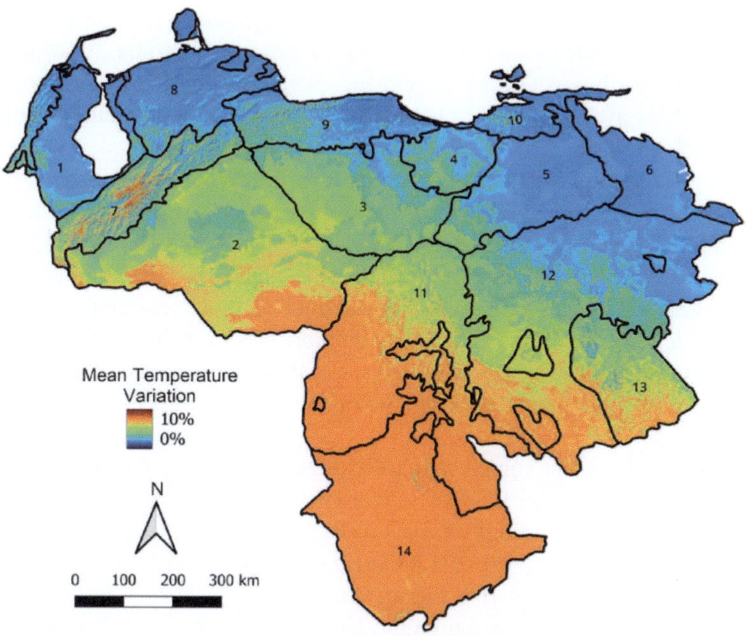

Fig. 5.1 Percent of variation of the mean annual temperature between 1970–2000 and 2041–2060 in Venezuela projected by the MPI-ESM 1–2-LR model under the SSP3–7.0 scenario according to Viloria et al. (2023). The boundaries of the physiographic regions numbered as in Table 5.1 are included

Table 5.2 Variation in mean annual *precipitation* between 1970–2000 and 2041–2060 by physiographic region projected by the MPI-ESM 1-2-LR model under the SSP3–7.0 scenario

Physiographic regions	1970–2000 (mm)	2041–2060 (mm)	Variation (mm)	Variation (%)
1 Lake Maracaibo depression	1429	1398	−31	−3
2 Western plains	1790	1758	−32	−2
3 Central plains	1278	1157	−121	−10
4 Unare depression	1070	961	−109	−10
5 Eastern plains	1174	1045	−129	−11
6 Orinoco-deltaic	1705	1511	−194	−12
7 Andes and Perijá	1357	1294	−63	−5
8 Falcón-Lara mountains	845	784	−61	−7
9 Central-coastal range	1137	1027	−110	−10
10 Eastern-coastal range	1141	1078	−63	−5
11 Intrusive Amazonian shield	2435	2359	−76	−3
12 Guayana shield	2021	1828	−193	−10
13 Ancient Roraima basin	2307	2076	−231	−10
14 Casiquiare shield	2956	2970	14	0
Mean	1618	1518	−100	−3

Conversely, the mean annual precipitation is projected to decline by 10%–12% (100–200 mm per year) in the east and north of Venezuela and by 3% or less in the south west and northwest (Table 5.2 and Fig. 5.2). The model indicates a potential slight increase in average annual precipitation in the Casiquiare shield (region 14) in the extreme south.

The projected annual precipitation changes exhibit significant internal variability across different regions, with some areas experiencing more pronounced variations than others. Notably the eastern plains, central plains, western plains, and the Andes are regions characterized by considerable spatial variability in precipitation patterns. These areas are of particular importance as they encompass a substantial portion of the country's agricultural production. The variability in precipitation could have profound implications for agricultural practices and productivity, necessitating adaptive strategies to manage potential impacts on crop yields and food security.

The study predicts that annual rainfall will decrease by more than 100 mm in the Orinoco-deltaic region, the Eastern plains Unare depression, and eastern central plains. This aligns with findings on meteorological droughts in agricultural areas of the central plains (Olivares et al. 2017; Cortez et al. 2018, 2019). Reduced precipitation will likely increase the risk of dry day sequences during the rainy season, impacting rainfed crops in these regions. Thus, in the northern half of the country, the eastern plains will experience more precipitation decreases but less temperature increases than the western plains. Moreover, the Lake Maracaibo depression in the country's northwest will also be affected by temperature increases and precipitation decreases, but with fewer changes than the other plains.

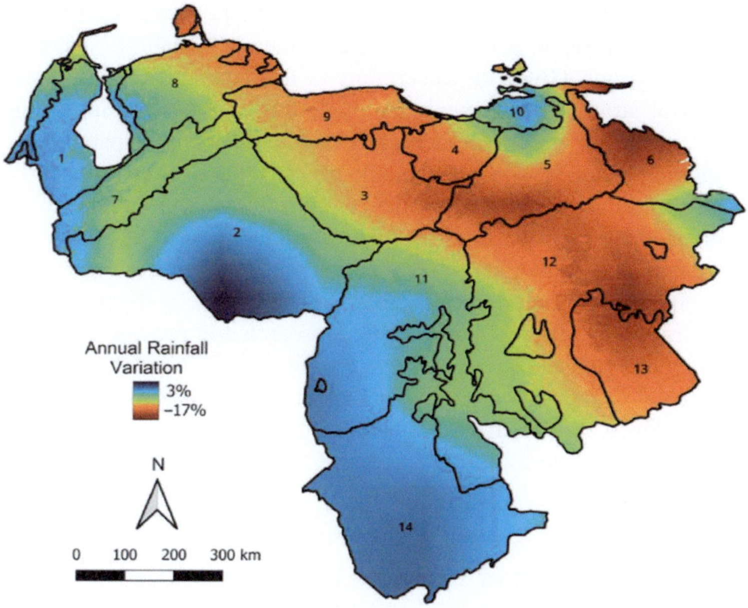

Fig. 5.2 Percent of variation of the mean annual precipitation between 1970–2000 and 2041–2060 in Venezuela projected by the MPI-ESM 1-2-LR model under the SSP3–7.0 scenario according to Viloria et al. (2023)

The annual mean temperature is anticipated to rise more significantly in the Andes and Central coast ranges compared to the Eastern-coastal range and the Falcón-Lara mountains. Concurrently, a more pronounced reduction in annual precipitation is expected in the Central-coastal range relative to other mountain regions. It is important to acknowledge that precipitation data from WorldClim carries greater uncertainty in mountainous areas due to spatial variability and interpolation imperfections (Fick and Hijmans 2017; Marchi et al. 2019). In the southern regions, the Amazonian shield and the Casiquiare shield are projected to experience higher temperature increases than other parts of the country. Similarly, the Guayana Shield and the Ancient Roraima basin are likely to face a more substantial decrease in precipitation. This reduction could significantly affect the Caroní River basin, potentially impact endemic species, and increase the risk of forest fires. Furthermore, it may severely influence the water inflow to the Lower Caroní hydroelectric reservoirs, which are crucial as they provide approximately 70% of the national electricity supply (Bautista 2012; Moreno-Pizani 2022).

The study highlights the necessity for region-specific strategies to address climate change in Venezuela, acknowledging the inherent variability across different physiographic regions. Given the uncertainty of the data, particularly in mountainous and coastal areas, it emphasizes the exploratory nature of the findings. This research is notable for its comprehensive examination of anticipated temperature and precipitation changes, offering valuable insights that could significantly impact Venezuelan agriculture adaptation strategies. The country's vulnerability to climate change underscores the importance of identifying the precise factors driving environmental changes to implement effective interventions. Further regional scientific studies are necessary to determine whether climate change is the primary driver or if other factors are also at play. Such detailed research will provide crucial information enabling policymakers to devise targeted solutions that address the root causes of environmental changes. Ultimately, these efforts are essential for crafting robust strategies to address the threat of climate change and ensure the sustainable management of natural resources.

5.3 Spatial Variation of Changes in Temperature and Precipitation Projected by Different GCMs

Most studies assessing agricultural land suitability under climate change scenarios rely on GCM data, which simulate future climate conditions based on assumptions about global greenhouse gas emissions and interactions among the atmosphere, oceans, cryosphere, and land (Karapetsas et al. 2024). However, these GCM projections are subject to significant uncertainties that arise from incomplete knowledge of the Earth's climate system and the challenges of accurately representing complex climate processes. The Coupled Model Intercomparison Projects (CMIPs) play a key role in systematically evaluating the performance of these models and the state of the art in general circulation modeling (O'Neill et al. 2016). The most recent iterations of CMIPs CMIP5 and CMIP6 have made significant contributions to the field of climate modeling.

In a study by Gómez and Pérez Godoy (Villamizar et al. 2018), various models from the CMIP5 project were evaluated identifying four models, HadGEM2-AO, GISS-E2-R, CCSM4, and MPI-ESM-LR that showed accuracy in replicating historical climate patterns and seasonal temperature and precipitation variations within Venezuela. In the subsequent CMIP6 project, the model CCSM4 was not directly used while the other models were replaced by updated and enhanced versions (HadGEM3 GISS-E2.1-G and MPI-ESM-MR). Based on Gómez and Pérez Godoy's insights, this analysis compares regional variations of future climate scenarios using projections of average annual temperature and precipitation from the GISS HadGEM and MPI models for the period 2041 to 2060.

5.3.1 Data Sources

A comparative analysis was performed utilizing both historical and projected future datasets to evaluate changes in monthly precipitation and maximum and minimum temperatures. The datasets were sourced from WorldClim 2.1 (Fick and Hijmans 2017) and consisted of 30" (approximately 1 km) resolution grids. From these datasets, annual mean temperature and precipitation values were calculated. The historical dataset spanning the period from 1970 to 2000 was developed through the interpolation of climatic station records. In contrast, future climate projections for the period 2041 to 2060 were derived from General Circulation Models (GCMs) under the CMIP6 framework (Meinshausen et al. 2017). These models included GISS-E2.1-G from the NASA Goddard Institute for Space Studies, USA (Rind et al. 2020), MPI-ESM-MR from the Max Planck Institute for Meteorology, Germany (Gutjahr et al. 2019), and HadGEM3 from the Met Office Hadley Centre, UK (Collins et al. 2008).

The climate projections were based on the SSP5–8.5 scenario, which prioritizes competitive markets, innovation, and participatory partnerships to drive rapid technological advancements. This scenario assumes a Radiative Forcing of 8.5 Wm^{-2} reflecting the additional energy retained by Earth due to greenhouse gas concentrations. Such a high level of radiative forcing is indicative of significant warming and pronounced climate impacts (Riahi et al. 2017).

5.3.2 Regional Variations

Figure 5.3 presents the regional variations in mean annual temperature changes across Venezuela, comparing the 1970–2000 period with the future projections for 2041–2060 based on the three selected GCMs. The projections collectively reveal that temperature increases are expected to be less significant in northern Venezuela, with a gradual intensification of warming trends moving both southward and westward. The GISS and HadGEM projections exhibit comparable spatial patterns of temperature changes. In contrast, the MPI model predicts a more pronounced warming in the southeast and relatively milder increases in the southwest of the country.

Although the spatial patterns of temperature changes across the three models share similarities, the magnitude of projected temperature increases varies considerably. According to the MPI projections, mean annual temperature is expected to rise by 6% to 10% equivalent to 1.5 °C to 2.2 °C. The GISS model anticipates an increase of 9% to 13% corresponding to 2.1 °C to 2.9 °C, while the HadGEM model predicts the most substantial rise, ranging from 10% to 18% equivalent to 2.4 °C to 4.1 °C (Table 5.3).

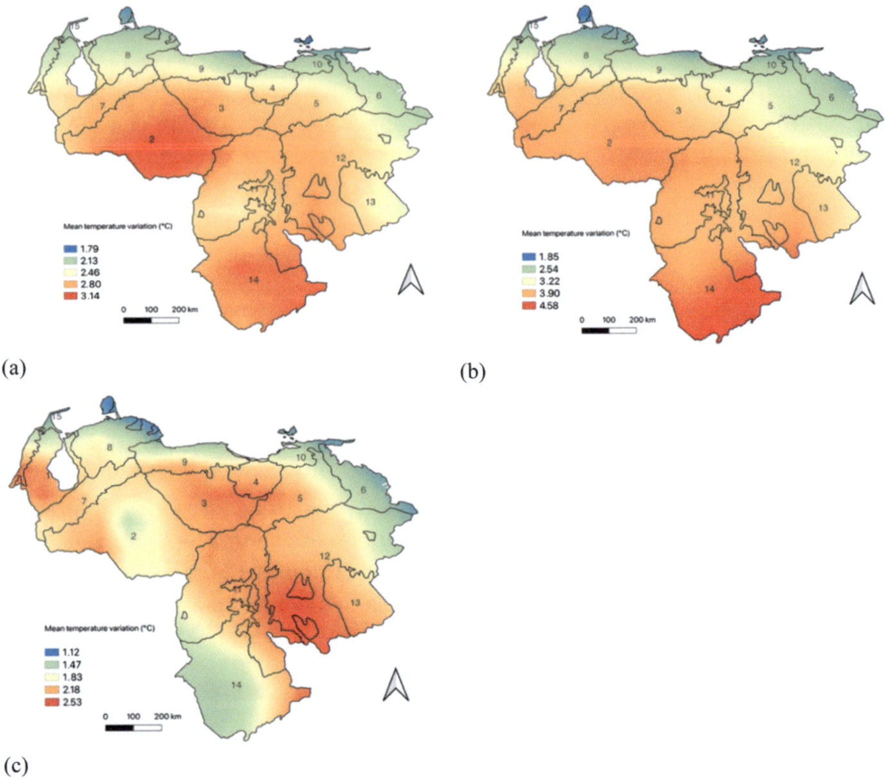

Fig. 5.3 Maps of changes in mean annual temperature (°C) from 1970–2000 to 2041–2060 projected by GISS (**a**), HadGEM (**b**), and MPI (**c**) with the SSP5–8.5 scenario in Venezuela. Each map includes the physiographic regions numbered as in Table 5.1

Discrepancies in spatial distribution patterns of future projections are larger for precipitation (Fig. 5.4) than for temperature (Fig. 5.3). This highlights the greater uncertainty in precipitation projections. All the applied models project a decrease in mean annual precipitation across Venezuela. On average, reductions of 156 mm (GISS), 178 mm (MPI), and 213 mm (HadGEM) are expected. While some regions, such as the Lake Maracaibo depression, show consistent precipitation, reduction patterns across model's other regions exhibit significant variability. For example, the GISS and HadGEM models align in four regions (2, 5, 7, and 12) while the MPI and HadGEM models align in three regions (4, 10, and 13). However, the proportions of precipitation reduction projected by these GCMs disagree in all the other physiographic regions (Table 5.4). Besides, Fig. 5.4 shows that there is little coincidence in the spatial variation of the changes projected by the compared GCMs.

The observed divergence between models emphasizes the need for caution when interpreting and utilizing these projections for policymaking or adaptation planning. Such discrepancies also underline the importance of employing a multi-model approach to better capture the range of potential outcomes and to enhance the robustness of climate impact assessments.

Table 5.3 Comparison of changes (%) on mean annual temperature from 1970–2000 to 2041–2060 as projected by the GCMs such as GISS, HadGEM, and MPI under the SSP5–8.5 scenario by physiographic regions in Venezuela

Physiographic regions	GISS	HadGEM	MPI
1 Lake Maracaibo depression	9	12	7
2 Western plains	11	14	7
3 Central plains	10	13	8
4 Unare depression	9	12	8
5 Eastern plains	10	12	8
6 Orinoco-deltaic	9	10	6
7 Andes and Perijá	13	18	10
8 Falcón-Lara mountains	9	11	7
9 Central-coastal range	10	12	8
10 Eastern-coastal range	9	10	7
11 Intrusive Amazonian shield	11	15	8
12 Guayana shield	10	14	8
13 Ancient Roraima basin	11	16	9
14 Casiquiare shield	11	16	7

Projections of temperature and precipitation changes generated by the applied GCMs vary regionally. However, the projections from these models show different spatial distribution patterns, which are most pronounced in the case of precipitation. This confirms the uncertainty in projections of future changes in these climate variables. Therefore, an appropriate strategy could be to analyze each region separately using several pre-tested GCMs as done in this study. Comparing the projections from these models would reveal various scenarios of possible future changes in temperature and precipitation. Examining these scenarios would allow for more robust decisions regarding climate change adaptation measures.

5.4 Potential Effects of the Changes in Temperature and Precipitation on the Suitability of Land for Maize Production in the Western Plains

Rising temperatures across the western and central plains present significant challenges to agriculture particularly in regions vital for food production. Projections indicate that maximum annual temperatures in these areas by the mid-twenty-first century could surpass the tolerance thresholds of key crops cultivated in Venezuela, including maize (*Zea mays* L.), common bean (*Phaseolus vulgaris* L.) and cocoa (*Theobroma cacao* L.) (Viloria et al. 2023). Maize stands out as especially critical due to its widespread cultivation its role in generating employment and its importance as a staple food source for both human and animal consumption (Segovia and Alfaro 2009).

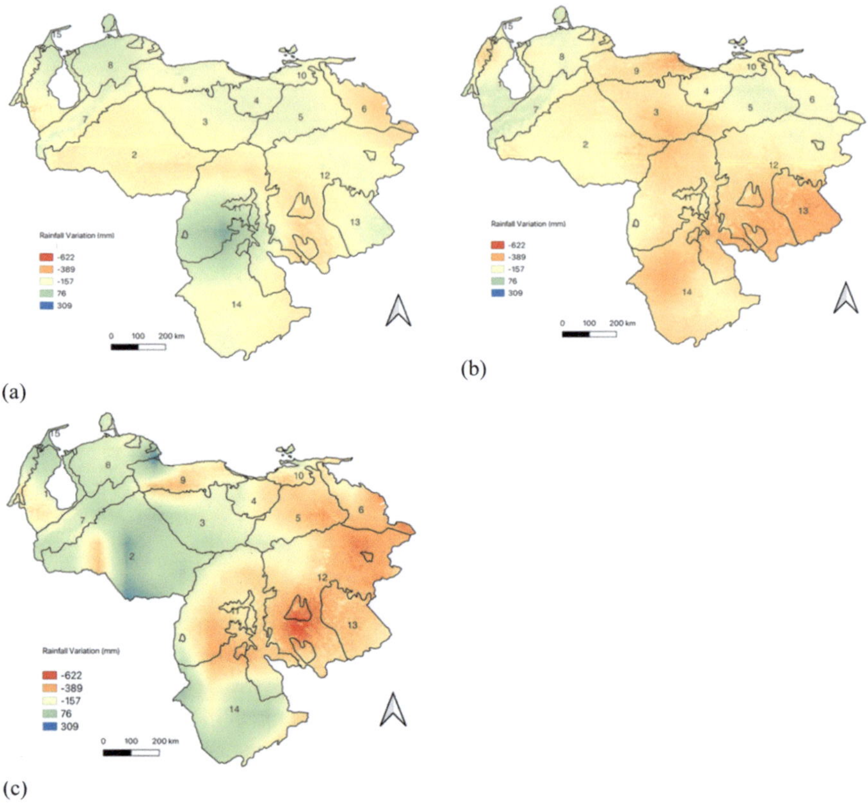

Fig. 5.4 Maps of changes on *annual mean* precipitation (mm) from 1970–2000 to 2041–2060 projected by GISS (**a**), HadGEM (**b**), and MPI (**c**) with the SSP5–8.5 scenario in Venezuela. Each map includes the physiographic regions numbered as in Table 5.1

Maize production in Venezuela employs various planting systems ranging from manual methods with minimal inputs for self-consumption or local markets to highly mechanized systems with significant input levels catering to agribusiness demands. The manual planting system is the most prevalent across the country. This traditional approach involves cultivating small plots with diverse crops and occasionally integrating livestock without relying on agricultural machinery or chemical inputs. In contrast, mechanized systems represent intensive agricultural practices characterized by large-scale operations advanced technologies and the application of synthetic fertilizers and biocides (Barrios and Bolotín 2012).

The land evaluation framework established by the Food and Agriculture Organization of the United Nations (FAO 2007) has been widely applied by researchers to assess the impacts of climate change on agricultural land suitability (Akpoti et al. 2019; Bilas et al. 2022; Karapetsas et al. 2024). Land suitability analysis (LSA) serves as a critical tool for identifying optimal areas for current and future agricultural activities by aligning the specific requirements of crops with the

Table 5.4 Comparison of changes (%) on mean annual precipitation from 2041–2060 to 1970–2000 as projected by the GCMs such as GISS, HadGEM, and MPI under the SSP5–8.5 scenario by physiographic regions in Venezuela

Physiographic regions	GISS	HADGEM	MPI
1 Lake Maracaibo depression	−10	−10	−8
2 Western plains	−11	−11	−1
3 Central Plains	−13	−20	−8
4 Unare depression	−12	−17	−18
5 Eastern plains	−13	−14	−22
6 Orinoco-deltaic	−14	−10	−17
7 Andes and Perijá	−11	−11	−6
8 Falcón-Lara mountains	−12	−19	−8
9 Central-coastal range	−13	−23	−15
10 Eastern-coastal range	−12	−16	−17
11 Intrusive Amazonian shield	−5	−11	−9
12 Guayana shield	−10	−12	−16
13 Ancient Roraima basin	−7	−14	−14
14 Casiquiare shield	−5	−9	−4

characteristics of the land (Akpoti et al. 2019). This approach offers valuable insights for sustainable agricultural planning in the context of climate change providing actionable recommendations to enhance resilience and productivity in affected regions.

Incorporating future climate projections into the LSA process enables the evaluation of climate change impacts on land suitability (Karapetsas et al. 2024). However, research in this area remains relatively limited (Akpoti et al. 2019). This section focuses on evaluating the suitability of the western plains of Venezuela for rainfed maize production using the FAO framework (FAO 2007). Furthermore, the study projects future changes in land suitability for maize cultivation by analyzing spatial data on precipitation and temperature derived from the HadGEM, GISS, and MPI models for the periods 2021–2040 and 2041–2060. Based on these projections, the research proposes strategies to adapt maize production to the anticipated impacts of mid-twenty-first-century climate change, ensuring greater resilience and sustainability in agricultural systems.

5.4.1 Projected Variation in Temperature and Precipitation for the Periods 2021–2040 and 2041–2060 in the Western Plains of Venezuela

The highest monthly maximum temperature in the western plains is recorded in March while the lowest values occur from June to August (Fig. 5.5). This suggests that rainfed maize cultivation in this region aligns with the less warm months of the

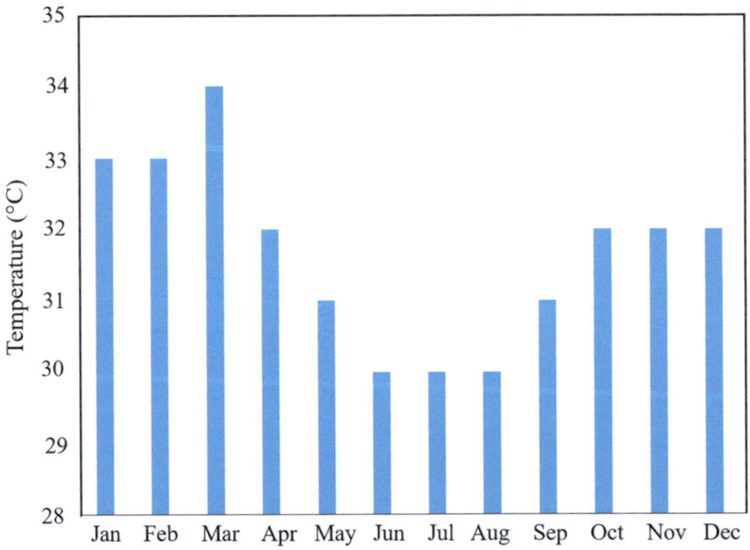

Fig. 5.5 Monthly variation of maximum temperature values (°C), data from 1950 to 1988 from the Acarigua climatological station, Portuguesa State, Venezuela

year as planting dates typically occur between May and June with a growth cycle lasting approximately 90 days. Therefore, the analysis of the potential impact of climate change on land suitability for rainfed maize production should focus on the months from May to September.

Figure 5.6a compares the average monthly precipitation across the study area estimated for the period 1970–2000 with projections from the MPI GISS and HadGEM models for 2021–2040. Overall, the variation is small (7% or less) except for the decreases projected by the GISS (12%) and HadGEM (19%) models for September in any case, the precipitation projected by any of the tested GCMs for three consecutive months exceeds the lower limit of water availability required for highly suitable land (class S1) which is defined as more than 500 mm of precipitation during the crop cycle.

The MPI model's projections for monthly average precipitation in the western plains during 2041–2060 indicate minimal differences with deviations of 3% or less compared to the 1970–2000 baseline (Fig. 5.6b). In contrast, the GiSS and HadGEM models forecast extensively reduced precipitation levels for May, August, and September relative to the same historical period with the HadGEM model predicting a substantial 36% decline in average precipitation for September. Nevertheless, the cumulative rainfall over three consecutive months is anticipated to remain above the minimum threshold required for class S1 of land suitability for maize (Table 5.5). Hence precipitation levels are not expected to pose a limitation to the viability of rainfed maize production in the western plains during 2041–2060.

Figure 5.6c illustrates that the projected increase in maximum monthly temperature by the MPI GISS and HadGEM models for 2021–2040 is relatively small (10%

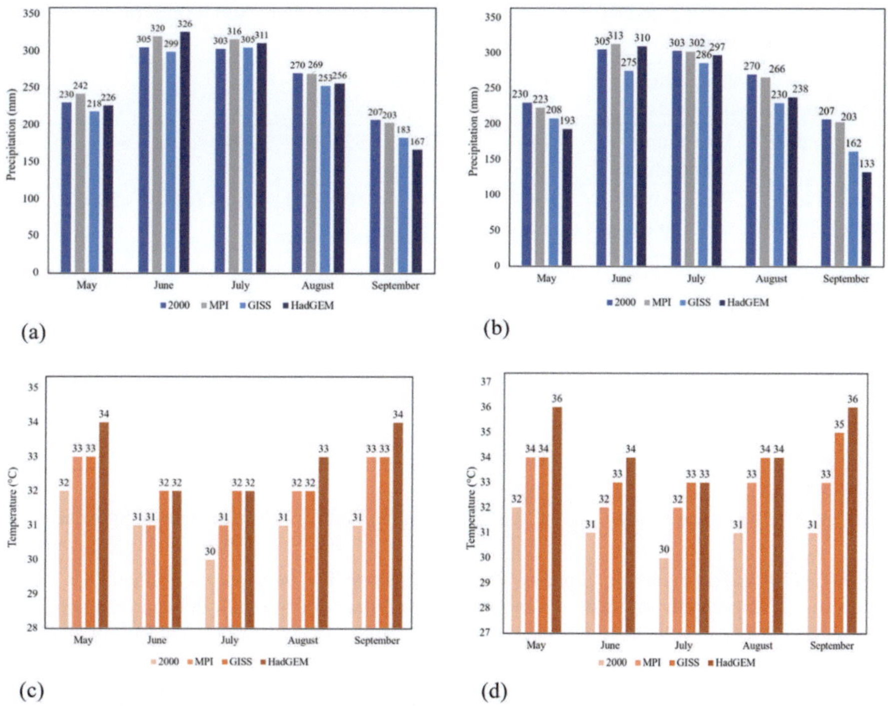

Fig. 5.6 (**a**) Variation of the average monthly precipitation projected by the MPI GISS and HadGEM models with the SSP5–8.5 scenario for the western plains of Venezuela period 2021–2040; (**b**) variation of the average monthly precipitation projected by the MPI GISS and HadGEM models with the SSP5–8.5 scenario for the western plains of Venezuela period 2041–2060; (**c**) variation in monthly maximum temperature projected by the MPI GISS and HadGEM models with the SSP5–8.5 scenario for the western plains of Venezuela period 2021–2040; (**d**) variation in monthly maximum temperature projected by the MPI GISS and HadGEM models with the SSP5–8.5 scenario for the western plains of Venezuela period 2041–2060

or less). This increase will not reduce the area of land suitable for rainfed maize as the highest projected temperature (34 °C) remains below the upper limit for moderately suitable land (S2) (Table 5.5). In general, all tested GCMs project that the difference in maximum temperature between 2021–2040 and 1970–2000 will be most pronounced in September compared to previous months. Notably for all months considered the maximum temperature projections by HadGEM are 1 °C higher than those of MPI. Additionally, the GISS projections align with MPI for May, August, and September and with HadGEM for June and July.

Figure 5.6d presents a comparison of regional averages of maximum temperature estimates for the historical period 1970–2000 and projections for 2041–2060. The percentage increases in temperature between these two periods vary across different models: 3–7% for MPI, 6–13% for GISS, and 13–16% for HadGEM. The projected maximum temperatures for 2041–2060 range as follows: 32–34 °C for MPI, 33–35 °C for GISS, and 33–36 °C for HadGEM. The highest projected

Table 5.5 Interpretive model for assessing rainfed maize land suitability in the western Venezuelan plains

Land quality	Diagnostic attributes	Unit	Aptitude classes			
			S1	S2	S3	N
Temperature	Maximum temperature	°C	<32	32–35	35–40	>40
Available water	Precipitation in the crop cycle	mm	>500	500–400	400–300	<300
	Soil profile texture					Coarse up to 1 m deep or more
Available oxygen	Drainage class	Classes	5–7	4	3	1–2
Available nutrients	Base saturation	%	≥80	50–80	35–50	≤35
Nutrient retention	Apparent CEC	Cmol/kg	>24	16–24	11–15	<11
Tillage	Granulometry	Classes	Medium	Fine	Expandable clays	Skeletal

Note: CEC = cation exchange capacity; drainage classes: 1 = very poorly drained; 2 = poorly drained; 3 = imperfectly drained; 4 = moderately well drained; 5 = well drained; 6 = somewhat excessively drained; 7 = excessively drained, adapted from Sys et al. (1993) and Ruiz et al. (2013)

temperatures are expected to occur in July for MPI, September for GISS, and both May and September for HadGEM.

Particular attention should be given to the projections from the HadGEM model as it indicates a maximum temperature of 36 °C during the 2041–2060 period. According to Waqas et al. (2021), such high temperatures could lead to a significant reduction in sucrose production. Furthermore, if maize is planted late and the grain-filling phase coincides with September, a month when elevated temperatures are projected, crop productivity could be severely impacted.

5.4.2 *Current Suitability of Land for Rainfed Maize Production*

The classification of land suitability for rainfed maize production in the western plains was divided into four classes: very suitable (S1), moderately suitable (S2) marginally suitable (S3), and not suitable (N). These classifications were applied to cartographic units on the soil map based on specific attributes that critically influence the viability of maize cultivation under rainfed conditions. To establish these classes, classification rules were formulated using available data on land characteristics supplemented by bibliographic research and expert consultations regarding the crop's agroecological requirements. These rules were integrated into an interpretive model (Table 5.5), which was systematically implemented across all soil map units to assess their suitability for rainfed maize cultivation under both low-input and high-input farming systems.

The study found no significant differences in land-use requirements between the low-input and high-input systems within the region. For example, constraints such as limited nutrient availability and reduced oxygen levels negatively impact crop yields in low-input systems. Conversely, addressing these constraints through fertilization and drainage practices in high-input systems leads to increased production costs. Additionally, challenges such as the presence of expandable clays or skeletal granulometry hinder cultivation regardless of whether manual or mechanical tools are employed. This suggests that certain inherent land characteristics pose universal limitations to rainfed maize production across different input systems.

Figure 5.7 shows the suitability of land for rainfed maize production in the western plains of Venezuela considering factors in Table 5.5. Due to the original soil map scale (1:100000) and soil variability, each unit includes two or three dominant soil classes possibly covering different land suitability classes. For practicality, the most frequent suitability class was assigned to each unit. The map shows that lands suitable for maize cultivation in the western plains are located above 7 degrees north latitude; soils conditions south of this latitude are unsuitable for this crop. Some suitable areas are small and isolated, but none are smaller than 500 hectares.

Table 5.6 provides an overview of the land suitability classifications for maize production in the western plains of Venezuela, highlighting that the total availability

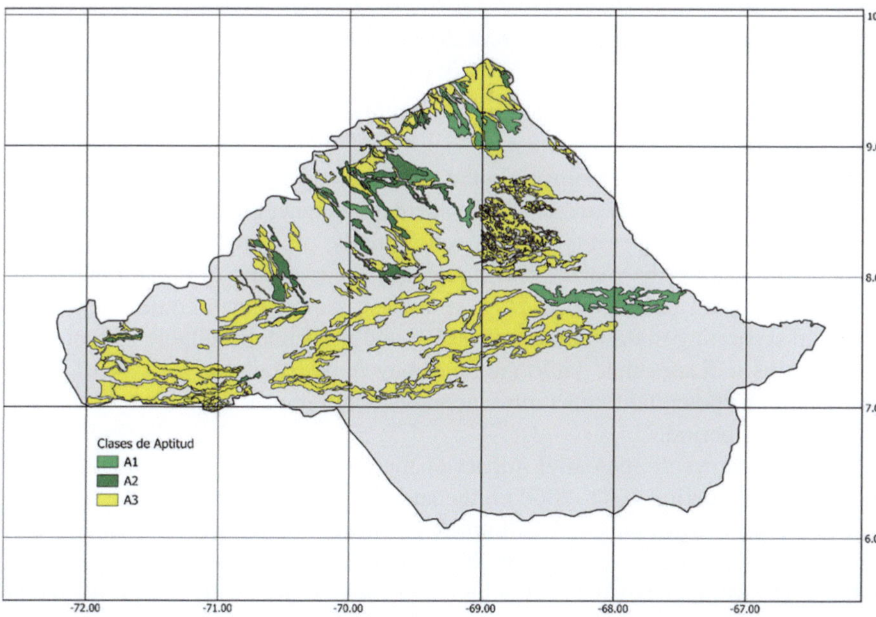

Fig. 5.7 Land suitable for rainfed maize production in the western plains of Venezuela. *S1* very suitable; *S2* moderately suitable; *S3* marginally suitable. Note: Gray areas are not suitable for this land use

Table 5.6 Land suitability for rainfed corn production in the western plains of Venezuela under climatic parameters from 1970 to 2000

		%	
Land suitability classes	Ha	Total area	Suitable area
S1: Highly suitable	325,500	3	12
S2: Moderately suitable	279,200	2	10
S3: Marginally suitable	2,159,700	17	78
N: Not suitable	9,857,800	78	0
Suitable area (S1 + S2 + S3)	2,764,400		100
Total area (S1 + S2 + S3 + N)	12,622,200	100	

of suitable land exceeds 2.7 million hectares. Among this total, only 605,000 hectares fall under the classes of highly or moderately suitable for maize cultivation. In contrast, over 2 million hectares are classified as marginally suitable with significant limitations that either hinder productivity or escalate production costs. As a result, these marginally suitable lands are unlikely to support profitable yields under high-input maize production systems making them impractical for such purposes. However, these lands may still be appropriate for low-input systems where reduced production costs could make their use more economically viable.

5.4.3 *Expected Suitability of Land for Rainfed Maize Production in the Period 2041–2060*

The analysis of the expected suitability of land for rainfed maize production during the period 2041–2060 assumes that no significant changes will occur in soil attributes relevant to maize suitability. On the other hand, projections of mean monthly precipitation suggest that rainfall will be adequate to meet the crop's water requirements throughout the growing season. Additionally, projected maximum monthly temperatures during the maize growing season are expected to remain below the threshold for land unsuitability. However, these temperatures could surpass the threshold separating marginally suitable (S3) and moderately suitable (S2) classifications. This indicates that while the land may remain suitable for rainfed maize production, the potential temperature increases could impact the degree of suitability during this period.

Table 5.7 shows the potential impact of the projected maximum monthly temperature for the period 2041–2060 on the suitability of land for rainfed maize production in the western plains. This impact is reflected in an increase in marginally suitable land (S3) at the expense of highly suitable and moderately suitable land (S1 and S2). The estimated extent of S3-class land varies significantly depending on the projections of different GCMs for May and September. Specifically, this area represents less than 5% of the suitable surface according to MPI projections 30% according to GISS and 90% according to HadGEM. The variation in the extent of

Table 5.7 Area of land suitability classes for rainfed maize estimated with monthly maximum temperature data projected by the GISS HadGEM and MPI models for the period 2041–2060

	GISS				HADGEM				MPI			
	S1 + S2		S3		S1 + S2		S3		S1 + S2		S3	
	km^2	%	km^2	%	km^2	%	km^2	%	km^2	%	km^2	%
May	22,747	82	8392	30	2349	8	25,350	92	27,539	99	161	1
June	27,497	99	202	1	25,258	91	2442	9	27,700	100	0	0
July	27,696	100	3	0	26,888	97	812	3	27,696	100	4	0
August	26,740	97	959	3	20,304	73	7396	27	27,324	99	376	1
Sept.	19,307	70	8392	30	1243	4	26,457	96	26,676	96	1024	4

marginally suitable land predicted by the different GCMs highlights a high degree of uncertainty in these projections. It is not possible to determine which GCM provides the most accurate predictions. However, all projections are likely relevant as interannual climate variability could result in some years being as warm as the HadGEM projections while others may align more closely with GISS or MPI. Considering this adopting the principle of "preparing for the worst while hoping for the best" appears to be a prudent approach.

Future increases in maximum temperatures in the study area could directly impact maize production suitability leading to a range of metabolic and morphological changes in plants that may reduce grain yields by 10 to 45%. These changes include reduced leaf growth, accelerated leaf senescence, shedding decreased pollen grain production and viability disruption of the fertilization process and a shortened grain-filling period (Waqas et al. 2021). Additionally indirect effects such as increased evapotranspiration, higher water demand for the crop, and greater incidence of pests and diseases may further exacerbate these challenges. However, this study does not account for the indirect effects.

The findings emphasize the urgent need for adaptive agricultural strategies to address the challenges posed by rising temperatures. Strategies such as adjusting planting schedules to avoid critical crop development stages during peak heat periods and adopting heat-resistant crop varieties are essential for sustaining agricultural productivity. Projections indicate that maize land suitability is less affected by maximum temperatures from June to August suggesting that planting between May and early June aligned with the rainy season is a viable adaptation measure. To optimize this, short planting window direct seeding and zero tillage practices are recommended. These methods not only enhance climate change adaptation but also serve as mitigation measures by improving soil carbon content reducing organic matter losses and enhancing soil structure (Kubar et al. 2018; Cooper et al. 2021; Liu et al. 2023; Amin et al. 2025; Joshi et al. 2025).

Additionally, the economic viability of utilizing marginally suitable land for maize production under high-input TUT systems is limited due to low yields. Therefore, technological advancements must focus on reducing costs for high-input systems while improving productivity in low-input systems. Integrating agronomic practices that enrich soil organic matter and biological activity is critical for sustainable adaptation and mitigation (Zingore and Njoroge 2022; Yang et al. 2024; Abbas et al. 2024). Furthermore, accelerated genetic improvement programs are essential for developing maize varieties resilient to high temperatures particularly during seedling emergence and grain-filling stages ensuring sustainable production amidst climate change.

5.5 Conclusions

This chapter summarizes a sequence of three studies aimed at predicting the spatial variation of potential temperature and precipitation changes by the mid-twenty-first century in Venezuela. Understanding the spatial distribution of temperature and precipitation changes will contribute to developing more effective adaptation strategies.

An initial exploration mapped the spatial variation of the temperature and precipitation changes projected by the MPI-ESM 1.2-LR model under the SSP3–7.0 scenario (Viloria et al. 2023). This study underscored the need for region-specific strategies to address climate change in Venezuela considering the variability across different physiographic regions. Acknowledging the exploratory nature of the findings due to data uncertainty particularly in mountainous and coastal areas, the research offered valuable insights that could shape Venezuelan adaptation strategies. The study highlighted the country's vulnerability to climate change and the importance of identifying the factors driving environmental changes to implement effective interventions. Therefore, further regional studies are crucial to provide essential information for developing robust strategies to address climate change and ensure sustainable management of natural resources.

A second analysis compared projections of the models MPI-ESM 1.2-LR GISS-E2.1-G and HadGEM3 with the scenario SSP5–8.5. The analysis of spatial variations in temperature and precipitation changes based on projections from these models highlighted notable regional differences for Venezuela. Temperature increases are generally more pronounced in southern and western regions with HadGEM projecting the highest rise while precipitation reductions are observed across all models albeit with significant variability in spatial patterns. The differences between the compared model results confirm the uncertainty of GCM projections of future climatic conditions. Discrepancies in precipitation projections are more pronounced compared to temperature emphasizing the inherent uncertainties in GCM outputs. This underscores the importance of employing a multi-model approach and region-specific analyses to enhance the reliability of climate impact assessments and inform adaptive strategies effectively.

The studies indicated that projected temperature increases could exceed the tolerance thresholds for key crops such as maize posing significant challenges for sustainable agricultural practices. This is a cause for concern in regions like the western plains of Venezuela given their dependence on agriculture for economic stability. Therefore, a third assessment used projections of the MPI GISS and HadGEM models under the SSP5–8.5 scenario to predict the impacts of projected temperature and precipitation changes on rainfed maize cultivation. While projected precipitation levels in Venezuela's western plains are expected to remain above the thresholds required for highly suitable land, the changes in temperature can significantly impact the suitability of land for rainfed maize production. Rising temperatures particularly those forecasted by the HadGEM model may lead to increased areas of marginally suitable land reducing the availability of highly and moderately suitable areas. This shift could be the result of reduced yields due to heat stress shortened grain-filling periods and increased water demands particularly during critical growth stages. Adaptive strategies such as adjusting planting schedules to avoid peak temperature periods adopting heat-resistant maize varieties and implementing sustainable practices like zero tillage are essential to mitigate these impacts and ensure agricultural resilience in the face of climate change. By implementing these measures, the agricultural systems in the western plains can build resilience to climate change safeguarding food security and livelihoods in the region.

These studies emphasize the value of GCM projections in estimating potential future temperature and precipitation patterns influenced by climate change. At the same time, they highlight the inherent uncertainties associated with these projections. Consequently, the establishment of a robust and well-distributed network of climate monitoring stations capable of delivering accurate and reliable data is essential for evaluating the reliability and applicability of these projections.

References

Abbas A, Naveed M, Shehzad Khan K, Ashraf M, Siddiqui MH, Abbas N, Mustafa A, Ali L (2024) The efficacy of organic amendments on maize productivity, soil properties and active fractions of soil carbon in organic-matter deficient soil. Span J Soil Sci 14:12814. https://doi.org/10.3389/sjss.2024.12814

Akpoti K, Kabo-bah AT, Zwart SJ (2019) Agricultural land suitability analysis: state-of-the-art and outlooks for integration of climate change analysis. Agric Syst 173:172–208. https://doi.org/10.1016/j.agsy.2019.02.013

Amin R, Jan SS, Butt M, Manan A, Khan M, Khan RA, Sadiq AB, Ali S (2025) Soil carbon sequestration through cover crops and reduced tillage: a climate-smart agricultural strategy for mitigating climate change and improving soil health. Kashf J Multidiscip Res 2(04):20–37. https://doi.org/70.71146/kjmr396

Babalik AA, Sarikaya O, Orucu OK (2021) The current and future compliance areas of kermes oak (Quercus coccifera L.) under climate change in Turkey. Fresenius Environ Bull 30(01):406–413. hppts://www.researchgate.net/publication/347953462_THE_CURRENT_AND_FUTURE_COMPLIANCE__AREAS_OF_KERMES_OAK_QUERCUS_COCCIFERA_L_UNDER_CLIMATE_CHANGE_TURKEY

Barrios M, Bolotín RE (2012) Análisis de la producción de maíz (*Zea mays* L.) en Venezuela entre 1999-2010. Rev Fac Agron. Universidad Central de Venezuela. Maracay.

Bautista S (2012) A sustainable scenario for Venezuelan power generation sector in 2050 and its costs. Energy Policy 44:331–340

Bilas G, Karapetsas N, Gobin A, Mesdanitis K, Toth G, Hermann T, Wang Y, Luo L, Koutsos TM, Moshou D, Alexandridis TK (2022) Land suitability analysis as a tool for evaluating soil-improving cropping systems. Land 11(12):2200. https://doi.org/10.3390/land11122200

Bunn C, Läderach P, Pérez Jimenez JG, Montagnon C, Schilling T (2015) Multiclass classification of agro-ecological zones for Arabica coffee: an improved understanding of the impacts of climate change. PLoS One 10(10):e0140490. https://doi.org/10.1371/journal.pone.0140490

Cabré MF, Solman S, Núñez M (2016) Regional climate change scenarios over southern South America for future climate (2080-2099) using the MM5 model. Mean, interannual variability and uncertainties. Atmosfera 29(1):35–60. https://doi.org/10.20937/ATM.2016.29.01.04

Collins WJ, Bellouin N, Doutriaux-Boucher M, Gedney N, Hinton T, Jones CD, Liddicoat S, Martin G, O'Connor F, Rae, J (2008) Evaluacion del modelo HadGEM2. Oficina Meteorológica de Exeter, Reino Unido. Vol.74. https://doi.org/10.5194/gmd-4-723-2011

Cooper HV, Sjögersten S, Lark RM, Girkin NT, Vane CH, Calonego JC, Rosolem C, Mooney SJ (2021) Long-term zero-tillage enhances the protection of soil carbon in tropical agriculture. Eur J Soil Sci 72(6):2477–2492. https://doi.org/10.1111/ejss.13111

Cortez A, Orlando Olivares B, Mayela Parra R, Lobo D, Rey BJC, Rodríguez MF (2019) Systematization of the calculation of the standardized precipitation index as a methodology to generate meteorological drought information. Rev Fac Agron 36:209–223. https://researchgate.net/publication/341109966_Systematization_of_the_calculation_of_the_standardized_precipitation_index_as_a_methodology_to_generate_meteorological_drought_information

Cortez A, Orlando Olivares B, Mayela Parra R, Lobo D, Rey JC, Rodríguez MF (2018) Event description of meteorological drought in locations central mountains, Venezuela. Cienc Ing Appl 1:22–44. https://doi.org/10.2166/wcc.2014.143

Dutta R (2014) Climate change and its impact on tea in Northeast India. J Water Clim Chang 5(4):625–632

FAO (2007) Land evaluation: towards a revised framework. Italy, Food Agriculture Organization of the United Nations, Rome. https://doi.org/10.1016/j.geoderma.2008.11.001

Fernández G, Milla D (2018) Validez de los datos de precipitación media mensual (1970-2000) generados por el modelo WorldClim V2. 0 para Venezuela. Terra Nueva Etapa 34(56). https://www.redalyc.org/journal/721/72157132004/html/

Fick SE, Hijmans RJ (2017) WorldClim 2: new 1-km spatial resolution climate surfaces for global land areas. Int J Climatol 37(12):4302–4315. https://doi.org/10.1002/joc.5086

Guenni L, Degryze E, Alvarado K (2008) Análisis de la tendencia y la estacionalidad de la precipitación mensual en Venezuela. Rev Colomb Estad 31(1):41–65. https://www.redalyc.org/articulo.oa?id=89912218003

Gutjahr O, Gutrasahan D, Lohmann K, Jungclaus JH, Bruggemann N, Haak H, Stossel A (2019) Max Planck Institute Earth System Model (MPI-ESM1.2) for the High-Resolution Model Intercomparison Project (HighResMIP), Geosci. Model Dev. 12, 3241–3281. https://doi.org/10.5194/gmd-12-3241-2019

Joshi DR, Clay DE, Alverson R, Clay SA, Westhoff S, Johnson JMF, Wang T, Sieverding H (2025) Tillage intensity reductions when combined with yield increases may slow soil carbon saturation in the Central United States. Sci Rep 15(1):10697. https://doi.org/10.1038/s41598-025-95388-x

Karapetsas N, Gobin A, Bilas G, Koutsos TM, Pavlidis V, Katragkou E, Alexandridis TK (2024) Analysis of land suitability for maize production under climate change and its mitigation potential through crop residue management. Land 13(1):63. https://doi.org/10.3390/land13010063

Karger DN, Wilson AM, Mahony C, Zimmermann NE, Jetz W (2021) Global daily 1 km land surface precipitation based on cloud cover-informed downscaling. Sci Data 8(1):307. https://doi.org/10.38/s4159597-021-01084-6

Kruskal WH, Wallis WA (1952) Use of ranks in one-criterion variance analysis. J Am Stat Assoc 47(260):583–621. https://doi.org/10.2307/2280779

Kubar KA, Huang L, Lu J, Li X, Xue B, Yin Z (2018) Integrative effects of no-tillage and straw returning on soil organic carbón and water stable aggregation under rice-rape rotation. Chil J Agric Res 78(2):205–215. https://doi.org/10.4067/S0718-58392018000200205

Läderach P, J R–V, Navarro-Racines C, Zelaya C, A M–V, Jarvis A (2017) Climate change adaptation of coffee production in space and time. Clim Chang 141(1):47–62. https://doi.org/10.1007/s10584-016-1788-9

Lescher Soto I, Villamizar A, Olivares BO, Gutiérrez M E, Nagy GJ (2024) Navigating the uncertain terrain: Venezuela's future using the Socioeconomic Pathways framework-A systematic review. Climate 12(7), 98. https://doi.org/10.3390/cli12070098

Liu X, Song X, Li S, Liang G, Wu X (2023) Understanding how conservation tillage promotes soil carbon accumulation: insights into extracellular enzyme activities and carbon flows between aggregate fractions. Sci Total Environ 897:165408. https://doi.org/10.1016/j.scitotenv.2023.165408

Marchi M, Sinjur I, Bozzano M, Westergren M (2019) Evaluating WorldClim version 1 (1961–1990) as the baseline for sustainable use of forest and environmental resources in a changing climate. Sustainability 11(11):3043. https://doi.org/10.3390/su11113043

Medina E, Velásquez G, Hernández I (2016) Impacto del calentamiento global y enriquecimiento atmosférico de CO2 sobre cultivos tropicales: la perspectiva para Venezuela. Rev Fac Agron UCV 42(1):25–37. https://saber.ucv.ve/ojs/index.php/rev_agro/article/view/11189

Meinshausen M, Vogel E, Nauels A, Lorbacher K, Meinshausen N, Etheridge DM, Fraser PJ, Montzka SA, Rayner PJ, Trudinger CM, Krummel PB, Beyerle U, Canadell JG, Daniel JS, Enting IG, Law RM, Lunder CR, O'Doherty S, Prinn RG, Reimann S, Rubino M, Velders GJM, Vollmer MK, Wang RHJ, Weiss R (2017) Historical greenhouse gas concentrations for climate modelling (CMIP6). Geosci. Model Dev. 10, 2057–2116. https://doi.org/10.5194/gmd-10-2057-2017

Moreno-Pizani MA (2022) Water management in agricultural production, the economy, and Venezuelan society. Front Sustain Food Syst 4:624066. https://doi.org/10.3389/fsufs.2020.624066

Olivares BO, Parra R, Cortez A (2017) Characterization of precipitation patterns in the Anzoategui state, Venezuela. ERIA-Rev Cuatrimest Geogr 37(3):353–365. https://doi.org/10.17811/er.3.2017.353-365

O'Neill BC, Tebaldi C, van Vuuren DP, Eyring V, Friedlingstein P, Hurtt G, Knutti R, Kriegler E, Lamarque J-F, Lowe J, Meehl GA, Moss R, Riahi K, Sanderson BM (2016) The scenario model Intercomparison project (ScenarioMIP) for CMIP6. Geosci Model Dev 9(9):3461–3482. https://doi.org/10.5194/gmd-9-3461-2016

Riahi K, van Vuuren director de fotografía, Kriegler E, Edmonds J, O'Neill BC, Fujimori S, Bauer N, Calvino K, Dellink R, Fricko O, Lutz W, Popp A, Cuaresma JC, Kc S, Leimbach M, Jiang L, Kram T, Rao S, Emmerlig J, Ebi K, Hasegawa T, Havlik P, Humpenoder F, Da Silva Luisiana, Smit, S, Stehfest E, Bosetti V, Eom J, Gernaat D, Masui T, Rogelj J, Strefler J, Drouet L, Krey V, Luderer G, Harmsen M, Takahashi K, Baumstark L, Doelman JC, Kainuma M, Klimont Z, Marangoni G, Lotze-Campen H, Obersteiner M, Tabeau A, Tavoni M (2017) Las trayectorias socioeconomicas compartidas y sus implicaciones en la enrgía, el uso del suelo y las emisiones de gases de efecto invernadero: una vision general. Cambio Ambiental Global 42, 153–168. https://doi.org/10.1016/jgoenvcha.2016.05.009

Rind D, Orbe C, Jonas J, Nazarenko L, Zhou T, Kelley M, Lacis A, Shindell D, Faluvegi G, Romanou A, Russell G, Tausnev N, Bauer M, Schmidt G, (2020) Un modelo climático optimizado para la atmósfera media: estructura del modelo, climatología, variabilidad y sensibilidad climática. JGR Atmospheres 125 (10), e2019JD032204. https://doi.org/10.1029/2019JD032204

Ruiz J, García G, Acuña I, Flores H, Ojeda G (2013) Requerimientos agroecológicos de cultivos, 2nd edn. Instituto Nacional de Investigaciones Forestales Agrícolas y Pecuarias-CIRPAC, Jalisco, México

Rodríguez MF, Cortez A, Olivares B, Rey JC, Parra R, Lobo D (2013) Análisis espacio temporal de la precipitación del estado Anzoátegui y sus alrededores. Agron Trop 63(1–2):57–65. https://n9.cl/14iow

Segovia VF, Alfaro YJ (2009) El maíz: un rubro estratégico para la soberanía agroalimentaria de los venezolanos. Agron Trop 59(3):237–247. https://ve.scielo.org/scielo.php?script=sci_arttext&pid=S0002-192X009000300001

Sondermann M, Chou SC, Lyra A, Latinovic D, Siqueira GC, Junior WC, Giornes E, Leite FP (2022) Climate change projections and impacts on the eucalyptus plantation around the

Doce River basin, in Minas Gerais, Brazil. Clim Serv 28:100327. https://doi.org/10.1016/j.cliser.2022.100327

Sys IC, Van Ranst B, Debaveye J, Beernaert F (1993) Land evaluation. Part III. Crop requirements. General Administration for Development Cooperation, Belgium. https://researchgate.net/publication/324330469_Land_Evaluation_Part3_Crop_requirements

Vila DA, De Goncalves LGG, Toll DL, Rozante JR (2009) Statistical evaluation of combined daily gauge observations and rainfall satellite estimates over continental South America. J Hydrometeorol 10(2):533–543. https://doi.org/10.1175/2008JHM1048.1

Villamizar A, Buroz Castillo E, Lairet Centeno R, Gómez JA (2018) Primer reporte académico de cambio climático de Venezuela (PRACC). Ediciones ACFIMAN—CITECI, Caracas, Venezuela

Viloria JA, Olivares BO, García P, Paredes-Trejo F, Rosales A (2023) Mapping projected variations of temperature and precipitation due to climate change in Venezuela. Hydrology 10(4):96. https://doi.org/10.3390/hydrology10040096

Waqas MA, Wang X, Zafar SA, Noor MA, Hussain HA, Azher Nawaz M, Farooq M (2021) Thermal stresses in maize: effects and management strategies. Plants 10(2):293. https://doi.org/10.3390/plants10020293

Yang L, Chen TY, Li ZY, Muhammad I, Chi YX, Zhou XB (2024) Straw incorporation and nitrogen fertilization regulate soil quality, enzyme activities and maize crop productivity in dual maize cropping system. BMC Plant Biol 24(1):729

Zhang Y, Fu B, Feng X, Pan N (2022) Response of ecohydrological variables to meteorological drought under climate change. Remote Sens 14(8):1920

Zheng P, Fang P, Liu P, Xu W, Wang G, Wang L (2022) Dominate tree species classification on large-scale mountainous areas using voting strategy-based ensemble learning method. In: China high resolution earth observation conference. Springer, pp 102–115. https://doi.org/10.1007/978-981-19-8202-6_10

Zingore S, Njoroge S (2022) Soil organic matter regulates maize productivity and fertilizer response in maize production. Grow Afr 1:8–11

Chapter 6
Climate Change Mitigation Strategies

6.1 Introduction

Climate change mitigation in the agricultural sector involves reducing greenhouse gas (GHG) emissions and enhancing the capacity of ecosystems to store carbon, especially in soils, forests, and agroforestry systems. Sánz-Sanchez and Uriarte (2020) highlight that land management affects key processes such as photosynthesis, respiration, organic matter decomposition, and fermentation in ruminants, all of which directly influence GHG emissions. These processes, driven by biological and physical activities (e.g., combustion and leaching), can either increase or reduce atmospheric CO_2 levels.

Agricultural ecosystems have a dual capacity: they emit and sequester greenhouse gases (mainly CO_2, methane, and nitrous oxide). When sequestration exceeds emissions, carbon accumulates in biomass and soil organic matter, contributing to mitigation. According to Lal et al. (2021), soils and biomass store approximately three times more carbon than the atmosphere. The balance between sequestration and net CO_2 emissions depends on ecosystem management, particularly agroecosystem practices. Given the wide range of agricultural activities, numerous actions can be implemented to reduce GHG emissions or enhance carbon sinks.

The main GHGs that can be emitted or sequestered through agricultural activities include carbon dioxide (CO_2), methane (CH_4), and nitrous oxide (N_2O). These gases may be emitted through activities such as deforestation or vegetation removal, the use of nitrogen-based fertilizers, burning of agricultural residues, enteric fermentation in animals, and manure management. Conversely, good agricultural practices can help sequester these gases, for instance, cover cropping, crop rotation, intercropping, agroforestry systems, soil conservation practices (such as reduced tillage), optimized fertilizer use, and the incorporation of organic matter, all of which contribute to atmospheric CO_2 removal. In addition to reducing emissions, these

A. Rosales Hernández et al., *Agriculture Under Climate Change*, The Latin American Studies Book Series, https://doi.org/10.1007/978-3-032-06368-7_6

practices also improve the productivity and sustainability of agricultural production units (Fig. 6.1).

This chapter is grounded in the analysis of 52 scientific studies published between 2018 and 2025 that evaluate a diverse range of mitigation strategies in Venezuela's agricultural sector. These studies are grouped into 12 strategic thematic clusters, each reflecting a different aspect of emission reduction or carbon sequestration potential. From renewable energy and soil conservation to agroecological practices and legal frameworks, the selected literature reveals the breadth and depth of Venezuela's emerging mitigation research agenda. The review integrates agronomic, technological, social, and institutional perspectives to provide a comprehensive understanding of the country's current capabilities and future needs.

In addition to presenting empirical results, this chapter emphasizes the value of spatial tools and ecosystem-level indicators, such as Priority Rank % of Areas of global significance for biodiversity conservation, carbon storage, and water provision, to guide mitigation efforts geographically. Areas with high ecosystem resilience, such as the Amazon and Guayana regions, offer opportunities for conservation-based mitigation, while degraded zones in the north require active restoration and land-use transformation. This spatial logic not only improves the efficiency of resource allocation but also reinforces the link between mitigation, adaptation, and biodiversity conservation.

Another key element discussed in this chapter is the integration of mitigation and adaptation strategies. Recognizing their synergies is essential for building climate-resilient agricultural systems that can simultaneously reduce emissions and strengthen rural development. Venezuela's policy frameworks, such as its Nationally

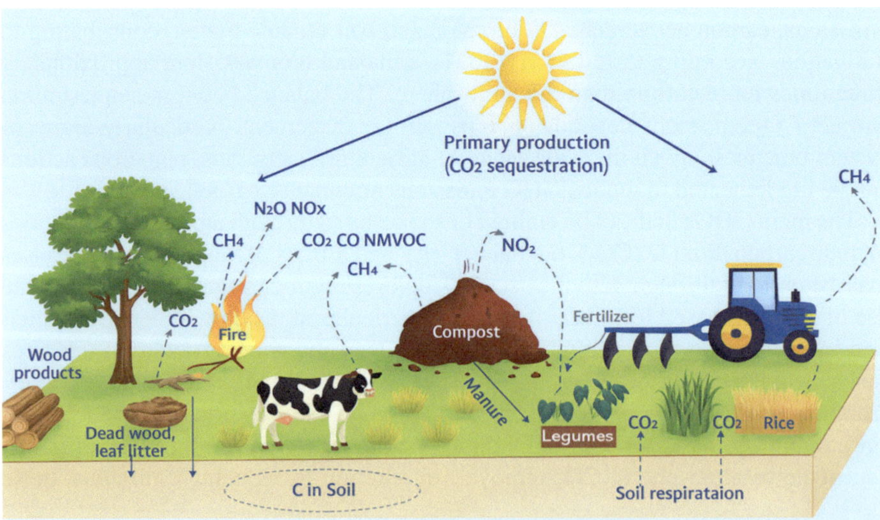

Fig. 6.1 The primary sources and sinks of GHGs in managed ecosystems (agroecosystems) Source: Own elaboration adapted from Sánz Sanchez and Uriarte (2020)

Determined Contributions (NDCs), already highlight priority actions in agroforestry and soil emissions management. However, the effective implementation of these strategies requires institutional support, technical training, and inclusive governance mechanisms that engage local communities and producers.

Ultimately, this chapter aims to provide an evidence-based roadmap for scaling up agricultural mitigation in Venezuela. By synthesizing scientific research with policy and territorial analysis, it outlines a path toward a climate-smart agriculture that not only contributes to emission reductions but also enhances food security, ecological health, and rural equity. The findings underscore that, despite structural and economic challenges, Venezuelan agriculture holds untapped potential to become a key factor in the country's climate response.

6.2 GHG Emissions from Agriculture in Venezuela

According to the IPCC (2019) global CO_2e emissions from agriculture, forestry and land use change (LUCF) are approximately 12–13 Gt per year (2010–2019). Agriculture alone accounts for about 5–6 Gt CO_2e annually, whichis appoximately 40–50% of LUCF sector emissions. This represents 10–12% of total glolal anthropogenic emissions. Agricultural emissions intensity represents a critical indicator for assessing the environmental footprint of food production systems (Fig. 6.2a). In Venezuela, historical trends from 1961 to 2021 reveal that livestock, particularly ruminants, exhibit the highest greenhouse gas (GHG) emissions per unit of product. Cattle meat, for instance, consistently shows the greatest emissions intensity, reaching approximately 73 kg CO_2 eq per kilogram of product in 2021. Goat and sheep meat also remain substantially high, at 57 and 41 kg CO_2 eq/kg, respectively. These elevated values are primarily attributed to enteric fermentation, manure management, and the extensive land requirements associated with ruminant production systems.

Conversely, non-ruminant animal products and crops demonstrate significantly lower emission intensities. Pork (6.6 kg CO_2 eq/kg), cow's milk (2.7 kg CO_2 eq/kg), and paddy rice (1.9 kg CO_2 eq/kg) present relatively moderate values, while chicken meat, eggs, and cereals excluding rice fall below 1 kg CO_2 eq/kg. The markedly lower intensity in poultry and crop-based products is linked to higher feed conversion efficiencies, shorter production cycles, and lower methane and nitrous oxide emissions. These differences underscore the mitigation potential of dietary shifts and production system diversification in achieving national climate goals.

The Venezuelan agricultural sector offers clear pathways for mitigation based on these intensity profiles. Transitioning from high-emission ruminant livestock to lower emission animal- and plant-based food systems could substantially reduce the sector's overall GHG footprint. Additionally, implementing best practices such as improved manure management, feed optimization, and sustainable rice cultivation techniques can further diminish emissions intensity. As such, emissions intensity metrics not only inform policy development and emission inventories but also provide strategic guidance for advancing climate-smart agriculture in Venezuela and other regions facing similar agroecological challenges.

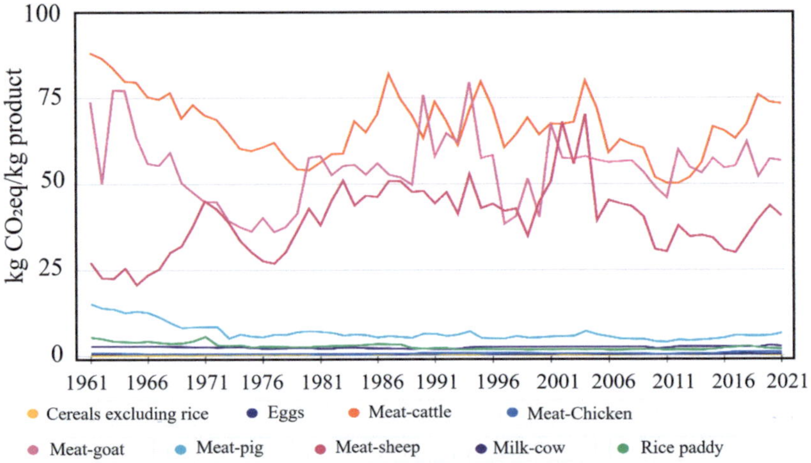

(a)

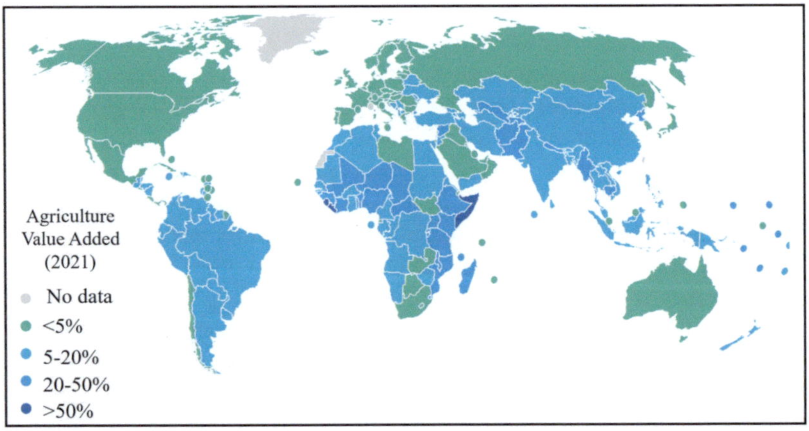

(b)

Fig. 6.2 (**a**) Historical agricultural emissions intensity (kg CO_2 eq per kg of product) in Venezuela from 1961 to 2021; (**b**) agriculture value added as a share of national GDP by country (2021), according to FAOSTAT (2024) available on https://www.climatewatchdata.org/

Based on the 2021 data presented by Climate Watch (Fig. 6.2b), Venezuela reports an agriculture value added of just 5.1% of its Gross Domestic Product (GDP). This places the country in the lowest category (<5%) globally, alongside most developed and some middle-income nations, indicating that agriculture plays a relatively minor role in the national economy. This contrasts starkly with many countries in Sub-Saharan Africa, where agriculture contributes over 30%, and in some cases, more than 50% to the national GDP, such as Liberia (66.9%) and Sierra Leone (57.5%).

The modest contribution of agriculture to Venezuela's economy suggests that while the sector may not be the dominant economic driver, it still holds strategic importance, particularly in terms of emissions intensity and mitigation potential. As seen in previous data, some Venezuelan agricultural products, notably ruminant livestock, contribute disproportionately to greenhouse gas (GHG) emissions per kilogram of product. Therefore, despite its relatively small economic share, the sector represents a key leverage point for climate action.

This highlights a dual imperative for Venezuela: on the one hand, to improve agricultural productivity and sustainability to enhance food security and rural livelihoods; and on the other, to target high-emission sub-sectors with mitigation strategies, such as sustainable livestock management, agroforestry, and low-emission rice cultivation. Such interventions could contribute not only to climate mitigation but also to the revitalization of a sector that, while economically marginal today, could become a cornerstone of a low-carbon development pathway.

6.3 Synergies Between Mitigation and Adaptation Measures in Agriculture

The agreements reached at COP21 in Paris (2015), under the framework of the United Nations Framework Convention on Climate Change (UNFCCC), marked a turning point in climate governance by explicitly recognizing the need for integrated policy approaches that combine mitigation and adaptation efforts (UNFCCC 2015). This paradigm shift enables countries to explore broader alternatives for reducing greenhouse gas (GHG) emissions while simultaneously addressing the increasing impacts of climate change. In sectors such as agriculture and forestry, this integration is particularly critical for designing sustainable strategies that not only reduce climate vulnerability but also promote carbon sequestration, enhance productivity, and preserve ecosystem services (UNFCCC 2015; FAO 2019).

From this perspective, framing climate change as an opportunity to strengthen the synergies between mitigation and adaptation, as emphasized by Vallejo et al. (2016), is essential for advancing toward truly sustainable development. These synergies offer multiple environmental and socioeconomic benefits across scales, while also facilitating the coordination of intersectoral public policies and the engagement of diverse stakeholders, ranging from rural communities to government agencies. However, their effective implementation requires institutional reforms, a robust research agenda, and a monitoring and evaluation system capable of identifying positive impacts, preventing unintended consequences, and tailoring strategies to local contexts (Locatelli et al. 2015). Ultimately, integrating mitigation and adaptation into agricultural policy in Venezuela is not merely a technical recommendation but a strategic imperative for achieving long-term climate resilience, equity, and sustainability.

In the agricultural and forestry sectors, it is essential to address the research agenda and implement institutional reforms that advance knowledge, regarding the benefits of synergies in different contexts and minimize their negative or compensatory impacts. The design and implementation of monitoring and evaluation systems are crucial components for ensuring long-term positive outcomes.

The agreements adopted by the COP21 on Climate Change acknowledge the need for new policy approaches that integrate adaptation and mitigation, enabling countries to generate and explore alternatives for reducing emissions and climate vulnerability.

6.4 Characteristics of Mitigation Measures

The IPCC Special Report on Climate Change and Land (2019) states that proper management of emission reduction and carbon sequestration measures in the agriculture, forestry, and other land use (AFOLU) sectors can play a significant role in climate change mitigation while also protecting many other ecosystem services, such as biodiversity. Conservation, restoration, and improved land management actions enhance carbon storage and/or reduce GHG emissions in agricultural lands.

Given the complexity and characteristics of the agricultural sector, mitigation expectations are global estimates based on assumptions that often overlook the potential intensification of climate change impacts on these ecosystems, as well as the environmental and socioeconomic barriers to implementing these measures across all geographical areas. Other authors, such as Sánz-Sanchez and Uriarte (2020), also emphasize the value of jointly developing mitigation and adaptation strategies in the agricultural sector at both national and regional levels.

The IPCC (2019) proposes measures that cover the entire food production and consumption chain: from on-farm agricultural production, transportation and logistics of harvested products, agro-industrial processing, and food consumption (including dietary changes and food waste reduction), to agricultural risk management. This comprehensive approach enables integrated progress on mitigation, adaptation, food security, and the interconnections with land degradation and desertification. Furthermore, it encourages the participation of multiple stakeholders (across different contexts and scales) in implementing climate solutions.

Pursuing a path that addresses both adaptation and mitigation measures contributes to increased productivity and resilience while also reducing GHG emissions in an integrated manner. Agricultural practices considered "climate-smart" help adjust agricultural production to the new realities imposed by climate change (see Chap. 7).

The IPCC (2019) also highlights the importance of integrating mitigation and adaptation strategies that both reduce emissions and improve agricultural resilience in the face of climate change. Climate-smart agricultural practices (see Chap. 7) represent a pathway to achieving these dual objectives, as they promote higher productivity, reduce GHG emissions, and enhance climate adaptability. Likewise, mitigation strategies in the agricultural and forestry sectors must include the establishment of effective monitoring and evaluation systems to ensure meaningful and measurable impacts in the medium and long term.

In line with the commitments made under COP21 on climate change, Vallejo et al. (2016) stress the need for new policies that integrate adaptation and mitigation efforts to reduce climate vulnerability. Joint implementation of both strategies can optimize results in the agricultural sector.

To contribute meaningfully to climate mitigation, it is essential in agriculture to implement sustainable practices tailored to local conditions. Conservation and restoration of agricultural lands increase their carbon sequestration capacity and reduce GHG emissions. However, mitigation expectations in this sector are constrained by climatic impacts, environmental and social barriers, and the complexities of global implementation. Nonetheless, agriculture can play a key role in combating climate change if institutional knowledge and capacity are strengthened to support sustainable management practices across the entire production process and diverse production systems.

Finally, effective mitigation in the agricultural sector requires careful management that considers both GHG emission reductions and the carbon sequestration potential of agroecosystems. A systemic approach that integrates both mitigation and adaptation is necessary to achieve resilient and sustainable agriculture, aligned with emission reduction goals and capable of delivering biodiversity and other ecosystem service benefits across all ecological levels.

While emissions are rising, countries are also stepping up their commitments to reduce emissions. As part of the Paris Agreement, 135 countries proposed agricultural mitigation measures (i.e., targets, policies, actions, and plans) and 171 countries proposed agricultural adaptation measures (i.e., targets, policies, actions, and plans). In Venezuela's Nationally Determined Contribution (NDC), two key sectoral mitigation actions have been explicitly identified: agroforestry and soil fertilization emissions management (Fig. 6.3a). These measures reflect the country's acknowledgment of agriculture and land use as strategic entry points for climate mitigation. Agroforestry systems, which integrate trees with crops and/or livestock, offer the potential to enhance carbon sequestration, restore degraded lands, and improve biodiversity while maintaining agricultural productivity. By incorporating such systems into national policy, Venezuela aligns itself with nature-based solutions that simultaneously address climate, ecological, and socioeconomic goals.

The inclusion of soil fertilization emissions highlights Venezuela's concern over nitrous oxide (N_2O) emissions, a potent greenhouse gas often overlooked in low-emission inventories. Excessive or inefficient use of synthetic fertilizers in agricultural soils is a major contributor to N_2O emissions globally. In Venezuela, targeting fertilization practices through precision agriculture, organic amendments, or optimized nitrogen management can play a significant role in reducing agricultural GHG emissions. These sectoral commitments within the NDC framework not only demonstrate national climate ambition but also provide a foundation for mobilizing international climate finance and technical cooperation focused on sustainable land management.

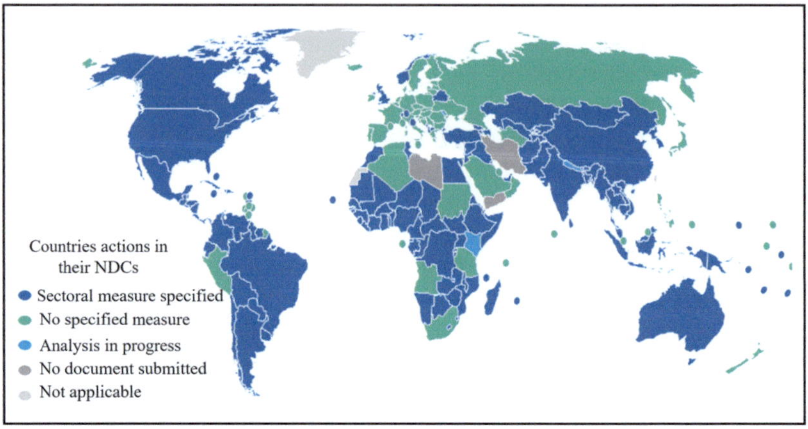

(a)

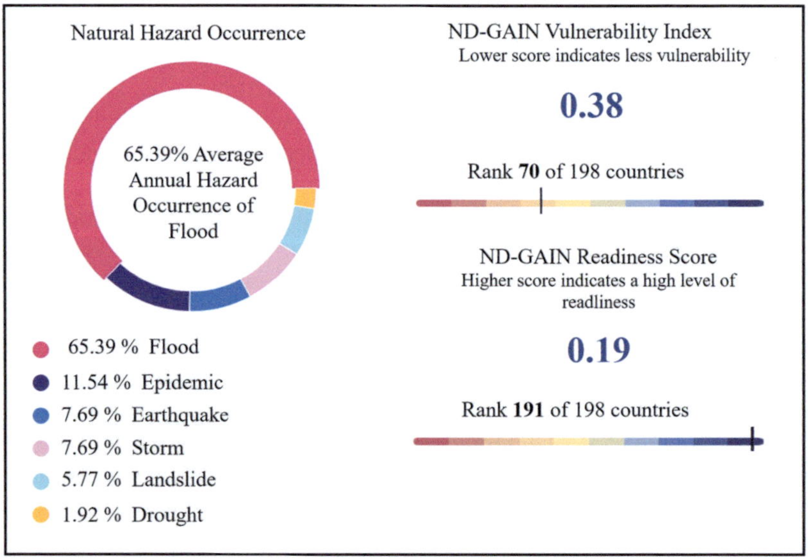

(b)

Fig. 6.3 (**a**) Global distribution of sectoral mitigation measures in Nationally Determined Contributions (NDCs); (**b**) ND-GAIN (Notre Dame Global Adaptation Initiative) vulnerability score for Venezuela (2021); ND-GAIN Readiness Score for Venezuela (2021); average annual occurrence of natural hazards in Venezuela (Floods, Epidemics, Earthquakes, and Others)

Despite being ranked 70th out of 198 countries in the ND-GAIN vulnerability index with a score of 0.38 (Fig 6.3b), Venezuela remains moderately exposed to the adverse impacts of climate change. This index reflects vulnerability across sectors such as water, food, health, infrastructure, and ecosystems. The relatively moderate score suggests that while Venezuela is not among the most climate-vulnerable countries globally, it still experiences significant systemic sensitivity to climatic stressors, particularly due to its socioeconomic fragility and environmental degradation.

In stark contrast, Venezuela's ND-GAIN readiness score is critically low, only 0.19 (Fig 6.3b), placing the country at 191st globally. This metric evaluates a nation's institutional, economic, and social readiness to plan for and implement climate adaptation and mitigation strategies. Venezuela's position highlights severe limitations in governance capacity, infrastructure, and investment climate, indicating a low ability to translate climate risk awareness into effective action. This readiness deficit substantially undermines the country's capacity to cope with both chronic climatic pressures and acute natural hazards.

Further compounding the risk is the frequency of natural hazards, with floods constituting the majority (65.4%) of events reported annually (Fig 6.3b). Other notable threats include epidemics (11.5%), earthquakes (7.7%), and storms (7.7%). The dominance of flood events underscores the urgent need to prioritize hydrometeorological risk management within national adaptation frameworks. Without significant improvements in institutional readiness and resilience planning, Venezuela is likely to face escalating climate-related losses, particularly in vulnerable sectors such as agriculture, public health, and urban infrastructure.

6.5 Spatial Patterns of Ecosystem Resilience and Implications for Climate Mitigation in Venezuela

The identification of priority conservation areas constitutes a strategic tool to guide climate change mitigation policies in Venezuela's agricultural sector, as it enables the generation of simultaneous co-benefits in biodiversity, carbon storage, and water provision. In this context, the Areas of Global Significance for Biodiversity Conservation, Carbon Storage, and Water Provision index, developed by UNEP-WCMC (2025), provides an analytical framework to identify critical regions for implementing Nature-based Solutions (NbS).

This index is expressed as a percentage priority rank (Priority Rank %) integrating global data on terrestrial biodiversity richness and vulnerability, carbon stocks in aboveground biomass and soils, and hydrological ecosystem services such as flow regulation and aquifer recharge. Its scale classifies areas from 1% (highest global priority) to 100% (lowest priority), highlighting territories where conservation can have a disproportionately high global impact.

In Venezuela, the available cartographic analysis shows that areas of highest global priority (percentile 1–25) are mainly located in Amazonas and southern Bolívar, where extensive forests of the Guyana Shield concentrate high densities of carbon in biomass and soils, along with elevated levels of endemism. Likewise, the Orinoco Delta presents wetlands, mangroves, and floodplain forests critical as carbon sinks and hydrological regulators.

Additionally, the Gran Sabana and the tepui massifs stand out for hosting relict ecosystems with unique species and essential hydrological functions. In contrast, large areas in the north of the country, including the central-northern coastal region and parts of the central and eastern Llanos, exhibit priority ranks between 50% and 100%. Although these areas are relevant for agricultural production, their combined global significance for biodiversity, carbon, and water is lower.

The application of this index is particularly relevant in Venezuela, given that numerous areas with high conservation priority overlap with regions of recent or historical agricultural expansion, generating land-use conflicts. This is especially critical in protected areas under agro-productive pressure, such as Yapacana National Park (Amazonas), Canaima National Park (Bolívar), or the Imataca Forest Reserve, which face pressures from subsistence agriculture (conuco), deforestation, illicit crops, and extensive livestock grazing; buffer zones of parks and reserves, where the agricultural frontier advances without ecological planning; and the Llanos and Andean piedmonts, which have high agricultural potential but are adjacent to remnants of tropical dry forests whose ecological functions are fundamental for regional hydrological stability.

This spatial prioritization approach enables the hierarchical application of agroecological measures, such as soil restoration, strategic reforestation, and agroforestry systems, maximizing the climate and ecological efficiency of interventions. However, the index also reveals that many high-priority areas overlap with regions of recent or historical agricultural expansion, generating land-use conflicts, especially in protected areas under agro-productive pressure, buffer zones, and Andean piedmonts.

Therefore, harmonizing agricultural policies with conservation and climate mitigation objectives requires instruments such as rural land-use planning, ecological-economic zoning, and strengthening legal frameworks to regulate land use in sensitive areas. The integrated global priorities index provides a robust technical basis for planning strategic interventions that articulate productive, climate, and ecological objectives, thus contributing to an agricultural climate policy consistent with Venezuela's international commitments (Fig. 6.4).

6.6 Studies Related to Mitigation Strategies in the Agricultural Sector of Venezuela

Mitigation strategies in the agricultural sector in response to Global Climate Change are numerous and diverse. For this chapter, a selection of 52 studies conducted in Venezuela during the 2018–2025 period was evaluated and compiled into 12 sets of strategies, as presented below. Each group includes a brief introductory overview of the topic, followed by a summary of each study, which is cited in the corresponding bibliographic section of this chapter.

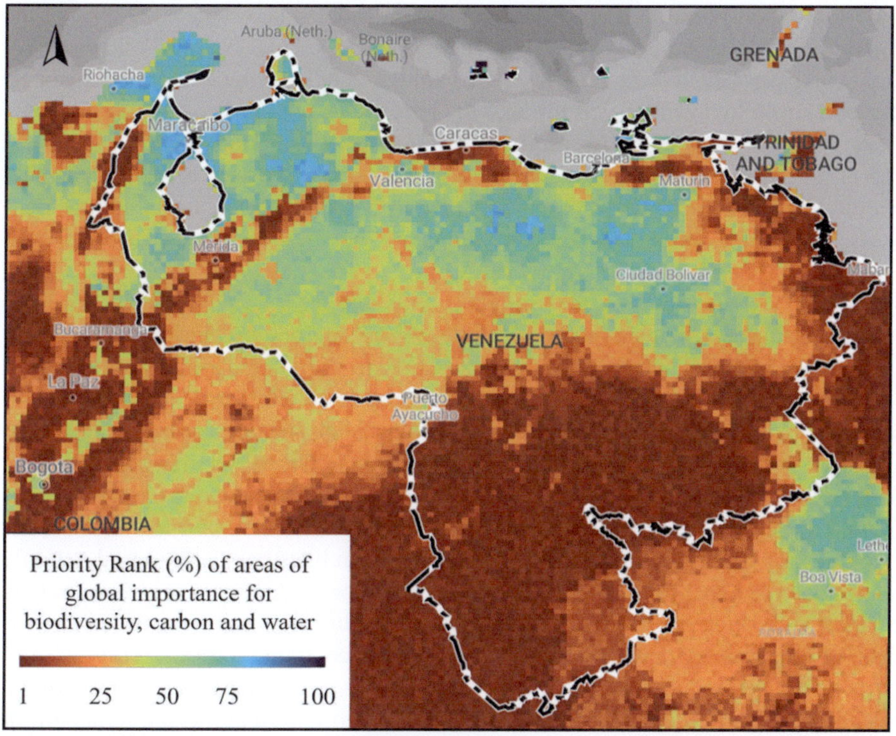

Fig. 6.4 Spatial priority ranking of areas of global importance for biodiversity, carbon storage, and water provision in Venezuela. Source: United Nations Environment Programme World Conservation Monitoring Centre, UNEP-WCMC (2025) available on https://n9.cl/uuzdz

6.6.1 Improving Energy Efficiency and Promoting Renewable Energy

Mitigation in agriculture is a crucial component of global efforts to address climate change, given the sector's significant contributions to greenhouse gas emissions and its vulnerability to environmental shifts. In Venezuela, some initiatives have emerged to support a transition toward cleaner, more resilient agricultural systems. The following case studies illustrate a small range of mitigation strategies, ranging from renewable energy generation using agricultural residues to the development of hybrid energy systems in rural areas. Each example highlights practical approaches with the potential to reduce emissions, enhance rural livelihoods, and contribute to national energy transition goals.

The study by Watson and Rao (2017) remains a valuable contribution to the use of renewable energy in Venezuelan agriculture. It highlights the socioeconomic impact of ETAVEN C.A., a company that patented a process to produce eco-friendly ethanol, known as Yarethanol, from cassava processing waste. This non-edible, non-toxic, non-polluting fuel is in high demand and can be produced at 50% of the market cost. Located in a cassava-producing region, ETAVEN sources low-quality cassava and waste at reduced prices. This approach helps reduce river and lake pollution caused by cyanide-rich runoff, cuts greenhouse gas emissions, and increases local farmers' income by up to 50%. Additionally, the initiative has created over 1500 jobs, fostering rural agricultural development. The study demonstrates how biofuel innovation can deliver significant environmental, social, and economic benefits, especially for small- and medium-scale cassava farmers across different regions of Venezuela.

Solano (2019) addresses the important issue of constructing biogas plants using agricultural and livestock waste in cattle-raising regions of Venezuela. The outcomes of the Conference of the Parties (COP) to the United Nations Framework Convention on Climate Change confirm that one of the most pressing concerns for signatory countries is the need for short-term decision-making regarding actions to mitigate the impacts associated with climate change. Global warming is one of the driving forces behind the transition from fossil fuels to cleaner energy sources aimed at reducing global carbon dioxide (CO_2) emissions. Within this context, the biodigestion of agricultural waste, particularly from livestock, is proposed not only as a renewable energy technology through biogas production, but also as a strategy to directly reduce methane (CH_4) emissions resulting from the natural decomposition of livestock manure. This study analyzes the economic pre-feasibility of constructing agro-industrial-scale biogas plants in livestock regions of the country, using various specialized software tools. It aims to contribute to the reduction of greenhouse gas emissions and assess the viability of such initiatives in supporting Venezuela's energy transition.

Ramírez and Méndez (2020) propose a solar and wind map for Venezuela as a tool to identify the most suitable locations for renewable energy generation parks across the country, particularly in rural areas, as the authors note. The aim is to support the reduction of electricity generation through the combustion of fossil fuels in thermal power plants. The variables studied for the map's development include wind speed, wind direction, and solar radiation. The research employed platforms such as Global Atlas, IRENA, and Google Earth to use downscaled satellite imagery provided by international agencies that maintain average data records spanning more than 10 years. The resulting information is intended to inform decision-making in the formulation of energy-related public policies and to set a precedent for Venezuela as a nation committed to meeting its CO_2 emissions reduction targets.

This study by Barroso Molina (2022) focuses on Toas Island, in northern Lake Maracaibo, Zulia State, though its findings apply to other isolated agricultural regions in Venezuela with limited access to electricity. Venezuela's energy system is heavily reliant on fossil fuels and large hydroelectric plants, leaving renewable sources underutilized. The study aims to evaluate the solar and wind energy potential of Toas Island and propose hybrid energy scenarios. It begins with a literature review, followed by data analysis using solar irradiation and wind speed information from Global Solar, NASA Power, and the Homer software. Based on these inputs, a viable microgrid project was developed. The results show that the most effective solution involves a hybrid system combining solar, wind, battery storage, and grid interconnection. This system not only ensures energy reliability and availability but also presents the potential to export surplus electricity to the national grid, enhancing sustainability and resilience.

6.6.2 Conservation Tillage as a Soil Preparation Practice

In the context of climate change mitigation in agriculture, direct seeding represents a key strategy for reducing emissions and enhancing the resilience of agroecosystems. This practice, which eliminates or minimizes soil tillage, helps preserve soil structure, reduce erosion, increase organic matter content, and promote carbon sequestration in agricultural soils. One of the main mitigation mechanisms associated with direct seeding (non-tillage) is the reduced use of agricultural machinery, which significantly lowers fossil fuel consumption and, consequently, greenhouse gas emissions. This section presents two studies conducted in Venezuela that assess the agronomic and environmental impacts of direct seeding in different productive contexts. One of these studies provides evidence supporting the potential of this practice as an effective tool for advancing toward more sustainable agriculture aligned with the country's climate commitments. The works are summarized below.

Barboza et al. (2019) conducted a study to compare energy and operating costs of mechanized agricultural operations in soybean cultivation, focusing on tillage and planting in Mesa de Guanipa, Anzoátegui State, Venezuela. The goal was to quantify potential savings from using a conservation system versus a conventional one, with the hypothesis that machinery energy costs could be reduced by over 10%. Using the "Energy and Operating Costs (CEE)" software, the researchers evaluated bulk density, soil moisture, resistance, fuel consumption, and energy costs across experimental plots. Results showed no significant differences in soil physical-mechanical properties across treatments, indicating no interference in implementation. The conservation system achieved the lowest energy and operating costs, as well as reduced fuel consumption, showing savings of 88.19% and 89.21%, respectively. Additionally, it reduced labor needs by lowering machinery operation time to 4 $h{\cdot}ha^{-1}$, thereby decreasing greenhouse gas emissions. This supports the conservation system as a more efficient and environmentally friendly agricultural approach.

In his undergraduate thesis, Blanco Pulido (2021) highlights the global importance of cowpea (*Vigna unguiculata*) as a widely produced and consumed legume. Successful cultivation typically requires soil preparation, making tillage systems crucial for proper crop establishment. This study aimed to evaluate conventional tillage versus no-till systems for Bayo cowpea yield in a production unit of La Cumbre C.A. Two plots were planted under commercial-like conditions. Germination and plant height showed significant differences between treatments ($p < 0.05$), while yield did not differ significantly ($p > 0.05$). However, conventional tillage achieved the highest yield at 906 $kg{\cdot}ha^{-1}$. It also presented the best cost-benefit ratio (1.41), making it the most viable option for this context, mainly due to higher herbicide costs in the no-till system. Although no-till did not negatively affect yield, the study recommends incorporating cover crops into no-till systems to reduce herbicide dependency and improve sustainability in future agricultural practices.

6.6.3 Reduction, Improvement of Chemical Fertilizer Use, and Alternative Fertilization

Reducing the use of synthetic nitrogen fertilizers significantly lowers nitrous oxide (N_2O) emissions, a potent greenhouse gas with a global warming potential nearly 300 times that of CO_2. It also reduces the carbon footprint of agricultural and industrial activities, as fertilizer production is energy-intensive and heavily reliant on fossil fuels. From an environmental standpoint, reducing the use of chemical fertilizers enhances soil health by encouraging the application of organic amendments, crop rotation, and nitrogen-fixing species—practices that improve soil carbon sequestration. Additionally, it helps minimize indirect greenhouse gas emissions associated with nutrient runoff and leaching, which contribute to water pollution and eutrophication. Between 2018 and 2024, several studies conducted in Venezuela explored this mitigation approach by evaluating alternatives to synthetic fertilization.

Blanco et al. (2018) conducted an interesting study in Falcón State to evaluate the biomass production of two cultivars of *Pennisetum purpureum* (Elephant Grass), using organic fertilization (cattle manure) and mineral fertilization (urea), with treatments involving different doses of both types of fertilizers. No significant differences were found among the treatments studied, indicating that through organic fertilization with cattle manure in the *P. purpureum* cultivars studied, it is possible to eliminate or reduce the use of chemical fertilizers and therefore decrease CO2-e emissions associated with the industrial production of urea, in addition to lowering the production costs of Elephant Grass.

Torres et al. (2018) conducted a greenhouse study similar to the previous one but with tomato (*Solanum lycopersicum*) to investigate the effects of different organic and inorganic fertilization treatments on tomato plant growth. An experiment was conducted comparing the effect of five treatments: (T0) control, (T1) inorganic fertilization, (T2) compost, (T3) vermicompost, and (T4) compost + vermicompost. During the crop cycle, vegetative variables were measured: plant height, number of leaves, stem diameter, and changes in some soil variables. The results show that the organic treatments (T2 and T3) promoted greater vegetative development, with significantly higher values for plant height, number of leaves, and stem diameter compared to the control (T0) and inorganic fertilization (T1). The greater plant development was due to the contribution of organic matter from the vermicompost, which increased nutrient availability, confirming the elimination or reduction of GHG emissions in the agricultural production process and generating a lower demand for chemical fertilizers, which produce large amounts of $CO_{2\text{-}e}$ in their manufacturing process.

Abreu Cruz et al. (2018) conducted a study to evaluate the effects of combining chemical fertilizers and earthworm humus on the growth and development of *Capsicum L.* (Ají Magistral). The chemical fertilizer used was 10-20-10, while the organic fertilizer was humus derived from cattle manure. Five treatments were tested: 100% chemical fertilizer, 75% chemical fertilizer combined with 2, 4, or 6 $t{\cdot}ha^{-1}$ of humus, and 4 $t{\cdot}ha^{-1}$ of humus alone. Growth, reproductive, and yield indicators were assessed during two harvests. In the first harvest, 100% chemical fertilizer produced the best results for plant height, number of leaves, flowers, fruits, and fruit weight. However, in the second harvest, combining 75% chemical fertilizer with 4 or 6 $t{\cdot}ha^{-1}$ of humus yielded similar results, indicating the positive effect of humus on plant productivity. The study demonstrates that earthworm humus can reduce the need for chemical fertilizers while maintaining crop yield, supporting more sustainable agricultural practices for *C. annuum.*

Acosta et al. (2019) studied the impact of adding organic residues to a soil in Santa Ana, Paraguaná Peninsula, Falcón State, Venezuela, aiming to evaluate carbon (C) and nitrogen (N) dynamics. Three organic residues were tested at a 2% application rate: sludge from an aerobic lagoon (L), compost from municipal sludge and goat manure (R), and compost from vegetable waste and goat manure (A). Treatments SL, SR, SA and a control soil (S, 0%) were evaluated at the time of addition and after 30 days. All treatments showed a significant initial increase in total C and N compared to the control. After 30 days, a decrease was observed due to organic matter mineralization. Despite this reduction, the study demonstrated that these organic materials can serve as alternative sources of organic matter, improving soil conditions in Venezuela's semi-arid regions. This highlights their potential for enhancing soil fertility and sustainability in similar agricultural contexts.

Coraspe and Lanza (2020) conducted a trial in José Vicente de Unda municipality, Portuguesa State, Venezuela, to evaluate the effect of organic and natural inorganic fertilizers on exchangeable aluminum in acidic soils cultivated with coffee, without using chemical fertilizers. The plantation is located at 1018 m above sea level. Organic fertilizers included vermicompost, bokashi, and decomposed coffee pulp, while inorganic treatments used agricultural lime and phosphorite. Results showed a highly significant effect on coffee productivity. Bokashi and vermicompost led to higher yields due to their richer supply of macro- and micronutrients. Both organic and inorganic fertilizers reduced exchangeable aluminum levels, helping prevent aluminum toxicity. This effect is attributed to the contribution of exchangeable bases, OH^- radical production, and organic matter acting as a complexing agent. The study confirms that organic fertilization, without chemical inputs, can enhance coffee cultivation on acidic soils in Venezuela, promoting sustainable agricultural practices in sensitive environments.

Along the same lines regarding the reduction or elimination of chemical fertilizers and the application of alternative options, Escalona-Sánchez et al. (2021) evaluated the effect of an organic-mineral fertilization plan on silage corn production. The trial was established in Duaca, Lara State, Venezuela, with six treatments: T1: Don Manuel compost (CDM); T2: Minco Fertil compost (CMF); T3: Phosphogypsum (FY); T4: CDM + FY; T5: CMF + FY; and a control without amendments. These were manually applied in the planting furrow, while organic fertilization with 16 products and mineral fertilization (urea) were applied through irrigation at 15 and 22 days after planting. The results showed that the Minco Fertil compost preparation increased growth, plant height, number of leaves, and leaf area. The highest silage corn yields and total dry matter were obtained with the Don Manuel and Minco Fertil compost preparations. Forage corn can be produced with organic applications, which have positive effects on crop growth and production.

Flores (2021) highlights corn's importance for Venezuelan food security and notes that its yield is affected by nutrient deficiencies. Given current difficulties in accessing inorganic fertilizers, the study explored biofertilizers as a cost-effective, environmentally friendly alternative with lower greenhouse gas emissions. Conducted at the La Salle Foundation of Natural Sciences in San Carlos, Cojedes State, Venezuela, the research tested the INIA-7 corn variety in two plots with differing fertility levels. Three treatments were applied: T1, inorganic chemical fertilizer (15–15-15 SP formula and prilled urea with 46% ammoniacal nitrogen at 400 kg/ha and 200 kg/ha, respectively); T2, chemical fertilizer plus biofertilizer; and T3, biofertilizer alone (beneficial microorganisms). Results showed no significant differences in yield between treatments or plots, suggesting that biofertilizers are a viable alternative to chemical fertilizers. The response of microbial communities may have been influenced by local soil conditions, including temperature, humidity, pH, and chemical composition, potentially affecting the biofertilizer's performance.

6.6.4 *Use of Mycorrhizae and Manure in Agricultural Productivity and the Reduction of N_2O Emissions from Soils*

The incorporation of mycorrhizae and properly managed cattle and sheep manure offers a sustainable strategy to reduce the carbon footprint of Venezuelan agriculture while enhancing climate resilience. Mycorrhizae improve nutrient and water uptake, reducing the need for synthetic fertilizers, while manure enriches soil organic matter and mitigates methane emissions when handled correctly. Together, these practices enhance carbon sequestration, improve soil structure, and reduce nitrous oxide emissions, particularly in degraded or acidified soils. In Venezuela, documented experiences in forest restoration and circular livestock systems highlight the potential of these approaches. The following section presents five studies on this topic: three on mycorrhizae, one on urban agriculture, and one on bioremediation of acid soils.

Jiménez et al. (2017) evaluated the effect of arbuscular mycorrhizal fungi (AMF) and bovine manure (BM) on early growth and pigment content in "Misterioso" chili pepper (*Capsicum frutescens*). Five treatments were tested: absolute control in sand (T1), T1 + BM (T2), and T1 with *Scutellospora fulgida* (T3), *Funneliformis mosseae* (T4), or *Glomus manihotis* (T5). After 55 days, variables measured included mycorrhizal colonization, plant height, number of leaves and nodes, root length, biomass, leaf area, leaf area index, and pigment content. T4 produced the best results in most variables except for leaf and node counts and leaf area. T3 also performed well, while T2 led in leaf area and index, and had high values for height, node and leaf numbers, biomass, and root length, though not for pigments. The study concluded that *Funneliformis mosseae* (58.47% colonization), *Scutellospora fulgida* (26.31%), and bovine manure significantly improved early plant development, highlighting AMF and organic fertilizers as effective growth enhancers.

Pírela-Almarza et al. (2018) studied the effects of arbuscular mycorrhizal fungi (AMF) and sheep manure on the early growth of Maradol Roja papaya (*Carica papaya* L.). Five treatments included inoculation with *Scutellospora fulgida* (SF), *Funneliformis mosseae* (FM), *Glomus manihotis* (GM), sheep manure (M), and an absolute control (sand only). Substrates were mixed at a 2:1 sand-to-AMF/manure ratio. After 60 days, the researchers measured AMF colonization, plant height, number of leaves and nodes, root length, stem diameter, shoot biomass, leaf area, leaf area index, and pigment content (chlorophyll a, b, total, and carotenoids). Statistical analysis revealed significant differences in colonization, plant height, pigments, and leaf measurements. The manure treatment achieved the highest levels of chlorophyll b, total chlorophyll, and carotenoids. The study concluded that both AMF and manure positively influenced early growth and pigment content in papaya, with organic treatments promoting plant development and potentially serving as sustainable alternatives to chemical fertilizers in early cultivation stages.

Márquez et al. (2019) evaluated mycorrhizal activity and soil quality in native and introduced potato plots in the Gavidia páramo, Mérida, along with a reference forest. Native potatoes (*S. tuberosum andigenum*) were grown traditionally; introduced potatoes (*S. tuberosum tuberosum*) used chemical fertilizers. Indicators included colonization by AMF and DSE, spore density, and soil properties. All sites showed AMF and DSE presence, but DSE was highest in introduced crop soils. Forest soils had better physical and biological quality, while crop plots had more nutrients. Multivariate analysis showed that native potato plots were more similar to forest soils, suggesting traditional practices have lower ecological impact. These findings point to the importance of preserving low-impact, culturally rooted farming methods in fragile ecosystems like the páramo.

Flores and Paredes (2020) examine how urban expansion contributes to environmental issues, particularly the Urban Heat Island (UHI) phenomenon, which impacts public health and thermal comfort. They also highlight the environmental degradation caused by industrial agriculture, including excessive use of fertilizers and pesticides that disrupt ecosystems, increase erosion and soil salinity, and lead to mineral loss. These practices are linked to serious human health effects, such as neurological disorders, genetic mutations, and infertility. As a response, the authors advocate for urban agriculture, particularly home gardens and small plots integrated into city landscapes, as a strategy to mitigate UHI effects. Urban agriculture can reduce local temperatures, enhance biodiversity, and limit greenhouse gas emissions such as CO_2 and N_2O. The study, documentary in nature and based on bibliographic sources, emphasizes the multifunctional benefits of urban agriculture. Final considerations highlight that urban agriculture is both ecologically and socially valuable, offering environmental services such as erosion reduction, emission control, and biodiversity conservation, while also fostering community well-being in densely populated urban areas.

Atacho et al. (2021) addressed the challenge of acid soils in Venezuela, which reduce productivity due to aluminum toxicity and limited nutrient availability. Soil pH also influences N_2O emissions by affecting microbial activity and nutrient dynamics. To rehabilitate acid soils in Sierra de San Luis, Falcón State, Venezuela, two vermicompost solutions, Uredo and El Pozón, were tested at 50%, 75%, and 100% concentrations. Treatments (Uredo 1–3 and Pozon 1–3) were incubated for 45 days at 28 °C and 80% moisture. Researchers measured pH, exchangeable acidity, organic matter, cation levels, base saturation, and basal respiration. All treatments increased soil pH above the initial 3.80 reaching values over 4.05. Pozón 2 and Uredo 2 showed the highest basal respiration rates (2.42 and 2.36 μg CO_2 g^{-1} h^{-1}). Uredo 2 and 3 effectively reduced exchangeable acidity (1.87–1.88 cmol kg^{-1}), demonstrating their effectiveness for early-stage rehabilitation. These results support vermicompost as a practical bioremediation strategy for improving acid soils and mitigating environmental stressors in agriculture.

6.6.5 *Agroforestry Systems*

Agroforestry systems including agrosilvopastoral systems represent sustainable land-use strategies that combine trees with crops and/or livestock in the same area, enhancing ecosystem services and increasing resilience to climate change. In the context of Venezuelan agriculture, these systems offer promising pathways to mitigate greenhouse gas emissions, improve soil health, conserve biodiversity, and support smallholder livelihoods. Their multifunctional nature aligns with both climate adaptation and mitigation goals, making them particularly relevant for regions vulnerable to environmental degradation and socioeconomic challenges. This section presents below a selection of seven research studies conducted in Venezuela that explore the potential and effectiveness of various agroforestry and agrosilvopastoral practices as climate-smart agricultural solutions.

Farreras (2018) explores shade-grown coffee as an agroforestry system supporting sustainable mountain agriculture in Venezuela. The study focuses on traditional systems in the Sucre municipality, Portuguesa state, using a descriptive-analytical and bibliographic methodology with a quantitative perspective. An analytical matrix was developed to highlight how shade-grown coffee can help mitigate problems in mountain watersheds and promote sustainability in social, economic, and ecological terms. The essay emphasizes that traditional coffee agroforestry systems not only support income generation for local farmers but also act as a systemic strategy to reduce greenhouse gas (GHG) emissions. This is achieved through the preservation of vegetated areas that serve as CO_2 sinks, making coffee cultivation both climate-friendly and economically viable. The findings reaffirm the value of integrating agroforestry with traditional practices to enhance environmental resilience in mountainous agricultural landscapes.

Continuing with the agroforestry coffee system (*Coffea arabica* L.) in Venezuela, Paolini-Gómez (2018) reports that coffee covers about 200,000 hectares, mostly in the Andes. Traditionally, it is cultivated under shade using leguminous trees with minimal external input. The study analyzed microbiological activity, microbial biomass, and ecophysiological parameters in surface soils (0–5 cm) on eight farms in Mérida and Trujillo, comparing traditional, conventional, and organic systems. Organic farms showed the highest values of organic carbon, basal respiration, and microbial biomass, indicating stronger conservation of soil resources. Conversely, conventional systems, using fertilizers and pesticides, negatively affected soil microbial communities and reduced their carbon-use efficiency. The findings highlight organic coffee farming as a more sustainable alternative for preserving soil health and environmental balance in mountainous Venezuelan regions.

Likewise, Fernández-Molina et al. (2019) analyzed a sustainable agro-productive strategy to diversify citrus monocultures in the tropical dry forest of Mesa de los Gabaldones, Trujillo State, Venezuela. Diversification enhances the sustainability of citrus systems, especially in regions where monocultures dominate. A key approach for sustainability was implementing agroforestry systems: (a) diversified cropping systems, (b) silvopastoral systems, and (c) home gardens. Innovative farmers rapidly adopted agroecosystem diversification through changes in agronomic practices and the introduction of technologies like silvopastoral systems and intercropping. Their adoption was facilitated by the participatory development of the strategy, involving producers from the outset. As a result, the number of diversified agroecosystems doubled, increasing from 25% at baseline to full adoption among the community. This shift reflects both a practical and cultural transformation in local production strategies, highlighting the importance of bottom-up approaches in fostering sustainable agriculture tailored to regional ecological and socioeconomic contexts.

For their part, Pardo-Plaza et al. (2019) explored the influence of coffee varietals and seasonal changes on microbiological indicators in agroforestry systems, emphasizing their role in sustainable soil management. In Venezuela, coffee cultivation commonly occurs under the canopy of Inga, Cedrela, or fruit trees, forming traditional multistrata systems that enhance biodiversity. At El Laurel station (UCV), soil samples were taken at 0–5 cm depth during both dry and rainy seasons to compare the varieties Caturra and Catuaí. Catuaí showed higher CO_2 emissions and metabolic quotient (qCO_2), especially in the dry season, while Caturra had a higher microbial quotient (qMic). Although $qCO_2 \cdot Corg^{-1}$ values did not differ significantly, they increased under drier conditions. These indicators proved useful to assess coffee systems' responses to environmental variability and crop management.

In response to shifting global markets, Rivera (2019) proposes expanding forest plantations in Venezuela to enhance carbon sinks and tap into the rising demand for renewable forest products. With fossil fuel demand expected to decline due to technological and climate pressures, the country must diversify its exports. Forest-based materials, such as wood, paper, bioplastics, and bioenergy, are projected to see demand double by 2050. Venezuela holds significant forestry potential, making reforestation a strategic move to meet environmental commitments while strengthening rural economies. The author argues that scaling up sustainable plantations can boost carbon sequestration, support biodiversity, and position Venezuela competitively in global green markets. To succeed, investments must be guided by long-term sustainability and robust governance frameworks.

Then, Camargo Roa et al. (2020) examined land-use conflicts in the Mejías sub-watershed (Mocotíes River basin, Mérida, Venezuela) to propose conservation-oriented strategies. The study aimed to restore degraded areas, reduce soil erosion contributing to river sedimentation, and prevent CO_2 loss. Using satellite images and a Digital Elevation Model (DEM), they classified land use and slope gradients for 2018. These data were integrated to develop a land potential use map. The study identified 2918.51 ha (25.37%) for reforestation, 1479.96 ha (12.86%) for agroforestry, and 1006.43 ha (8.75%) with agricultural potential. Additionally, they recommended pasture establishment on 131.11 ha (1.14%) and erosion control on 10.75 ha (0.09%). This research offers practical tools for sustainable land management in watersheds with high ecological vulnerability. It underscores the importance of spatial planning, particularly in Andean zones where topographic and climatic conditions heighten environmental risks, and where land use must align with both conservation and livelihood goals.

In their examination of land degradation in Barinas State, Venezuela, Cañas et al. (2020) focused on the environmental impacts of deforestation in Quebrada Negra, situated in the middle basin of the Masparro River, Venezuela. Deforestation accelerated over the last two decades due to unplanned rural settlements, mainly from migrants pursuing agro-livestock activities. These practices led to declining land productivity, increased erosion, organic matter loss, and elevated greenhouse gas emissions from tree felling and burning. The essay takes a transdisciplinary approach to assess the impact of this land-use transformation and proposes agroforestry systems as a solution aligned with sustainable development. The adoption of agroforestry could mitigate land degradation, restore ecosystem services, and support rural livelihoods. By integrating ecological, social, and economic dimensions, the proposed model promotes long-term landscape restoration and resilience. The study highlights the urgent need for institutional and community-level strategies to halt further environmental damage and foster sustainability in vulnerable rural territories affected by unsustainable agricultural expansion.

6.6.6 *Pastoral System and Livestock Farming*

Mitigating climate change in agriculture is a key challenge for sustainable development, with traditional and extensive livestock systems playing a major role. Although these systems generate significant greenhouse gas (GHG) emissions, they also offer opportunities for mitigation and resilience. This section reviews four studies on climate mitigation in pastoral and livestock systems. Two are literature reviews that explore global strategies, policy tools, and the importance of incentives and technical support for sustainable livestock. The other two focus on Venezuelan research, highlighting silvopastoral practices, rotational grazing, improved feeding, and manure management. These strategies reduce emissions, enhance carbon storage, and improve productivity. The summaries below illustrate the potential for transforming livestock systems toward climate-resilient models.

The review by González et al. (2015) remains a key reference for greenhouse gas (GHG) mitigation in Latin American livestock systems, including Venezuela. The paper, titled "Limitations for the Implementation of GHG Emission Mitigation Actions in Livestock Systems in Latin America," outlines strategies applicable to the Venezuelan context. It highlights that livestock contributes 18% of global GHG emissions, while tropical deforestation adds 20–25%. In Latin America, cattle production has a high environmental impact. Mechanisms such as the Clean Development Mechanism (CDM), NAMAs, and REDD support mitigation efforts. The authors emphasize Life Cycle Assessment (LCA) as essential for identifying emissions. They conclude that most countries lack fully integrated mitigation policies in agriculture.

Similar to the previously cited review, Caballero (2017) analyzes the challenge of climate change in Latin America, highlighting its economic, social, and environmental impacts. The study presents climate change as a global negative externality linked to the prevailing development model, thus requiring deep structural transformations. To address this challenge, a set of public policies is needed to correct market failures and promote a low-carbon economy. This involves changing current patterns of production and consumption and developing infrastructure aligned with sustainable growth. The evidence presented shows that no single public policy can simultaneously address climate change and achieve sustainable development. Therefore, policy packages are proposed, adapted to national or regional specificities, combining common elements such as incentives, regulations, and mechanisms for adaptation and mitigation in the productive sector.

In her study, Laya (2019) assesses the sustainability of dairy production and processing systems in rural plots of the Rómulo Gallegos municipality, Cojedes State, Venezuela, using the MESMIS methodology. This tool evaluates economic viability, ecological suitability, and the conservation of natural resources. The research, carried out on 650 plots, began with a diagnostic phase that identified unsustainable practices. In response, an agroecological model was proposed based on silvopastoral and agrosilvopastoral systems, pasture rotation, and technological improvements in forage cultivation. Post-harvest conservation included acidified silage made with whey and molasses, stored without refrigeration. The integration of tree and shrub species as protein sources for animal feed, particularly useful during droughts, was also part of the strategy. Overall, the proposal aims to reduce agrochemical use, enhance system resilience, and support local agricultural sustainability under circular economy principles.

As part of their exploration of agroecological alternatives, Ojeda-Falcón and Domínguez-Quintero (2020b) implemented a modified Voisin Rational Grazing (VRG) system on acidic sandy soils in Barinas, Venezuela. This approach avoids agrotoxins and tillage while promoting intensive, clean food production. Conducted on a 27-hectare pasture of *Urochloa* spp., the study tested four treatments: asphalt emulsion, phosphate rock, their combination, and a control using only VRG. Treatments with asphalt emulsion produced the highest dry matter yields and increased stocking rates from 1.2 to 4.3 LU/ha within 2 years. VRG also enhanced soil phosphorus fertility, addressing a major constraint in savanna ecosystems. The method enabled significant pasture productivity gains without synthetic fertilizers, tillage, or pesticides. This led to lower production costs, reduced greenhouse gas emissions, and eliminated dependence on agro-industrial inputs, making it a highly relevant model for environmentally and economically sustainable livestock systems.

6.6.7 Land-Use Change and Vegetation Cover

The connections between land-use change, vegetation cover, and climate change mitigation are particularly relevant in the Venezuelan context, considering that forested areas act as natural carbon sinks. These ecosystems store greenhouse gases and help reduce the impact of climate change, while deforestation threatens their ability to absorb carbon and affects biodiversity. Understanding and protecting natural forest and shrubland ecosystems from deforestation driven by agricultural expansion are essential to promote sustainable practices that contribute to climate change mitigation in Venezuela.

Olivares et al. (2019) analyzed land-use changes between 2001 and 2013 in the agricultural community of Kashaama, located in Anzoátegui State, Venezuela. The authors used digital images derived from a Moderate Resolution Imaging Spectroradiometer (MODIS), which provide information on different types of land use and vegetation cover for the years 2001 and 2013. The study concluded that the indigenous community experienced a significant transformation of vegetation cover into anthropogenic land uses, putting at risk the ecosystem goods and services available to the population.

For the same agricultural community and period, Olivares and López (2019) analyzed the behavior of the Normalized Difference Vegetation Index (NDVI) using MODIS sensor data (MOD13Q1), which includes this index. The results showed that the average NDVI behavior is directly associated with the region's main physiographic and climatic features. The authors determined that the method facilitated the creation of maps which, together with computerized satellite image processing, reliably distinguished vegetation types and conditions, such as cultivated versus natural savanna. The study confirmed that multi-temporal satellite image analysis is a useful tool to monitor the evolution of plant communities and crops, particularly in vulnerable areas.

In addition, Morales Montero et al. (2021) conducted a spatiotemporal analysis of land cover in the northern Orinoco River region, Venezuela, between 1988 and 2010. This area has experienced extensive environmental degradation due to deforestation, land-use changes, agricultural and livestock expansion, and urban growth.

To analyze spatial dynamics, the authors produced two land cover maps based on national vegetation maps from 1988 and 2010. Using Geographic Information Systems (GIS) and cross-tabulation, they generated a matrix to identify changes in vegetation. The results revealed a sharp increase in human-altered areas, from 21.9% in 1988 to 43.5% in 2010, mainly at the expense of natural vegetation. Forests experienced the most significant loss, declining from 36.8% to 22.3% over the study period. The authors concluded that the region's natural vegetation is undergoing severe degradation and may face complete disappearance if current trends continue. The study highlights the urgent need for conservation strategies to prevent further ecological loss.

Similarly, Pino Suarez et al. (2021) analyzed forest successional dynamics 57 years after enrichment strip interventions in the "El Caimital" University Forest, Barinas State, Venezuela. They compared floristic composition, structure, and dynamics between a primary forest (control) and a forest with enrichment strips. Field assessments included all individuals with a diameter at breast height (DBH) >10 cm. In the non-intervened forest (NIF), 49 species across 24 families were identified; in the intervened forest (IF), 45 species in 23 families were recorded. Key families in both areas were Arecaceae, Leguminosae, Moraceae, Malvaceae, and Euphorbiaceae. Based on the Importance Value Index (IVI) and Expanded Importance Index (EII), dominant species in the NIF were *Trophis racemosa*, *Attalea butyracea*, and *Lonchocarpus pictus*, while the IF was dominated by *Attalea butyracea*, *Hura crepitans*, and *Lonchocarpus pictus*. Both forests showed high biodiversity, with Margalef Diversity Index values exceeding 5.0, indicating long-term ecological richness. The study highlights the potential of enrichment strategies to support natural regeneration and carbon sequestration. These findings are relevant for forest-based climate change mitigation efforts in Venezuela.

6.6.8 Agroecological Sustainability

Agroecological sustainability is emerging as a key pathway toward climate change mitigation and adaptation in Venezuela's agricultural sector. In a country marked by ecological diversity and socioeconomic challenges, agroecological systems offer a promising alternative to conventional agriculture by integrating ecological principles with local knowledge, social equity, and productive resilience.

This section presents two empirical studies: one conducted in the endorheic basin of Lake Valencia, and another in the Venezuelan Andes. The evidence highlights the potential and challenges of transitioning to more sustainable agricultural models.

The research conducted by Silva-Laya et al. (2018) evaluated sustainability indicators in two agroecological systems producing vegetables and medicinal plants in the Venezuelan Andes: a peri-urban zone (NUDES-La Limonera) and a rural agricultural area (FIRP). Over 11 months (2014–2015), 16 indicators were assessed across ecological, economic, and sociocultural dimensions.

NUDES-La Limonera, a national government project, showed weaker ecological and economic performance, such as limited soil life conservation, dependence on external inputs, and poor food self-sufficiency. Sociocultural strengths included community ties, but housing conditions were inadequate due to lack of land ownership. In contrast, FIRP, driven by organized farmers promoting crop diversification, demonstrated stronger sustainability across all dimensions. Productivity, profitability, and land tenure were especially favorable. The study emphasizes how synergies within production systems, driven by ownership, organization, and diversification, can support comprehensive sustainability in rural and peri-urban contexts.

Likewise, Ruiz-Dagger and Paolini (2021) studied biological indicators in lacustrine and alluvial soils of Venezuela under different land uses, demonstrating that land use can affect microbial activity and alter biochemical transformations in soils. The authors evaluated basal respiration and ecophysiological coefficients in soils of the Lake Valencia basin under three land uses: natural vegetation, sugarcane monoculture, and banana monoculture. Total organic carbon (C_org), water-soluble carbon, and microbial activity significantly decreased in agricultural soils compared to those under native vegetation, both in alluvial and lacustrine types. The metabolic quotient (qCO_2) was highest in banana agroecosystems, followed by natural vegetation, and lowest in sugarcane fields. On the other hand, the metabolic efficiency quotient (qCO_2/C org) was significantly lower in soils with natural vegetation, illustrating the relationship between more efficient carbon use and the nature of the usable organic matter in those soils. The findings reveal that both crop type and agricultural practices exert a substantial influence on soil biological activity.

6.6.9 State of Ecosystems and their Conservation

The condition of agricultural and natural ecosystems directly affects their ability to contribute to climate change mitigation. Healthy and functioning ecosystems can act as carbon sinks, regulate microclimates, conserve biodiversity, and support agricultural productivity. In contrast, degraded ecosystems lose these key functions and, in many cases, become net sources of greenhouse gas emissions. This section focuses on diagnosing the current state of agroecosystems and natural ecosystems linked to agriculture as a fundamental basis for designing management, conservation, and restoration strategies that enhance their contribution to climate change mitigation.

In this session, five published studies are presented that address various dimensions of climate change mitigation, focusing on the conservation of natural ecosystems in Venezuela. From integrated watershed management to reforestation as a strategy to conserve water sources, restore carbon sinks, and enhance ecosystem resilience, these studies underscore the importance of adopting systemic and participatory approaches. Summaries of the works are presented below.

Zambrano (2018) examined key environmental challenges affecting Venezuelan watersheds. He attributes these issues to climate change-related natural disasters, rapid environmental degradation, and significant alterations to the hydrological cycle, resulting in decreased water quality, quantity, and availability. The study focuses on the middle Santo Domingo River Basin (Mérida–Barinas states), Venezuela, and emphasizes the importance of integrated watershed management. From ecological, economic, and social perspectives, Zambrano proposes principles for the sustainable use of water resources that can balance environmental protection, economic development, and social equity. His essay underscores the need for inclusive approaches that involve the active participation of local communities in planning and implementation. This inclusive model seeks to ensure that watershed management not only meets technical and environmental standards but also responds to the real needs of the population, ultimately promoting long-term sustainability in regions vulnerable to degradation.

Molina Pereira (2019) implemented a reforestation project near the Masparro River dam in Barinas State, Venezuela, with the goal of using reforestation as a conservation strategy for water bodies. Through a quantitative research approach, the study followed five phases: diagnosis, feasibility analysis, design, execution, and evaluation. Thirty local residents participated, providing data via a validated Likert-scale questionnaire. Bar charts were used for analysis. The project emphasized the importance of reforesting key water sources to guarantee water availability for human use, agriculture, and livestock, especially in light of climate change. A total of 250 bamboo trees were planted. The results reflected strong community engagement and highlighted the critical need to protect and restore ecosystems that support water supply. Molina Pereira concludes that reforestation efforts rooted in community participation offer an effective environmental strategy for mitigating degradation, conserving natural resources, and promoting awareness in areas affected by deforestation and water scarcity.

Pacheco et al. (2019) applied geotechnologies to estimate the C factor of the RUSLE model in the Carache River Basin, Trujillo, Venezuela. The C factor reflects vegetation cover and land use influence on soil erosion. The team used corrected multispectral images from the Miranda VRSS-1 satellite to calculate the Normalized Difference Vegetation Index (NDVI) and generate a land-use classification. C values were assigned theoretically, and a strong correlation ($R^2 = 0.99$) was found between NDVI and C factor, represented by a fifth-order polynomial function. Around 30% of the basin had high C values (0.5–1.0), indicating intense erosive processes on sloped agricultural lands with poor vegetation cover. The study underscores the importance of Geographic Information Technologies (GIT) in environmental diagnostics and decision-making. It recommends protective measures and improved agricultural practices to prevent soil erosion and promote sustainability, especially in erosion-prone areas with fragile agroecosystems and high human pressure.

Peña-Colmenares and Gordon-Colón (2019) studied water and soil characteristics and vegetation structure in a morichal of the upper Tigre River Basin, Anzoátegui, Venezuela. Data were collected from eight stations along the river between May 2000 and May 2001, including bimonthly measurements of water pH and conductivity and composite soil sampling. Vegetation analysis involved identifying tree species ≥2 m tall and recording diameter and canopy height. Structural variables and species composition were analyzed through correspondence analysis. Soils were mineral, with sandy-to-sandy clay loam textures and high spatial variability due to flooding and lateral river dynamics. The floristic survey recorded 218 species, with *Mauritia flexuosa* (moriche palm) dominating the tree stratum. Other abundant species included *Virola surinamensis* and *Calophyllum brasiliense*. The findings revealed a mosaic of plant communities shaped by elevation gradients, sedimentation, and hydrological variability. The study highlights the ecological complexity and conservation value of palm swamp forests in tropical riverine systems.

Méndez et al. (2020) developed indicators to assess rainfall erosivity in the Chichiriviche River Basin, Vargas State, Venezuela, a mountainous watershed where agriculture drives soil erosion. The study analyzed the spatial and temporal distribution of rainfall and calculated several indices: the Annual and Modified Fournier Index (AFI, MFI), Precipitation Concentration Index (PCI), total erosivity, and rainfall erosive potential. Results showed that rainfall was influenced by altitude and concentrated toward year-end. Peak 1-h rainfall intensities for 100-year events reached up to 81.95 mm/h. While most indices classified erosive potential as low to moderate, the IMF2 yielded high values in some cases. Extreme rainfall events, though isolated, played a significant role in triggering erosion. The study identified IMF1, IMF2, ICP1, and ICP2 as the most effective indicators. However, the lack of hourly rainfall data limited detailed analysis. Findings emphasize the need to monitor short-duration, high-intensity events, which are key drivers of erosion in tropical mountainous areas.

6.6.10 Land Capability and Territorial Planning

Land suitability, which means the soil's capacity to support specific agricultural uses sustainably, and territorial planning and zoning, that refers to organizing those uses spatially, are key to reducing GHG emissions and protecting carbon sinks in Venezuela. Identifying suitable areas for cropping, silvopastoral systems, or conservation helps prevent agricultural expansion into high-carbon ecosystems, protects fragile soils, and promotes carbon-sequestering practices. Proper land-use planning enables production on high-potential soils with lower input use, supports conservation zones as CO_2 sinks, and encourages multifunctional landscapes with agroforestry and recharge zones. It also helps direct rural infrastructure to already modified areas and facilitates policies like payments for ecosystem services and green loans aligned with soil capacity. In Venezuela, where plains, mountains, and Amazonian forests coexist, linking land suitability with territorial planning is crucial for achieving mitigation targets under the Paris Agreement. This synergy maximizes agriculture's climate benefits while safeguarding productivity, food security, and the ecosystem services essential for the country's sustainable development. Below are two published works in this area.

Ferrer et al. (2018) conducted a detailed land evaluation of a 76.85 ha tabletop plateau in eastern Venezuela (Monagas State), aiming to guide land-use planning. The soils Ultisols with low fertility, kandic horizons, kaolinitic clays, and an ustic-isohyperthermic regime present major fertility constraint. Of the total area, 62.94 ha fall under Capability Class III due to severe fertility limitations; 42.91 ha of this also show coarse texture constraints. Another 13.91 ha belong to Class II with moderate fertility limitations. Class III land supports moderately managed crops, pastures, or forestry, while Class II allows intensive cultivation. According to the Agricultural Vocational Classification, 73 ha are Class V, suited for livestock and wildlife uses. Only 3.85 ha are Vocational Class III, where cultivation of fruit, cereals, oilseeds, tubers, and tropical plantations (e.g., coffee, cacao) is feasible under conservation-oriented management.

Martens (2018) explores territorial planning in southern Venezuela, focusing on the Pemon territory within the Upper Caroní River Basin, Venezuela, a region vital for hydroelectric power and rich in Protected Areas (ABRAE). These areas act as key natural CO_2 sinks. Combining legal analysis of Venezuela's Territorial Planning Law and its proposed reforms with ethnographic fieldwork (2007–2010), the study exposes enduring sociocultural conflicts. The findings reflect tensions between indigenous governance traditions and state-driven land-use models, especially in regions with a long history of cultural resistance. Martens argues that territorial planning in such contexts must integrate environmental protection with indigenous land rights and identity. The case highlights how overlapping legal, cultural, and ecological frameworks shape policy conflicts. It calls for participatory planning approaches that respect traditional knowledge and the ecological significance of ABRAEs, ensuring social inclusion and climate resilience in Venezuela's most bio-diverse and culturally sensitive landscapes.

6.6.11 The Soil Conservation, Restoration, and Management Improvement

Soil plays a crucial role in mitigating climate change by storing atmospheric CO_2. However, degradation, through erosion, compaction, salinization, or organic matter loss, threatens this function. This section highlights soil degradation in agriculture and explores strategies for conservation, restoration, and carbon sequestration. Below are seven summaries from papers published in this area.

Batista and Catarí (2018) carried out a comprehensive study of landslide susceptibility and erosive processes in the Mamo River basin and its El Tigre sub-basin, Vargas State, Venezuela. Using a 60-year inventory of active landslides, a Digital Elevation Model (DEM), remote sensing, geological data, and precipitation records, they identified the most vulnerable zones. Their models showed that lower, densely populated areas of the basin are more prone to mass movements than upper and middle sections. Erosive dynamics were further characterized using indices such as total sediment production (TSP), unit sediment production (USP), and shear stress (SSP), which highlighted severe erosion in the lower basin and throughout El Tigre. The findings underline the need for preventive land management in high-risk regions.

Rivas and Mogollón (2018) analyzed wind erosion as a major desertification driver in the Paraguaná Peninsula, Falcón State, Venezuela. They assessed current and potential erosion levels by evaluating four key degradation factors: climate, soil texture, topography, and vegetation cover. The results revealed average current soil losses of 258 t/ha/year, while potential losses could reach 512 t/ha/year. Over 50% of the territory was at very strong to extreme erosion risk, with potential risk extending to 80%. Spatial distribution patterns were heavily influenced by vegetation and soil characteristics. The study emphasizes that wind erosion depletes organic matter and soil carbon, severely threatening ecosystem resilience. Vegetation emerged as the most crucial factor for reducing surface wind erosion in this arid environment.

Guillén and Elizalde (2020) examined the spatial distribution of erosion scars in the upper Guárico River Basin, Guárico State, Venezuela, where mass movements are a common environmental hazard. Using photointerpretation and GIS-based digitalization, they identified over 650 individual and 1600 grouped scars in the Caramacate River sub-basin. Scar density was calculated for lithogeomorphological and landscape units, revealing a greater concentration of grouped scars on slopes between 20° and 35°. The study found that slope gradient, rock disintegration, and vegetation cover were the primary conditioning factors, while precipitation and seismicity served as triggers. These results highlight the need for erosion monitoring in mountainous zones and emphasize the interaction between topographic vulnerability and anthropogenic pressures on fragile landscapes.

Ojeda-Falcón and Domínguez-Quintero (2020a) addressed water-induced soil erosion in agricultural lands within Guaraira Repano National Park, Vargas State, Venezuela, an illegal hillside farming area and extreme rainfall events. Using gamma spectroscopy to measure ^{137}Cs radionuclide activity, they estimated erosion rates via Proportional and Mass Balance I models. Gross and net soil loss ranged from 24.7 to 39.7 and 19.8 to 29.3 Mg ha^{-1} $yr.^{-1}$, respectively. These results demonstrate accelerated degradation in mountain agroecosystems, exacerbated by poor conservation practices and climate variability. The study highlights the value of nuclear techniques to evaluate erosion, soil redistribution, and productivity vulnerability, particularly under saturated moisture conditions. Such analyses are vital for guiding sustainable management in fragile agricultural zones.

Acevedo et al. (2021) investigated soil degradation in two horticultural systems in the Quíbor Valley, Lara State, Venezuela, through multivariate analysis. The traditional system focused on onion crops with furrow irrigation, while a newer system cultivated super sweet corn using conventional techniques. Soil samples were collected in a 10 × 10 grid at three depths and analyzed for physical, chemical, and biological attributes. Principal component analysis revealed significant degradation in both systems, influenced by intrinsic soil properties and management intensity. Indicators such as low porosity, poor aggregate stability, and reduced microbial activity pointed to deteriorating soil health. The study warns that intensified cultivation in the region could further degrade already vulnerable soils, undermining productivity and long-term sustainability.

Infante (2021) offered a conceptual analysis of remediation strategies for hydrocarbon-contaminated soils in Venezuela, with emphasis on how soil texture influences contaminant behavior. In clay-rich soils, hydrocarbons are strongly adsorbed and harder to remove, while in sandy soils, contamination spreads deeper but remediation is more feasible. The author compared regional regulatory thresholds and highlighted the need to revise Venezuelan standards to align with toxicity and bioavailability. Among remediation techniques, bioremediation with biostimulation stands out as widely used both nationally and internationally. Infante argues for a national review of current limits (e.g., 1% oil and grease) to ensure effective environmental recovery, especially considering the variability in soil type, crude oil composition, and weathering effects after a spill. Bioremediation with biostimulation, as an effective technique, can play a key role in restoring contaminated soils and reintegrating them into agricultural use. Moreover, adjusting Venezuela's permissible limits based on actual toxicity and bioavailability would contribute to a more accurate assessment of the agricultural suitability of remediated land.

6.6.12 Legal Frameworks and Environmental Education

Between 2018 and 2024, only three studies explored the relationship between legal frameworks, environmental education, and climate change mitigation in Venezuelan agriculture. Despite limited research, both dimensions are essential for advancing sustainable, climate-smart systems. Legal instruments guide policies to reduce GHG emissions and protect carbon sinks, while education strengthens farmers' adaptive capacity and climate awareness. Although not solely focused on mitigation, the studies offer valuable conceptual tools. This section summarizes their contributions to agricultural sustainability and environmental governance.

Farfán Tigre et al. (2018) present a case study on Venezuelan wetlands to highlight environmental education's role in climate mitigation. The study promotes field-based learning as a means to foster ecological awareness and encourage sustainable behavior. It integrates ecological, economic, and educational perspectives, proposing strategies that stimulate critical thinking and environmental responsibility. The authors advocate incorporating wetland-based education at all academic levels, framing wetlands as ideal platforms to teach sustainability, ethics, and adaptation. These experiences support the development of environmentally responsible agricultural practices and reinforce the role of education in building climate-resilient communities.

Abache Carvajal (2019) examines Venezuela's green tax framework and the "polluter pays" principle, emphasizing their theoretical value for reducing environmental impacts. However, economic barriers limit their application among smallholders. Yánes and Caula (2020), using Contingent Valuation Method, assess public support for conserving La Bocaina wetland. Though not agriculture-focused, 94% favored education on bird conservation, and 35% were willing to contribute financially. Results highlight the link between education, income, and conservation behavior. Together, these studies offer insights into regulatory and educational tools relevant for enhancing climate mitigation and environmental stewardship in agriculture.

Yánes and Caula (2020) applied the contingent valuation method to assess public support for conserving La Bocaina wetland, which hosts 33% of Venezuela's aquatic bird species, including the endemic *Rallus wetmorei*. Although not focused on agriculture, the study offers insights relevant to mitigation efforts. Results showed strong support for conservation: 94% favored educational programs, 80% backed the initiative, and 35% were willing to contribute financially. Willingness to pay increased with income and decreased with household size and distance. The findings underscore the value of environmental education in shaping conservation attitudes and fostering community involvement in climate action.

6.7 Conclusions

Venezuelan scientific production on climate change mitigation in agriculture showed significant progress in the diversity of approaches and strategies evaluated. Although the total number of studies remains limited given the sector's structural challenges, the analyzed works reflect growing concern with reducing greenhouse gas (GHG) emissions while simultaneously enhancing carbon sequestration capacity through practices adapted to local conditions.

The most frequently studied strategies include energy efficiency, the use of organic and biofertilizers, and the adoption of agroecological practices such as no-till farming, crop rotation, and the utilization of organic residues. Agroforestry and agrosilvopastoral systems also stand out as sustainable alternatives that integrate production with ecosystem conservation, especially in mountainous areas and fragile soils.

Additionally, the use of tools such as geospatial technologies and land-use evaluation models shows methodological progress in identifying priority areas for conservation, reforestation, or sustainable agricultural development. Although limited, research on environmental education and legal frameworks underscores the essential role of regulatory and pedagogical instruments in consolidating climate-smart agriculture.

A recurring theme across the studies is the need to integrate technical dimensions with social, legal, and educational components to achieve effective and sustainable outcomes. Most studies suggest that the positive impacts of mitigation practices are enhanced when supported by training, community organization, and public policies that recognize the diversity of agrarian contexts in the country.

Moreover, the studies highlight the urgency of preventing agricultural expansion into natural ecosystems with high-carbon storage capacity, such as forests, palm swamps (*morichales*), and water recharge zones. Preserving these ecosystems is not only a mitigation strategy but also a fundamental requirement for maintaining the ecosystem services essential to agriculture and rural communities.

The reviewed evidence confirms that Venezuelan agriculture holds significant potential to contribute to national and international climate mitigation goals. However, realizing this potential requires overcoming structural barriers such as limited access to sustainable technologies, insufficient investment in research, and poor coordination among institutional actors. Only through an integrated approach—combining science, public policy, and social participation—is that the country can achieve a more resilient, productive, and environmentally aligned agricultural sector.

References

Abache Carvajal S (2019) Principios y regulación de los tributos verdes en Venezuela. In: En: Revista de Derecho Tributario. Asociación Venezolana de Derecho Tributario, pp 67–82. Caracas, 164 (octubre-noviembre-diciembre). https://n9.cl/di2wt

Abreu Cruz E, Araujo Camacho E, Rodríguez Jiménez SL, Valdivia Ávila AL, Fuentes Alfonso L, Pérez Hernández Y (2018) Efecto de la aplicación combinada de fertilizante químico y humus de lombriz en Capsicum annuum. Centro Agrícola 45(1):52–61. https://n9.cl/9c58o9

Acosta Y, Prieto A, Rosales Y (2019) Dinámica de carbono y nitrógeno en suelos tratados con residuos orgánicos. Ingenium et Potentia 1(1):4–22. https://n9.cl/m14b4

Acevedo I, Sánchez A, Mendoza B (2021) Evaluación del nivel de degradación del suelo en dos sistemas productivos en la depresión de Quíbor. I. Análisis multivariado. Bioagro 33(1):59–66. https://revistas.uclave.org/index.php/bioagro/article/view/3023

Atacho P, Rodríguez N, Mogollón JP, Torres D, Yendis H, López M (2021) Uso de soluciones de Vermicompost para la biorremediación de un suelo ácido en la sierra de San Luis, Venezuela. Agroindustria, Sociedad y Ambiente 1(8):1–22. https://revistas.uclave.org/index.php/asa/article/view/3393/2118

Barboza N, Hernández-Alfonso P, Paneque-Rondón P, Gómez-Águila M, Miranda-Caballero A (2019) Labranza convencional y de conservación en el cultivo de la soya (Glycinemax (L) Merril). Ing Agrí 9(3). https://revistas.unah.edu.cu/index.php/IAgric/article/view/1142

Barroso Molina AA (2022) Potencial solar y eólico de Venezuela, escenarios de sistemas híbridos de suministro de energía eléctrica para la Islas de Toas, Edo. Zulia. Publ Cienc Tecnol 16(1):16–26. https://doi.org/10.5281/zenodo.6841805

Batista JR, Catarí XB (2018) Aplicación de modelo de susceptibilidad a deslizamientos superficiales e índices de erosión en la cuenca del río Mamo y subcuenca El Tigre. Edo. Vargas. Venezuela. Terra Nueva Etapa 34(55). https://www.redalyc.org/journal/721/72156172017/72156172017.pdf

Blanco Pulido G R (2021) Efecto de la labranza convencional y la siembra directa sobre el rendimiento del frijol (Vigna unguiculata (l.) Walp.) en la Finca Cumbres de Arona (Agrícola La Cumbre CA), Municipio Pedro María Freites, Edo. Anzoátegui, Venezuela. Tesis de grado Universidad Nacional Experimental del Táchira-UNET

Blanco JLG, Paneque OSG, Puertas A, Castillo JM, Cabrera IM (2018) Rendimiento en dos cultivares de Pennisetum purpureum Schumach a diferentes dosis de fertilización orgánica y mineral. Rev Arbitr Interdiscip Koinonía 3(6):96–108. https://dialnet.unirioja.es/servlet/articulo?codigo=7062680

Cañas K, Zambrano M, Velázquez M (2020) Sistemas agroforestales: una visión transdisciplinaria en el proceso de deforestación. Agroforestry systems: a transdisciplinary vision in the deforestation process. Revista Ambientellanía 3(1):128–138. http://revistas.unellez.edu.ve/index.php/ambientellania/article/view/898

Caballero K (2017) Políticas públicas sectoriales para el cambio climático en América Latina: una aproximación. Comisión Económica para América Latina y el Caribe (CEPAL). Unidad de Cambio Climático de la División de Desarrollo Sostenible y Asentamientos Humanos. https://n9.cl/y6975

Camargo Roa C, Calderón-Dávila AG, Lobo-Márquez JA, Ovalles-de-Cabezas Y (2020) Identificación de conflictos de uso de la tierra y propuesta de uso conservacionista en la subcuenta quebrada Mejías, municipio Antonio Pinto Salinas, Edo. Mérida, Venezuela. Rev Topografía AZIMUT 11(1):46–65. https://revistas.udistrital.edu.co/index.php/azimut/article/view/15979

Coraspe J, Lanza RG (2020) Efecto de fertilizantes orgánicos e inorgánicos naturales sobre el aluminio intercambiable en suelos ácidos cafetaleros. Effect of natural organic and inorganic fertilizers on interchangeable aluminum in soils cafetaleros acids. Revista Unellez de Ciencia y Tecnología 38:18–24. http://revistas.unellez.edu.ve/index.php/ruct/article/view/1727/1541

Escalona-Sánchez A, Gavilanez-Buñay T, Francisco-Yépez A, Ramírez-Guerrero HO (2021) Uso de enmiendas en la producción de maíz para ensilaje con riego orgánico mineral. Agron Costarric 45(1):177–192. https://doi.org/10.15517/rac.v45i1.45769

FAO (2019) Climate-smart agriculture and the sustainable development goals: mapping interlinkages, synergies and trade-offs and guidelines for integrated implementation. FAO, Rome. https://n9.cl/bos9ti

FAO (2024) FAOSTAT Macro Indicators. License: CC BY-NC-SA 3.0 IGO. Accessed 29 April 2024. https://www.climatewatchdata.org

Farreras JA (2018) El café bajo sombra como sistema agroforestal. Rev Equidad 2:23–30. http://revistas.unellez.edu.ve/index.php/Revequidad/article/view/477

Fernández-Molina F, Daboín L, Medina M, Tirado-Delfin E (2019) La diversiicación de agroecosistemas como estrategia de promoción del desarrollo local. Caso Mesa de los Gabaldones, Trujillo, Venezuela. Biblioteca Digital REVENCYT Academia 18(42). Julio-Diciembre (2019) Disponible en: https://bdigital2.ula.ve:8080/xmlui/bitstream/handle/654321/3938/5%20LA%20DIVERSIFICACION%20DE%20AGROSISTEMAS.pdf?sequence=1&isAllo

Locatelli B, Pavageau C, Pramova E, Di Gregorio M (2015) Integrating climate change mitigation and adaptation in agriculture and forestry: opportunities and trade-offs. WIREs Clim Change 6(6):585–598. https://doi.org/10.1002/wcc.357

Farfán Tigre AE, Montilla Pacheco A, Pinargote Sánchez J (2018) Educación ambiental en escenarios reales de aprendizaje. Caso humedales venezolanos. RefCalE: Revista Electrónica Formación y Calidad Educativa 6(2):111–122. ISSN 1390-9010. http://refcale.uleam.edu.ec/index.php/refcale/article/view/1600

Ferrer J, Valera A, Flores B (2018) Capacidad y vocación de uso agropecuario en un sector de altiplanicie de mesa. Revista UNERG Agro-Científica Año 1(1):13–13. https://n9.cl/51x2qx

Flores YE, Romero AJ, Torres AM, Briceño FA, García AJ (2021) Efecto de abonos biológicos y fertilizantes químicos en el cultivo de maíz, FLASA Cojedes Venezuela. Effect of biological fertilizers and chemical fertilizers in corn cultivation, FLASA Cojedes Venezuela. Cienc Tecnol Agropecuaria 6(1):21–27. https://dialnet.unirioja.es/servlet/articulo?codigo=9492419

Flores P, Paredes F (2020) Agricultura urbana: una alternativa mitigante sobre las islas térmicas urbanas. Revista Ambientis Occidentales 1. http://revistas.unellez.edu.ve/index.php/ambientis/article/view/1016

González R, Sánchez MS, Chirinda N, Arango J, Bolívar DM, Escobar D, Tapasco J, Barahona Rosales R (2015) Limitaciones para la implementación de acciones de mitigación de emisiones de gases de efecto de invernadero (GEI) en sistemas ganaderos en Latinoamérica. Livest Res Rural Dev 27:12. https://cgspace.cgiar.org/items/68da0f89-c3de-47f5-9c2e-93f30f21f6d0

Guillén C, Elizalde G (2020) Diagnóstico de cicatrices de erosión en masa, cuenca del rio Caramacate, Venezuela//Diagnosis of mass landslides cicatrices in the basin of the Caramacate River, Venezuela. Rev Geogr Venez 61(1):72–88. https://n9.cl/ankt1

Infante C (2021) Criterios y tecnologías de remediación en suelos contaminados con hidrocarburos. Criteria and remediation technologies in hydrocarbon contaminated soils. Bol Acad Cienc Fís Mat Nat. Vol. LXXXI, Numero 1, Paginas 29–34. Retrieved from: https://acfiman.org/boletines_articulos/criterios-y-tecnologias-de-remediacion-en-suelos-contaminados-con-hidrocarburos/

IPCC (2019) Climate change and land: an IPCC special report on climate change, desertification, land degradation, sustainable land management, food security, and greenhouse gas fluxes in terrestrial ecosystems. Intergubernamental Panel on Climate Change. https://n9.cl/fpbqp

Jiménez IJ, Ramírez M, Petit B, Colmenares C, Parra I (2017) Efecto de hongos micorrízicos arbusculares y estiércol de bovino en el crecimiento inicial y pigmentación en Capsicum frutescens L. (Ají). Bioagro 29(2):137–144. https://n9.cl/u9873

Lal R, Monger C, Nave L, Smith P (2021) The role of soil in regulation of climate. Phil Trans R Soc B376: 2020084. https://royalsocietypublishing.org/doi/10.1098/rstb.2021.0084

Laya Y (2019) Proyecto de evaluación de sustentabilidad de sistemas de produccion y procesamiento de leche vacuna. Rev Cienc Tecnol Agrollanía 17. Enero-Diciembre (2019). Universidad Nacional Experimental de los Llanos Occidentales Ezequiel Zamora. http://www.postgradovipi.50webs.com/archivos/agrollania/2019/Articulo4.pdf

Márquez M, Cáceres A, Hernández-Valencia I, Romero L (2019) Parámetros micorrízicos y calidad del suelo en cultivos de papas nativas e introducidas (Mérida, Venezuela). Acta Biol Venez 39(1):91–105. https://n9.cl/3ejfo

Martens R (2018) La ordenación del territorio en Venezuela y su impacto en las comunidades indígenas del Municipio Gran Sabana-Edo. Bolívar. Bol Antropol 36(96):274–306. https://www.redalyc.org/journal/712/71257885002/71257885002.pdf

Méndez W, Pacheco H, Landaeta L, Menéndez E, Pérez M, Parra G, Fuentes J (2020) Indicadores de erosividad de la lluvia en una cuenca de la Serranía del Litoral Central de Venezuela. Rev Geogr Norte Gd 76:279–301. https://doi.org/10.4067/S0718-34022020000200279

Molina Pereira YA (2019) La Reforestación como Estrategia Ambiental para la Conservación de ríos y quebradas. Rev Sci 4(13):182–199. https://www.redalyc.org/journal/5636/563659492010/563659492010.pdf

Morales Montero OF, Albarrán Torres AJ, Gutiérrez Camargo JJ (2021) Estudio espacio temporal de la cobertura de la tierra: Norte del río Orinoco-Venezuela, periodo 1988-2010. GeoFocus, Revista Internacional de Ciencia y Tecnología de la Información Geográfica 28:95–119. https://geofocus.org/index.php/geofocus/article/view/686/520

Ojeda-Falcón AD, Domínguez-Quintero O (2020a) Determination of net soil erosion rate using 137Cs in a slope agroecosystem. Rev Cienc Téc Agropecu 29(1) http://scielo.sld.cu/scielo.php?script=sci_arttext&pid=S2071-00542020000100004

Ojeda-Falcón AD, Domínguez-Quintero O (2020b) Pastoreo Racional Voisin, ruta agroecológica sustentable en suelos ácidos-arenosos de baja fertilidad natural. Centro Agrícola 47(2):41–53. http://scielo.sld.cu/scielo.php?pid=S0253578520200002000041&script=sci_arttext&tlngpt

Olivares B, López M (2019) Normalized difference vegetation index (NDVI) applied to the agricultural indigenous territory of Kashaama, Venezuela. UNED Research Journal 11(2):112–121. https://doi.org/10.22458/urj.v11i2.2299

Olivares B, López-Beltrán M, Lobo-Luján D (2019) Changes in land use and vegetation in the agrarian community Kashaama, Anzoátegui, Venezuela: 2001-2013. Rev Geogr Am Cent 2(63):269–291. https://doi.org/10.15359/rgac.63-2.10

Pacheco HA, Cevallos RX, Vinces CJ (2019) Cálculo del factor C de la RUSLE, en la cuenca del río Carache, Trujillo-Venezuela usando imágenes del Satélite Miranda VRSS-1. Rev Espac 40(03) http://www.revistaespacios.com/a19v40n03/19400306.html

Pardo-Plaza YJ, Paolini-Gómez JE, Cantero-Guevara ME (2019) Biomasa microbiana y respiración basal del suelo bajo sistemas agroforestales con cultivos de café. Rev UDCA Actual Divulg Cient 22(1):e1144. https://revistas.udca.edu.co/index.php/ruadc/article/view/1144/1740

Paolini-Gómez JE (2018) Actividad microbiológica y biomasa microbiana en suelos cafetaleros de los Andes venezolanos. Terra Latinoam 36(1):13–22. https://www.terralatinoamericana.org.mx/index.php/terra/article/view/257/329

Peña-Colmenares C, Gordon-Colón E (2019) Morichal de la cuenca alta del río Tigre Anzoátegui, Venezuela: aguas, suelos y vegetación. Acta Biol Venez 39:2. https://www.researchgate.net/publication/350610640_N

Pírela-Almarza AY, Aguirre-Serpa OE, Ramírez-Villalobos MDC, Petit B, Bracho B, Parra I (2018) Efecto de hongos micorrízicos arbusculares y del estiércol de ovino en el desarrollo inicial de la lechosa (bL.) var. Maradol roja. Bioagro 30(1):79–86. https://ve.scielo.org/scielo.php?script=sci_arttext&pid=S1316-33612018000100008

Pino Suarez MV, Rojas Uzcátegui YJ, Salcedo P, Rangel Vásquez E, Gómez Mejías AE (2021) Dinámica sucesional del bosque luego de establecer fajas de enriquecimiento, El Caimital, Barinas-Venezuela. Investig Geogr 61:99–115. https://investigacionesgeograficas.uchile.cl/index.php/IG/article/view/61075/67864

Ramírez L, Méndez Y (2020) Recurso eólico y solar del territorio venezolano. Tekhné 23(1):94–121. https://revistasenlinea.saber.ucab.edu.ve/index.php/tekhne/article/view/4589/3826

Rivera R (2019) El cultivo forestal: análisis comparativo y una propuesta técnica para Venezuela. In: Comparación de las experiencias de Brasil, Chile y Uruguay. Academia. Edu. https://www.academia.edu/40308683/

Rivas W, Mogollón JP (2018) Erosión eólica en la península de Paraguaná, edo. Falcón, Venezuela. La. Rev For Venez 62(1):9–29. https://www.researchgate.net/profile/Jose-Mogollon/publication/341685458

Ruiz-Dagger M, Paolini J (2021) Indicadores biológicos de suelos lacustres y aluviales de Venezuela bajo diferentes usos. Parte 1. Actividad microbiana y coeficientes ecofisiológicos. Terra Latinoamericana 39: e922. https://www.terralatinoamericana.org.mx/index.php/terra/article/view/922

Sánz-Sanchez MJ, Uriarte IR (2020) Cambio climático, agricultura y uso de la tierra, vol 163. Papeles de Economía Española, Madrid Iss, pp 84–204. https://www.proquest.com/docview/2399206709?sourcetype=Scholarly%20Journals

Silva-Laya SJ, Pérez-Martínez S, Álvarez-Del-Castillo J (2018) Evaluación de sostenibilidad agroecológica de dos experiencias de producción con énfasis en hortalizas, Venezuela. Rev Colomb Cienc Hortíc 12(3):632–645. https://www.researchgate.net/publication/337572608

Solano J A (2019) Pre-factibilidad de generación de energía en zonas ganaderas mediante la biodigestión de residuos agropecuarios como apoyo a la transición energética en Venezuela. Tekhné, 22(3). https://revistasenlinea.saber.ucab.edu.ve/index.php/tekhne/article/view/4091

Torres D, Mendoza B, Gomes C, Almao L, Hernández W, Carrero L, Escalona A (2018) Efecto de la fertilización orgánica e inorgánica sobre el crecimiento de tomate (Solanum lycopersicum) en ambientes protegidos. Revista Científica ASA:4–18. http://bdigital2.ula.ve:8080/xmlui/bitstream/handle/654321/4705/articulo1.pdf?sequence=1&isAllowed=y

UNEP-WCMC (United Nations Environment Programme World Conservation Monitoring Centre) (2025) Index of priority rank % of areas of global significance for biodiversity conservation, carbon storage, and water provision [map]. UN Biodiversity Lab. https://n9.cl/tpwxgu

UNFCCC (United Nations Framework Convention on Climate Change) (2015) Adoption of the Paris agreement: proposal by the president. FCCC/CP/2015/L.9/rev.1. In: Conference of the Parties, Twenty-First Session, Paris, December 12. https://unfccc.int/resource/docs/2015/cop21/eng/l09r01.pdf

Vallejo C, Chacón M, Cifuentes M (2016) Sinergias entre adaptación y mitigación del cambio climático (SAM) en los sectores agrícola y forestal. Concepto y propuesta de acción. USAID Programa Regional de Cambio Climático. Serie técnica, Boletíntécnico No. 79. https://hal.science/cirad-01426726v1/document

Watson P, Rao KC (2017) Chapter 6. Emerging fuel technologies from agro-waste. In: Bio-ethanol from cassava waste (ETAVEN, Carabobo, Venezuela). Section II: energy recovery from organic waste, pp 286–295. https://n9.cl/26fct

Yánes GG, Caula SA (2020) Assessment of ecological management project in a wetland at San Esteban National Park, Caribbean Coast, Venezuela. Ecosistemas 29(3), 2010. https://doi.org/10.7818/ECOS.2010

Zambrano M (2018) Gestión ambiental: una mirada sistémica en la cuenca del rio Santo Domingo, Barinas, Venezuela. Rev Ambientellanía 1(2) http://revistas.unellez.edu.ve/index.php/ambientellania/article/view/637

Chapter 7
Agricultural Adaptation Strategies to Climate Change

7.1 Introduction

Adapting agriculture to climate change consists of designing and implementing agricultural practices, either individually or as production systems, to overcome the current or potential detrimental effects of climate change. The objective is to eliminate or reduce such effects and contribute to increasing or maintaining the economic, social, and environmental sustainability of the agricultural production process (Pezo et al. 2019).

This chapter presents the challenges faced by agricultural research in Venezuela, particularly in adaptation to climate change. The effects of climate change in Venezuela, as discussed in previous chapters (such as in the Chap. 3, Chap. 4, and Chap. 5), which can be summarized in water deficit for crops, because of a decrease in rainfall, especially in the eastern part of the country, and an increase in temperatures in the west and the central region. The most important agricultural regions are located in eastern, central, and western Venezuela. Agriculture in these regions will undoubtedly be affected by the aforementioned impacts.

To address these challenges, it is essential to implement both mitigation measures, which reduce or prevent greenhouse gas emissions (Chap. 6), and adaptation strategies to climate change. These measures work synergistically, enabling farmers who adopt them to adjust to the adverse effects of climate change while ensuring the production of food and other agricultural goods. Such measures typically stem from agricultural research, whose findings are transferred to farmers through public or private agricultural extension and technology transfer programs.

Adaptation measures in agriculture are crucial not only for farmers, as they help protect their economic livelihoods and reduce the risks associated with extreme weather events such as droughts and floods, but also for society as a whole. These measures contribute to agricultural production, national food security, sectoral economic stability, and the conservation of natural resources such as soil and water.

A. Rosales Hernández et al., *Agriculture Under Climate Change*, The Latin American Studies Book Series, https://doi.org/10.1007/978-3-032-06368-7_7

Therefore, generating knowledge on agricultural adaptation strategies at local, regional, and national levels is of utmost importance. The agricultural adaptation measures developed in Venezuela during the 2018–2025 period, as reported in this chapter, are limited, with only 15 research papers (Sects. 7.2 and 7.3) available, of which 10 focus on irrigation. This reflects not only the absence of clear and effective public policies on climate change, particularly in agriculture, but also a scarcity of agricultural research on the subject. Additionally, Mercado et al. (2023) highlight that Venezuelan scientific journals are facing a difficult situation due to the sharp decline in local research and the total loss of state support. A significant number of journals have come to a standstill or have fallen considerably behind. An important aspect to highlight is the institutional crisis facing Venezuela, which has had an impact on the country's ability to address the challenges of climate change in agriculture. As a result, important opportunities to develop adaptation strategies based on scientific evidence have been missed.

In this chapter, the objective is to address these identified gaps and challenges and to critically analyze the available scientific production on adaptation practices for the Venezuelan agriculture. By examining the existing published literature in the period 2018–2024, it is feasible to identify the most up-to-date areas of knowledge, revealing those where more research is needed.

7.2 Agricultural Irrigation Adaptation Measures in Venezuela

The distribution of agricultural land in Venezuela reveals that rainfed agriculture dominates national production. Although official statistics on areas under rainfed and irrigated agriculture are unavailable, there is a disparity of opinion regarding this distribution. In any case, there is a consensus that irrigated land is currently limited, and therefore, agricultural production is markedly dominated by rainfed agriculture. This situation underscores a strong dependence on rainfall for agricultural productivity, a factor that increases the sector's vulnerability to the impacts of climate variability and climate change.

Rainfed systems, although widespread, are inherently riskier in the face of reduced precipitation and increased temperature variability, as projected for Venezuela. The limited proportion of irrigated land reflects constraints in infrastructure development, investment capacity, and management of water resources. It also suggests that efforts to expand or improve irrigation systems could substantially enhance resilience and productivity in the agricultural sector.

Strengthening irrigation coverage and efficiency would be a strategic adaptation measure, especially in regions already experiencing frequent droughts or where rainfall patterns are expected to shift significantly. Investment in irrigation technologies, such as controlled deficit irrigation or precision irrigation, could optimize water use, stabilize yields, and contribute to food security in the context of ongoing climatic changes.

The distribution of crop production by water risk in Venezuela highlights significant exposure to different types of hydrological stress (Fig. 7.1a). Notably, a large proportion of both irrigated and rainfed systems are subject to seasonal variability and interannual variability in water availability, indicating that water supply instability is a pervasive risk factor across the country's agricultural landscape. In irrigated areas, production is highly sensitive to both interannual variability and the risk of coastal salinization (coastal eutrophication potential), reflecting challenges associated with the proximity of some agricultural zones to coastal ecosystems. These risks are further exacerbated by the degradation of irrigation infrastructure and limited water management capacity, which can amplify the effects of climate-driven stresses.

Mitigating water-related risks will require a dual approach: reinforcing infrastructure for irrigation and drainage in vulnerable areas, and promoting water-smart agricultural practices such as drought-tolerant crop varieties and improved soil moisture conservation techniques. The figures emphasize the urgent need for national adaptation planning to integrate water risk management as a core strategy for sustainable agricultural development.

The analysis of drought risk across rainfed crop production systems in Venezuela shows that a substantial portion of agricultural output is already subject to moderate to high drought risk (Fig. 7.1b). A relatively small fraction of production areas is classified as low-risk, emphasizing the widespread exposure of the sector to drought events. The predominance of moderate to high drought risk in rainfed agriculture

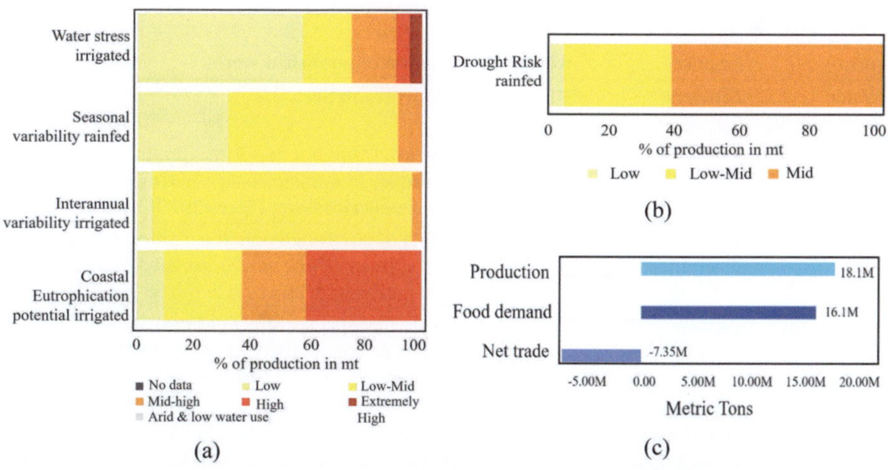

Fig. 7.1 (**a**) Percentage of total crop production exposed to different categories of water-related risk in Venezuela; (**b**) percentage of crop production classified by drought risk levels in Venezuela under rainfed conditions; (**c**) comparison of the volume of crop production, food demand, and net trade balance in Venezuela. Source: World Resources Institute (WRI 2024) available in https://www.wri.org/applications/aqueduct/food/#/?basemap=hydro&country=VEN

reflects the natural climatic characteristics of Venezuela, particularly in semi-arid and dry sub-humid zones. With climate change expected to exacerbate drought frequency and intensity, these vulnerabilities could become even more pronounced, severely impacting food security and rural livelihoods.

Addressing drought risk will require not only the expansion of irrigation infrastructure but also the adoption of climate-resilient practices, such as the diversification of crops, the use of drought-tolerant varieties, conservation tillage, and integrated soil fertility management. Strategic public policies and investments are critical to ensuring that agricultural systems are better equipped to cope with increasing hydrological variability.

The comparison between the volume of crop production and food demand in Venezuela reveals a precarious balance (Fig. 7.1c). In 2023, total crop production reached approximately 18.1 million metric tons, while food demand was close behind at 16.1 million metric tons. This narrow margin leaves the country vulnerable to fluctuations in production caused by climate variability, pests, or socioeconomic factors. The net trade figure shows a significant negative balance of −7.55 million metric tons, indicating that domestic production is insufficient to fully meet national food needs, and the country remains heavily reliant on food imports. This deficient context is reaffirmed by the data in Table 7.1 (Estimate of crop production for the years 2018–2019 according to Fedeagro 2019) of Chap. 1, where the percentages (%) of self-supply of 12 main crops are shown, and none

Table 7.1 Status of research in Venezuela (No. published works in 2018–2024) on agricultural adaptation measures carried out in the study period

Agricultural adaptation measure	Specific agricultural practice	No. of published works	Citation of published works
Water management	Efficient irrigation	10	See works in Sect. 7.2
Crop selection	Crops tolerant to climate stress	1	Cassava (*Manihot esculenta*) production and resilience to climate change in Cojedes state of Venezuela (Romero, Flores 2020).
	Crop diversification	1	Effect of controlled deficit irrigation on the yield of coriander (*Coriandrum sativum*) and its association with chives (Carrero et al. 2020)[1]
Agronomic practices	Agroforestry	1	Shade tree species affect gas exchange and hydraulic conductivity of cacao cultivars in an agroforestry system (Avila–Lovera et al. 2021)
	Integrated pest and disease management	1	Use of biocontrollers and biostimulation in Soybean Cultivation (*Glycine max*) in Venezuela (Goyo and Zambrano 2017)
	Technology transfer	1	Agricultural adaptation measures to climate change in the Pao River basin, Carabobo, Venezuela (Romero et al. 2014).
Total		15	

(1) It is also included in water management (efficient irrigation)

exceeds 21%. This dependency introduces additional risks, particularly given global market fluctuations and the potential for supply chain disruptions under climate change scenarios.

Improving agricultural self-sufficiency by enhancing productivity and climate resilience is a critical national priority (Olivares Campos 2023). Strategies such as improving irrigation efficiency, promoting sustainable intensification, rehabilitating degraded lands, and supporting farmer access to technologies and credit are vital to increasing domestic food production and reducing dependence on external food sources.

Figure 7.2 illustrates the spatial distribution of irrigated and rainfed agricultural areas across Venezuela, along with the corresponding food demand for various crops, expressed in 1000 metric tons. Particularly, the states of Zulia, Lara, Portuguesa, and Barinas exhibit the highest density of crop production areas. A diversity of crops is represented, including cereals such as maize and rice, pulses like beans, and oilseed crops such as sunflower and soybean. These regions coincide with Venezuela's historically most fertile plains, emphasizing the importance of water resources in sustaining agricultural productivity under the climate change scenarios.

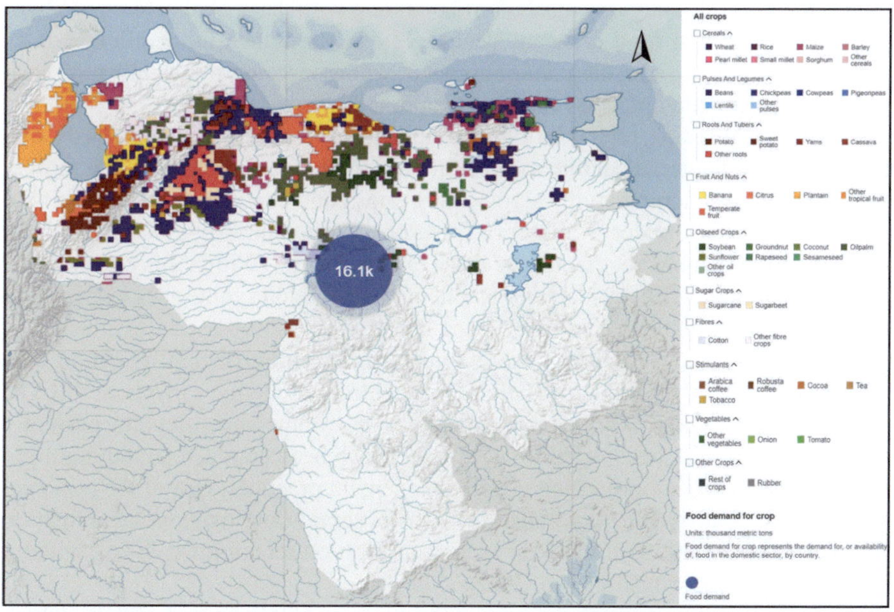

Fig. 7.2 Irrigated and rainfed crop producing areas and food demand for crop in Venezuela. The map highlights major agricultural zones by crop type, distinguishing between cereals, pulses, oilseeds, roots and tubers, fruits, vegetables, and stimulants. Food demand, represented in 1000 metric tons, is concentrated in key productive regions, underscoring the strategic importance of these areas for national food security under climate change scenarios, according to World Resources Institute (WRI 2024) available in https://www.wri.org/applications/aqueduct/food/#/?basemap=hydro&country=VEN

Figure 7.2 also highlights significant regional specialization in crop types. For instance, areas devoted to cereals are heavily concentrated in the Llanos region, whereas fruit and vegetable production shows greater concentration along the Andes region. Rainfed agriculture remains predominant in the majority of agricultural landscapes, particularly for extensive crops like maize and sorghum, while irrigated farming appears more localized and associated with higher-value crops such as vegetables, fruits, and sugarcane. This reflects the underlying constraints in irrigation infrastructure and the dependence on seasonal rainfall for a substantial portion of national agricultural output.

Furthermore, the blue circle indicating a food demand of 16.1 thousand metric tons underscores the concentrated domestic demand for staple crops in certain regions. The aggregation of food demand in these zones highlights the strategic role they play in national food security. However, it also reveals potential vulnerabilities, as climatic variability, particularly changes in precipitation patterns, could disproportionately impact the areas with the highest production and demand, thereby exacerbating food insecurity risks under climate change scenarios.

The spatial patterns depicted in Fig. 7.2 underscore the critical need for targeted adaptation strategies. Enhancing irrigation efficiency, promoting climate-resilient crop varieties, and investing in sustainable land and water management practices are essential actions. By aligning agricultural planning with the spatial heterogeneity of crop production and food demand, Venezuela can strengthen the resilience of its agri-food systems to climate change impacts while ensuring the stability of food supply for its population, and future exportation.

Due to the ongoing effects of climate change in Venezuela, as previously discussed in this chapter, a decline in precipitation, particularly in the eastern region, and an increase in temperatures across the western and central areas (Chap. 5) are projected to reduce water availability for crops. Agricultural production in these regions will inevitably be affected by these climatic shifts. Within this context, irrigation emerges as a critical adaptation strategy. This section synthesizes ten research papers, primarily published in Venezuelan scientific journals that address this adaptation measure. The majority of these studies focus on enhancing irrigation efficiency through the implementation of controlled deficit irrigation (CDI) methodologies. These works emphasize the severe moisture deficits currently impacting semi-arid areas, where most studies have been conducted, and highlight the urgent necessity of improving water management practices. One of the principal objectives of controlled deficit irrigation is to conserve water by reducing irrigation volumes to below 100% of the crop evapotranspiration requirement (ETc), without substantially compromising crop yields.

A notable example of such research is the study by Villafañe et al. (2018), which evaluated water productivity improvements in the cultivation of stevia (*Stevia rebaudiana* Bertoni) under different deficit irrigation treatments in Aragua State. The authors assessed the effects of applying 25%, 50%, 75%, and 100% of ETc, combined with non-deficit irrigation, on soil water depletion, crop yield, and water use efficiency. Their findings revealed that the 50% ETc treatment effectively prevented water stress while optimizing water use, thus achieving considerable water savings without compromising yield, illustrating the potential of CDI as an adaptation strategy under Venezuelan conditions.

Complementary to these findings, Bello et al. (2019) conducted a study in Colina Municipality, Falcón State, where three irrigation schedules were evaluated in aloe vera (*Aloe vera* L.) cultivation. The experimental treatments included no irrigation, irrigation at 8:00 a.m., 3:00 p.m., and 8:00 p.m., utilizing micro-sprinklers to satisfy crop water requirements. The results demonstrated that night-time irrigation at 8:00 p.m. produced the greatest leaf and gel yields, significantly enhancing water use efficiency. This study supports night irrigation as a feasible and effective adaptation practice in arid environments like northern Falcón, where water scarcity constitutes a critical constraint to agricultural productivity.

In addition to technical evaluations, Leroy (2019) explored the perceptions and adaptation responses of farmers in high-altitude Páramo ecosystems of Colombia and Venezuela regarding water scarcity under climate change. Their research demonstrated that farmers' adaptation strategies were influenced not only by their awareness of climate change but also by socioeconomic pressures, such as the expansion of cultivated land, population increases, aging irrigation systems, and management inefficiencies. Venezuelan farmers, in particular, responded by adopting more efficient irrigation technologies, rehabilitating degraded infrastructure, and implementing conservation measures for wetlands, highlighting the socioecological dimensions of agricultural adaptation.

Also, Rodríguez-Yzquierdo et al. (2020) examined the use of controlled deficit irrigation (CDI) in the cultivation of passion fruit (*Passiflora edulis* Sims) in Aragua State. By applying a 66% reduction in reference evapotranspiration (ETo) during non-critical growth phases, the authors assessed the impacts on plant nutrition, productivity, and fruit quality. The results indicated that CDI did not negatively affect plant nutritional status, yields, or fruit quality, while promoting significant improvements in water use efficiency. This evidence confirms CDI as a viable and sustainable water management strategy for mitigating climate change impacts on high-value horticultural crops in Venezuela.

Further research conducted by Carrero et al. (2020) evaluated the response of coriander (*Coriandrum sativum*) to controlled deficit irrigation under both monoculture and intercropped systems with chives (*Allium fistulosum* L.) in Zulia State, Venezuela. The study was motivated by the limited availability of water and agricultural inputs, which encouraged producers to diversify their crops through associations. The results demonstrated that coriander cultivated in monoculture and irrigated at 100% of the water volume corresponding to the substrate's saturation capacity (composed of 75% soil and 25% organic matter) achieved the highest yield. In the intercropping system, chives benefited from irrigation at replacement rates equal to or exceeding 80%, which enhanced their development and productivity. These findings underline the potential of crop association and irrigation management as synergistic strategies to improve water use efficiency and resilience under resource-scarce conditions.

Building on this body of knowledge, Rodríguez Padrón (2020) conducted a bibliographic review focusing on irrigation management practices for the cultivation of green pepper (*Capsicum annuum* L.), specifically examining controlled deficit irrigation under both shade netting and open field conditions. In the context of projected water scarcity exacerbated by climate change, the insights gathered in this review are highly relevant, particularly for the horticultural production zones in Mérida State. The analysis highlights the need to optimize irrigation practices to maintain productivity while conserving increasingly limited water resources, reinforcing the role of deficit irrigation as a vital component of adaptation strategies.

In addition to these studies, Gil Marín et al. (2020) evaluated the effects of different irrigation regimes on the yield, fruit production, quality, and water use efficiency of the long purple eggplant cultivar (*Solanum melongena* L.) under field conditions in Maturín, Monagas State. The authors tested four irrigation levels and concluded that an irrigation application equivalent to 120% of the crop evapotranspiration (ETc) yielded the best results. This corresponded to an average daily water application of 4 mm/day, or approximately 1.32 liters per plant during the flowering and fruiting stages. Their findings emphasize that, for certain crops and climatic conditions, moderate water surpluses can optimize production and quality outcomes without necessarily compromising water resource sustainability.

Similarly, Gil-Marín et al. (2021) extended their research to evaluate the response of the zucchini cultivar Shiraz (*Cucurbita pepo* L.) to different irrigation water depths under field conditions. The study applied 4 irrigation criteria, replacing 60%, 80%, 100%, and 120% of ETc. As in the case of eggplant, the 120% ETc treatment resulted in the highest yields and fruit quality, suggesting that certain vegetable crops under specific climatic regimes might benefit from slightly increased water applications to maximize both productivity and resource use efficiency.

Moreover, Blanco et al. (2021) investigated the effects of induced water deficit and biofertilization on the growth dynamics and bulb yield of the F1 2000 onion hybrid (*Allium cepa* L.) in a semi-arid environment in Mérida State. Two irrigation treatments were compared: daily watering and watering at three-day intervals. The authors recommend a lower irrigation frequency, combined with full (100%) NPK fertilization as the most sustainable strategy. Alternatively, a 50% NPK fertilization regime without water stress also demonstrated promising results. These agro-sustainable practices offer viable pathways to improve onion production under water-scarce conditions typical of Neotropical semi-arid zones.

Finally, Pereyra (2012) conducted an evaluation of five white corn hybrids developed by DANAC (Polar Companies Foundation, Venezuela) to assess their responses to variable water supplies. Among the hybrids tested, Danac-223 exhibited superior performance in terms of water use efficiency and yield, particularly under the conditions of the western plains' region of Venezuela. This research highlights the critical role of selecting and promoting crop varieties with greater tolerance to water stress, an essential component of long-term adaptation strategies to climate change in the Venezuelan agricultural sector.

7.3 Farming Practices as Agricultural Adaptation Measures in Venezuela

In addition to irrigation-focused strategies, agricultural adaptation to climate change encompasses a wide array of practices to improve resilience and sustainability at the farm level. Although the volume of research on these alternative practices remains limited in Venezuela during the 2018–2024 period, a few studies have explored key areas such as crop management, biological innovation, and sustainable production systems. The following analysis summarizes the main findings of five research papers addressing adaptation measures beyond irrigation, highlighting both the progress made and the existing gaps in this critical field. A detailed analysis of the research gaps and needs detected from this review will be presented in Sect. 7.5 of this chapter.

Among the works reviewed, Romero and Flores (2020) conducted a documentary analysis on cassava (*Manihot esculenta*) production and its resilience to climate change in Cojedes State, Venezuela. The authors proposed a series of agronomic improvements necessary to maintain cassava as a sustainable crop under future climatic conditions. These practices include seed management through cuttings or the use of vitro plants, the promotion of intercropping systems, the incorporation of organic matter and mycorrhiza inoculation into soils as suggested, the active involvement of families in cassava production, and the dissemination of technical knowledge with appropriate support systems as suggested by Mora et al. (2019).

Similarly, the study by Mora Guerrero (2022) analyzed the management practices adopted by Venezuelan agricultural producers, emphasizing continuous improvement and social responsibility. The research identified eight emerging categories that define these practices: economic sustainability, technological innovation, social capital development, management transparency, governance quality, environmental sustainability, and organizational contribution. Within this framework, the concepts of profitability, productivity, and the formation of agricultural partnerships were highlighted as critical factors in building more resilient agricultural enterprises to face climate change.

Additionally, Goyo and Zambrano (2017) conducted a review focused on the application of biocontrol agents and biostimulants in soybean (*Glycine max*) cultivation. Their study explored the role of biological agents in mitigating biotic and abiotic stresses in crops, highlighting the effectiveness of biocontrol in reducing pesticide use. Techniques such as seed microbiolization, using Rhizobium species to promote nodulation and Trichoderma as a biocontrol agent, alongside the application of organic fertilizers, were identified as promising strategies to enhance plant growth by stimulating the production of hormones, siderophores, secondary metabolites, vitamins, and amino acids.

Romero et al. (2014) examined agricultural adaptation measures in the Pao River basin in Carabobo State, Venezuela. In their analysis, nine adaptation measures were selected, categorized into five thematic areas: water management, soil conservation, adjustment of agricultural practices, application of biological control methods, and the adoption of sustainable ecosystem practices (Morales-Márquez et al. 2022). This comprehensive approach underscores the importance of integrating multiple dimensions of agricultural management to effectively respond to the multifaceted challenges posed by climate change.

Finally, Avila–Lovera et al. (2021), analyzed shade tolerance as a widespread strategy of rainforest understory plants. Cacao commonly grows in the shaded understory, and recent emphasis has been placed on diversifying the types of trees used to shade cacao plants to achieve additional ecosystem services. They studied three agricultural cacao cultivars growing in the shade of four timber species (*Cedrela odorata* L., *Cordia thaisiana* Agostini, *Swietenia macrophylla* King, and *Tabebuia rosea* (Bertol) A.D.C.) in an agroforestry system. The purpose was to (i) evaluate the timber species for their effect on the physiological performance of three cacao cultivars; (ii) assess the role of green stems on the carbon economy of cacao; and (iii) examine coordination between stem hydraulic conductivity and stem photosynthesis in cacao. The authors conclude that different shade regimes determined by timber species and the interaction with cacao cultivar had an important effect on most of the physiological traits and growth variables of three cacao cultivars growing in an agroforestry system. Results suggested that *C. odorata* is the best timber species to provide partial shade for cacao cultivars in the Barlovento region in Venezuela, regardless of cultivar. It is important to take into account that cocoa is one of the crops with the greatest export potential, and that Venezuela has a considerable area of land suitable for this crop under agroforestry system (Comerma et al. 1999).

7.4 Climate-Smart Agricultural Practices Suitable to Venezuela

Climate-smart Agricultural Practices (CSAP) is an approach that the FAO (2010) proposed to develop technical, policy, and investment frameworks that enable sustainable agricultural development in the face of climate change. Climate change's significant and widespread impacts on agriculture highlight the urgent need to integrate climate considerations into national agricultural planning and programs. This approach identifies and promotes sustainable agricultural development within the explicit climate change framework.

The implementation of climate-smart agricultural practices contributes to achieving internationally recognized objectives, such as the United Nations Sustainable Development Goals (SDGs) and the commitments outlined in the Paris Agreement (COP 21). CSAP aims to address three key objectives: (i) sustainably increasing agricultural productivity and income, (ii) enhancing resilience and adaptation to climate change, and (iii) reducing and, when possible, eliminating greenhouse gas emissions.

A critical aspect of CSAP is its implementation at the field level, working directly with farmers while continuously evaluating the feasibility and adaptability of these practices. Understanding the diverse needs and priorities of farmers is essential, making direct collaboration with them a fundamental component (FAO 2021).

Below is a summary of six agricultural practices proposed by Venezuelan agronomists in the climate-smart agricultural practices training manual for Belize (Mercado et al. 2018), developed by the Tropical Agricultural Research and Higher Education Center (CATIE) in Costa Rica. These practices serve as a bridge between agricultural research on climate change adaptation and its practical implementation by farmers through knowledge and technology transfer processes.

7.4.1 Rainwater Collection and Storage

Among the various agricultural adaptation measures available to confront the challenges of climate change, rainwater collection and storage represent a particularly simple and promising option for Venezuela. Given the country's projected decrease in precipitation and increasing climate variability, developing decentralized, cost-effective water management solutions has become a critical need. The following section analyzes the concept, potential benefits, and current gaps in research and implementation of rainwater harvesting practices in the Venezuelan context.

Rainwater harvesting refers to the capture, accumulation, and storage of rainwater runoff for reuse. By employing various structures and storage technologies, rainwater is collected before it reaches natural water bodies (e.g., rivers, lakes) or aquifers. The primary objective of this practice is to maximize the use of rainfall during wet periods and supplement water availability during dry seasons. A more stable water supply reduces crop stress, enhances product quality, and increases net income. The most significant benefits of this adaptation strategy include water conservation, diversification, and production stability (FAO 2013; FAO 2014).

Despite the recognized advantages of rainwater harvesting globally, scientific production specifically addressing this topic within Venezuela remains notably scarce. The limited number of studies, reports, and empirical evaluations highlights a significant gap in the national research agenda concerning climate change adaptation. This is particularly concerning given Venezuela's increasing exposure to erratic rainfall patterns and prolonged droughts, especially in its semi-arid and sub-humid regions, where rainwater harvesting could play a transformative role in strengthening agricultural resilience.

Incorporating rainwater harvesting into Venezuelan agricultural practices would offer multiple benefits beyond simple water provision. It would support smallholder farmers by reducing their dependence on increasingly unreliable traditional water sources, enhance the viability of rainfed farming systems, and enable more diversified, sustainable agricultural production. Moreover, widespread adoption of rainwater collection could mitigate some of the socioeconomic vulnerabilities exacerbated by water scarcity, such as rural poverty, food insecurity, and migration pressures.

Given Venezuela's natural variability in rainfall distribution and the ongoing institutional and infrastructural weaknesses, promoting rainwater harvesting technologies emerges as a strategic priority. Public policies, research funding, and technical public and private extension services should actively encourage the development, evaluation, and dissemination of context-appropriate rainwater collection systems. Such initiatives would not only align with climate change adaptation objectives but would also contribute to the revitalization of the national agricultural sector through more efficient resource management.

Ultimately, addressing the current research deficit on rainwater harvesting in Venezuela is critical for designing evidence-based adaptation strategies. Integrating local knowledge with technical innovation could facilitate the rapid scaling of rainwater management solutions, making agriculture more resilient to future climatic stresses. Therefore, fostering interdisciplinary scientific research, farmer-centered pilot programs, and knowledge exchange platforms will be essential for unlocking the full potential of rainwater harvesting as an effective adaptation measure in Venezuela.

7.4.2 *Irrigation Schedule*

Irrigation scheduling is more than a method or practice tailored to a specific crop; it is a comprehensive well known approach that defines both the timing and the amount of water to be applied, regardless of the irrigation technology utilized (Gu et al. 2020). By accurately regulating water application, irrigation scheduling minimizes the risk of crop failure caused by water stress, enhances irrigation efficiency, and increases water productivity, particularly when water resources are limited (Datta, Taghvaeian 2023). Additionally, this practice contributes to reducing energy consumption and water pumping costs, thereby improving the overall return on financial investments. Properly managed irrigation scheduling ensures optimal crop yields and product quality while promoting the efficient use of scarce water and energy resources, an aspect of critical importance in regions facing growing climate pressures.

Despite the proven benefits of irrigation scheduling in optimizing resource use and improving agricultural resilience, the scientific literature in Venezuela offers very limited research specifically addressing the development, validation, or implementation of irrigation scheduling strategies. Most studies have focused predominantly on controlled deficit irrigation, often without integrating a systematic scheduling component tailored to local climatic conditions, crop needs, and soil characteristics. This gap underscores the need for more focused research to adapt irrigation scheduling practices to Venezuela's diverse agroecological zones.

The lack of scientific attention to irrigation scheduling is particularly concerning given the country's increasing exposure to water scarcity and energy instability. In agricultural regions where access to reliable electricity or fuel for water pumping is intermittent, optimizing irrigation frequency and volume could dramatically improve system sustainability and crop performance. Moreover, irrigation scheduling, when combined with water-saving technologies such as drip or low-energy precision application (LEPA) systems, could reduce operational costs for farmers, making agricultural production more viable even under adverse climatic conditions.

Incorporating irrigation scheduling practices into Venezuela's agricultural adaptation strategies would bring multiple benefits beyond water conservation. It would promote more predictable agricultural outputs, enhance farmers' decision-making capabilities, and reduce the risk of soil degradation from over-irrigation or prolonged drought stress. Furthermore, training programs focused on could build technical capacity among producers and extension agents, fostering a more climate-resilient agricultural sector. Gu et al. 2020 provide in their review, the four most common methods of operation in irrigation scheduling: (i) evapotranspiration and water balance (ET-WB), (ii) soil moisture status, (iii) plant water status, and (iv) models, along with their pros and cons for comparison.

Given the urgency imposed by climate change and the evident research gap, prioritizing the development and transfer of irrigation scheduling technologies in Venezuela is crucial. Applied research, field trials, and participatory approaches that involve farmers directly in testing and refining irrigation calendars would facilitate faster adoption and local adaptation. Strengthening this area of innovation represents a strategic opportunity to improve agricultural water management efficiency, enhance rural livelihoods, and secure food production in an increasingly uncertain climatic future.

7.4.3 Low-Energy Precision Irrigation (LEPA)

The LEPA system is an irrigation method that delivers water to plants at specific points in the field, optimizing absorption. It combines a mechanical device with automatic movement (such as a central pivot or a lateral movement system) with water distribution procedures. This system applies water directly and close to the soil surface through drop tubes equipped with low-pressure application devices. Its design allows irrigation water to be delivered in the area closest to the crops or above the crop canopy, minimizing losses due to wind drift, interception, and evaporation, thereby improving irrigation efficiency and water productivity. Higher productivity is closely associated with this system due to its ability to significantly reduce water losses. Additionally, it enables frequent and small applications of water and fertilizers (*fertigation*), complementing rainfall in most cases. By requiring less water and energy, this system optimizes water resource use and ensures efficient delivery directly to the plants (Peters et al. 2016; Senninger Agriculture Irrigation 2025).

Beyond improving water and fertilizer use efficiency, LEPA systems also contribute to enhancing soil health by minimizing surface runoff and deep percolation losses. This preservation of soil structure supports better root development and nutrient absorption, ultimately leading to improved crop resilience against abiotic stresses such as drought and heat. Moreover, by reducing the operational pressure required for irrigation, LEPA systems lower energy demands, helping to make agricultural operations environmentally and economically viable, especially in regions where energy costs are a limiting factor for farmers.

Nevertheless, despite its advantages, the implementation of LEPA systems presents several challenges. The initial investment costs associated with installing pivot or lateral movement systems can be substantial, particularly for smallholder farmers. Furthermore, the effective operation of LEPA requires careful field leveling and the maintenance of appropriate emitter spacing to ensure uniform water distribution. Without adequate training and technical support, farmers may face difficulties in managing these systems efficiently, potentially compromising the expected gains in water productivity.

In the Venezuelan context, the adoption of low-energy precision irrigation systems could represent a strategic response to increasing climate pressures. Given the country's projected decline in precipitation and the existing inefficiencies in conventional irrigation practices, the deployment of LEPA technologies could significantly enhance agricultural sustainability, particularly in semi-arid regions such as Falcón, Lara, and parts of the central plains. However, widespread adoption will require not only financial incentives but also targeted research, pilot projects, and capacity-building initiatives to adapt these technologies to local conditions.

Considering the scarce national research on precision irrigation technologies, it is essential to promote studies evaluating the technical, economic, and environmental feasibility of LEPA systems under Venezuelan agroecological conditions. Integrating such technologies into national climate change adaptation strategies could help secure water availability, stabilize food production, and reduce the vulnerability of rural communities. Furthermore, encouraging public–private partnerships for the dissemination and financing of precision irrigation solutions would foster a more resilient and efficient agricultural sector, aligned with sustainable development goals.

7.4.4 Agricultural Drainage

Agricultural land drainage systems comprise a set of strategies, techniques, and hydraulic structures designed to remove excess water and/or salts from the soil in the root zone. This process creates a well-oxygenated environment that enables roots to develop properly. Maintaining an appropriate balance of water and air content, aligned with the physiological needs of crops, ensures soil sustainability for long-term productivity (Brouwer et al. 1985). The objectives of this practice are: (i) to enhance crop yields in poorly drained soils, (ii) to improve soil productivity by reducing production costs, (iii) to increase efficiency by allowing more time for other agricultural activities, and (iv) to reduce vulnerability to extreme weather events that cause water accumulation in the soil. By implementing drainage systems, excess water, which hinders root respiration and damages root tissues, is effectively removed. This practice also promotes microbial decomposition, as oxygen availability in the soil is essential for the growth of beneficial microorganisms. Additionally, drainage expands the range of crops that can be cultivated by improving conditions in moisture-prone soils and mitigating the risk of salinization caused by rising sea levels, salt intrusion, and groundwater contamination (Gurovich and Oyarce 2015).

Despite the well-documented importance of agricultural drainage systems for maintaining soil health and crop productivity, there is a conspicuous lack of scientific research on this subject in Venezuela. A review of the national literature reveals that very few studies have systematically addressed the design, implementation, or evaluation of drainage strategies adapted to Venezuela's diverse agroecological regions. This gap is particularly significant given the increasing occurrence of extreme weather events, such as prolonged rainy seasons and flooding, which amplify the need for efficient soil drainage management as part of climate change adaptation measures.

The absence of research is even more pronounced in specific regions that are highly vulnerable to waterlogging and soil salinization. Areas such as the lower central-western plains (Llanos), parts of Zulia, and low-lying coastal agricultural zones, where excessive soil moisture threatens agricultural viability, have received little to no scientific attention regarding drainage solutions. In these zones, inadequate drainage not only reduces immediate crop yields but also leads to long-term soil degradation processes, such as compaction, salinity build-up, and loss of microbial biodiversity, ultimately compromising the sustainability of agricultural systems.

The consequences of this research void are profound. Without locally adapted knowledge on drainage practices, Venezuelan farmers are left without technical guidelines to manage excess soil moisture effectively, especially in clayey soils. This situation exacerbates crop losses during periods of heavy rainfall, increases production costs due to reduced soil workability, and limits the diversification of cropping systems. Furthermore, the lack of data impedes policymakers from designing evidence-based interventions to support agricultural resilience, ultimately deepening the vulnerability of rural communities to climate-induced waterlogging and soil degradation.

Several factors explain Venezuela's current lack of scientific production on agricultural drainage. The progressive deterioration of research institutions, the drastic reduction in public and private funding for agricultural science, and the prioritization of short-term emergency responses over long-term land management strategies have all contributed to this vacuum. Additionally, the weakening of academic collaboration networks, the limited access to international research funding, and the migration of highly qualified researchers have further hindered the generation of knowledge in critical areas such as soil and water management. Reversing this situation requires a concerted national effort to rebuild research capacities, foster interdisciplinary studies, and prioritize sustainable land and water management within climate adaptation agendas.

7.4.5 Integrated Soil Fertility Management

It is a set of practices aimed at improving soil fertility by promoting organic inputs and genetically improved crops while minimizing chemical fertilizers. This approach optimizes crop productivity by maximizing the benefits of different fertilization methods, including organic inputs. It integrates agronomic principles, such as organic amendments, and genetically modified germplasm, while reducing chemical fertilizer use, thereby lowering greenhouse gas emissions from industrial production. This set of practices enhances climate change adaptation by increasing soil water retention, helping to stabilize crop yields during drought periods. It also reduces runoff and soil erosion while fostering soil organisms, which are crucial in maintaining agricultural productivity. Furthermore, minimizing chemical fertilizer use lowers production costs and provides environmental benefits, including reduced soil and water pollution (Roobroeck et al. 2016).

Although integrated soil fertility management (ISFM) is internationally recognized as a key component of sustainable agriculture and climate change adaptation, research efforts focused on this area in Venezuela remain extremely limited (Toro and Andrade 2021; Lobo et al. 2022). A very few empirical studies have addressed the adaptation of ISFM practices to Venezuelan soils and crops; this gap is alarming considering the critical role of soil health in maintaining agricultural productivity under increasingly variable climatic conditions.

The lack of research is particularly evident in regions where soil degradation is most pronounced, such as the central Llanos, western Zulia, and portions of the Andean mountains and piedmont (Márquez et al. 2019). In these areas, decades of intensive cultivation, coupled with poor fertilization practices, have led to declines in soil organic matter, increased erosion, and reduced nutrient availability. Yet, there is an almost complete absence of field-based research on organic amendments, biofertilizers, conservation tillage, or other integrated soil fertility strategies that could help reverse these trends and build long-term resilience (Gómez and Pulido 2019; Mogollón, Rivas 2021).

The consequences of this scientific gap are profound. Farmers continue to rely heavily on chemical fertilizers, often applying them inefficiently and without soil nutrient monitoring, which exacerbates environmental problems such as eutrophication, soil acidification, and contamination of water bodies. Furthermore, the lack of knowledge transfer on ISFM practices limits farmers' ability to adapt to drought, improve soil water retention, and sustain crop yields under changing climatic conditions. In the absence of robust scientific backing, national policies also fail to promote sustainable soil fertility management as a cornerstone of agricultural resilience.

Several structural factors explain the absence of significant research efforts on integrated soil fertility management in Venezuela. These include the severe contraction of agricultural research institutions, the chronic underfunding of soil science programs, the loss of experienced researchers due to brain drains, and the minimal interaction between universities, extension services, and farming communities. Rebuilding research capacity, fostering interdisciplinary collaborations, and promoting pilot projects that integrate agronomic, environmental, and socioeconomic dimensions are crucial steps to fill this gap. Strengthening ISFM research would not only enhance productivity and food security but also contribute significantly to Venezuela's climate change adaptation strategies.

7.4.6 Incorporation of Organic Matter into the Soil

It involves incorporating plant and animal residues in various stages of decomposition into the soil surface. As these residues break down, they release nutrients that become available to crops. These materials include leaves, dead roots, exudates, manure, urine, feathers, hair, bones, dead animals, and microorganisms (such as bacteria, fungi, and nematodes), which contribute organic substances to the soil or release their cellular components upon death (Peña et al. 2025). Additionally, organic matter plays a crucial role in carbon sequestration in cultivated areas, storing carbon in the upper soil horizons and enhancing soil water retention.

Increasing the organic matter content in the soil improves its adaptive capacity, fertility, water-holding capacity, and structure. This, in turn, boosts crop production and productivity, leading to higher farmer incomes and contributing to nutrition and food security (Lozano et al. 2018). The increase in crop yields results from the nutrients provided by organic matter and the formation of stable soil aggregates, which enhance water, air movement and improve nutrient uptake by plant roots. Furthermore, incorporating organic matter reduces the need for chemical fertilizers, thereby lowering greenhouse gas emissions associated with their production (FAO 2025).

The practice of organic matter incorporation is particularly valuable in tropical and subtropical environments, where high temperatures and intense rainfall can accelerate organic matter decomposition and leaching. In such contexts, maintaining a continuous supply of organic inputs is crucial to sustaining soil fertility over time. Moreover, organic matter acts as a buffer against soil acidification and helps retain essential nutrients, such as phosphorus and nitrogen, that might otherwise be lost through runoff or volatilization, particularly in highly weathered soils common in many agricultural zones of Venezuela.

Failure to systematically incorporate organic residues into agricultural soils can lead to progressive soil degradation, characterized by declining fertility, reduced water-holding capacity, compaction, and diminished biological activity. These conditions not only decrease crop productivity but also increase the vulnerability of farming systems to climate variability, pests, and diseases. The loss of soil organic matter is often irreversible on human timescales, highlighting the urgent need to prioritize organic matter management as a central pillar of sustainable agriculture (Rondon et al. 2021).

Despite its importance, the widespread adoption of organic matter incorporation practices faces several barriers. Among them are the limited availability of organic inputs in some regions, the logistical challenges associated with transporting and applying bulky organic materials, and the lack of technical knowledge among producers regarding the management of organic amendments. Overcoming these barriers requires concerted efforts in farmer training, the development of decentralized organic fertilizer production units, and the integration of organic matter management into broader soil fertility and climate adaptation programs.

The need to advance soil quality studies in Venezuela has become increasingly urgent given the growing environmental and agricultural challenges facing the country (Olivares et al. 2020). Soil degradation, including nutrient depletion, erosion, salinization, and loss of organic matter, threatens the long-term productivity of key agricultural regions such as the Llanos, the Andean, and the coastal plains. Comprehensive and systematic soil quality assessments are essential not only for diagnosing current levels of degradation but also for designing evidence-based interventions to restore and manage soils sustainably (González et al. 2021; Olivares et al. 2022; Pardo-Plaza et al. 2019).

Furthermore, understanding soil quality dynamics under changing climatic conditions is crucial for developing climate-resilient agricultural systems. Without a robust national framework for soil quality monitoring and research, Venezuela risks undermining its food security, compromising ecosystem services, and limiting its capacity to adapt to future environmental stresses.

7.5 Agricultural Adaptation Measures: A Viewpoint by Latin American and Caribbean Regional Organizations

Several international organizations, such as the Inter-American Development Bank (IDB) and the Economic Commission for Latin America and the Caribbean (ECLAC), along with non-governmental organizations, Latin American scientific journals and book publishers, have demonstrated increasing interest in addressing climate change in the Latin American and Caribbean region, particularly focusing on the adaptation of regional agriculture to these changes. Although the documents published by these entities are not strictly scientific research studies, some are research reviews; their content is highly valuable for informing agricultural adaptation policy development and for shaping implementation strategies at the national level and in critical agroecological regions such as the South American Andes, including Venezuela. In this context, the following analysis synthesizes the main contributions of six relevant papers that address agricultural adaptation measures to climate change.

One of these studies, conducted by Llambí and Garcés (2021), provides an analysis of the knowledge gaps and work priorities related to climate change adaptation in the Andes. It aims to guide future management efforts in this field and is particularly pertinent for the Venezuelan Andes, given that the principal author is a Venezuelan researcher. Closely aligned with this, Lozano-Povis et al. (2021) offer a systematic review emphasizing that agriculture in the Andes is highly sensitive to climate change. Their findings document key processes already underway, such as soil erosion, glacier retreat, vegetation loss, increased rainfall intensity, changes in thermal regimes, and alterations in staple crop dynamics—including potatoes, maize, coffee, and quinoa. The authors advocate for intensifying the adoption of innovative subsistence technologies and establishing research institutions focused on resilient agricultural systems, with specific references to the Venezuelan Andes alongside other Andean nations.

From a broader Latin American perspective, Cristiani (2023) addresses the opportunities and challenges of climate change for the extensive agricultural sector, stressing the importance of increasing public and private investments in agricultural research and development. The study identifies priority research topics, such as developing crop varieties adapted to new temperature, soil, and water conditions, evaluating alternative species for diversification, and refining soil and water conservation techniques. It also recommends the preservation of traditional crop varieties with valuable genetic traits compatible with anticipated climatic changes, despite their generally lower yields compared to commercially dominant varieties.

In a complementary approach, Nicholls et al. (2015) and Escalante and Messa (2023) argue that the most enduring benefits for climate resilience are likely to emerge from more transformative agroecological measures. These include diversification through polycultures, agroforestry systems, and integrated agriculture-livestock systems combined with organic soil management, water harvesting, and the enhancement of agrobiodiversity. The authors underscore that traditional farming systems offer a rich repository of practices capable of strengthening the resilience of modern agricultural systems to climatic extremes, many of which remain highly relevant for Venezuela.

Furthermore, Almeida et al. (2020) present a framework of five key adaptation measures for agriculture in Latin America and the Caribbean, which are highly applicable to Venezuela's current situation. These measures include prioritizing the adoption of agricultural adaptation policies, promoting research and development of crop varieties less susceptible to climate stresses, increasing the sustainable use of water in agriculture, restoring degraded lands while preventing deforestation, and implementing technologies and practices that simultaneously enhance productivity and reduce vulnerability. These strategic actions, if adopted systematically, could significantly improve the resilience of Venezuelan agriculture to the complex challenges posed by climate change (Fig. 7.3).

7.6 A Look at Research on Agricultural Adaptation in Venezuela

To achieve a general appreciation of research on agriculture adaptation measures to climate change in Venezuela, Table 7.1 provides straightforward data that contributes to appreciating the state of research on this topic. Table 7.1 includes, in its first two columns, some agricultural practices usually implemented in countries with advanced and sustainable agriculture. The purpose is to realize the number of adaptation measures proposed in the research carried out and published in Venezuela, whose summaries are included in Sect. 7.2 and 7.3.

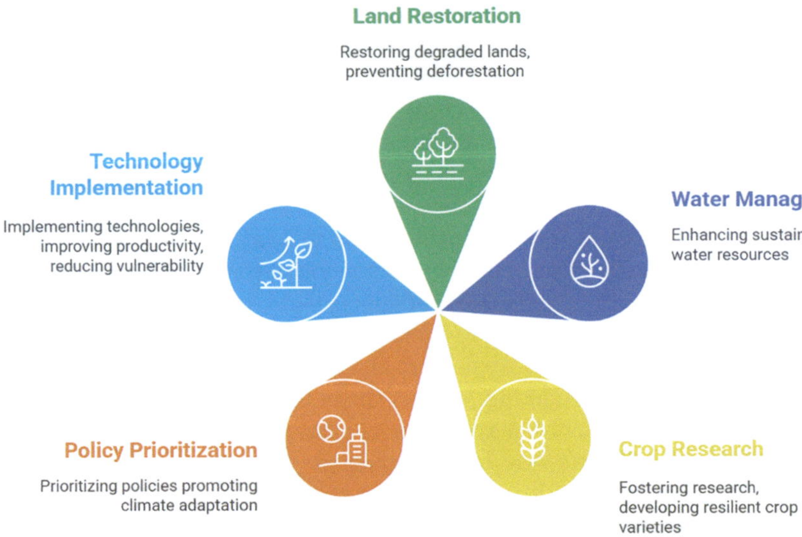

Fig. 7.3 Five key adaptation measures for Latin America and the Caribbean. Adapted from Almeida et al. (2020)

Although this study does not aim to provide an exhaustive assessment of agricultural adaptation research in Venezuela, the data in Table 7.1 highlight, on the one hand, quantitative aspects, such as the limited number of research papers published, only 15 in total. Additionally, studies are absent in areas considered priorities within the framework of a new agricultural approach in response to ongoing climate change in Venezuela. Exporting agricultural products represents a viable pathway for post-rentier development in the country. However, many developed countries, importers of agricultural products, such as those of the European Union, require certifications of sustainable agricultural production, whose practices are closely tied to adapting to climate change.

During the analyzed period, 66.6% (10/15) of the published studies focused on irrigation management, particularly the controlled deficit irrigation method (Sect. 7.2), aimed at improving irrigation use efficiency. This practice is especially beneficial in critical areas facing water shortages, such as northern Venezuela, as well as in regions expected to experience increasing water scarcity due to climate change, such as the eastern region, as mentioned in Chap. 5 of this book.

Only five studies were identified on other agricultural adaptation practices beyond irrigation: climate-stress-tolerant crops, crop diversification, integrated pest and disease management, and a combination of adaptation measures. These studies account for 33.3% of the total research on adaptation. Additionally, 46.6% (7/15) of the specific adaptation practices listed in Table 7.1 showed no published studies during the evaluated period.

It is also important to highlight some notable gaps in the research. Only one study on agroforestry was identified in the literature review, even though this approach is crucial in an intertropical country like Venezuela. Agroforestry has been recognized as a key strategy not only for climate change adaptation but also for biodiversity conservation. Unfortunately, only very limited attention was given to this area during the analyzed period, even though Venezuela has vast land areas with high and moderate suitability for tropical perennial crops that can be integrated into agroforestry systems, such as shade-grown coffee and cacao (Ávila-Lovera et al. 2021) which also have competitive export potential (Comerma et al. 1999, Corrales et al. 2023).

The absence of research on the specific adaptation practice of "developing new crop varieties" is also noteworthy. More than just an isolated measure, crop genetic improvement (genetic engineering) is a complex but highly effective research process. When directed toward climate change adaptation, such as developing crops with greater tolerance to water deficits and rising temperatures, its results can be crucial for ensuring food security and agro-exports. However, these initiatives require significant institutional efforts, legal updated, investments, advanced knowledge, innovative technologies, and the training of highly skilled researchers and assistants, all of which demand time to yield results. In Venezuela, institutional degradation has hindered progress in crop and livestock genetic improvement. Nevertheless, while the institutional crisis presents a challenge, it is not insurmountable. Strengthening this area is a strategic and decisive action for the future of Venezuelan agriculture.

7.7 Conclusions and Future Outlook

The findings indicate that research on agricultural adaptation to climate change in Venezuela has been limited. Only 15 papers were published in scientific journals during this period. This situation is attributed, among other factors, to the institutional crisis affecting research in Venezuela, particularly in the agricultural sector, as well as to the real lack of interest in climate change from the national government and society.

However, research was conducted on agricultural adaptation measures to climate change, focusing on improving irrigation water use efficiency (10 out of 15 papers). As discussed in Chap. 5 of this book, water use efficiency in agriculture is crucial in some regions of the country, such as the eastern plains, where precipitation is expected to decrease due to climate change. These studies, published primarily in national journals, explored "controlled deficit irrigation" as a key technique to address water scarcity in semi-arid and arid areas. Crops such as aloe vera (*Aloe vera*), passion fruit (*Passiflora edulis*), coriander (*Coriandrum sativum*), and stevia (*Stevia rebaudiana*) were evaluated. The findings indicated that controlled deficit irrigation enhances irrigation water use efficiency without compromising productivity or the quality of harvested agricultural products. Additionally, the studies analyzed different irrigation frequencies and water management strategies for crops such as bell pepper (*Capsicum annuum*), zucchini (*Cucurbita pepo*), and eggplant (*Solanum melongena*), recommending optimal irrigation rates to maximize yield and water use efficiency.

Outside the field of irrigation, few studies have been identified on general agricultural practices (5 out of 15 papers), lacking a coherent approach such as that seen in research on water deficit/irrigation. As evidenced in this Chapter (Table 7.1), research on agricultural adaptation to climate change in Venezuela reveals a knowledge gap, at least for the study period, in key adaptation measures and specific practices associated with them. These measures and practices have typically been developed and implemented by tropical countries with an advanced agricultural sector and export-oriented production targeting developed markets. One such example is the genetic improvement of crops. In contrast, no scientific studies on this topic were reported in Venezuela during the study period, despite its critical importance in ensuring the development of crop varieties resistant or tolerant to changes in parameters linked to climate change, as well as to the pests and diseases that arise because of these changes. Similarly, agroforestry, as an effective agricultural system for climate change adaptation, has been an under-researched area in the study period (1/15 papers). Especially for permanent tropical crops such as coffee and cocoa, as well as for cattle farming, which are key sectors with the potential to contribute to Venezuelan exports. Agroforestry is a crucial field for a mega-biodiverse country like Venezuela.

The institutional crisis has led to reduced investment in agricultural research and a significant loss of human talent, resulting in missed opportunities to develop effective strategies for adapting to ongoing climate change. However, research on agrarian adaptation has made progress in optimizing irrigation and water management, an urgent priority for Venezuelan agriculture, which is increasingly affected by climate change.

Similarly, agroforestry, as a practical agricultural system for adapting to climate change, remains an underexplored area of research in the study period. This is especially true for permanent tropical crops such as coffee and cocoa, as well as for cattle farming, sectors that have significant potential to contribute to Venezuela's vital export economy. Agroforestry represents a crucial opportunity for a mega-biodiverse country like Venezuela. On a positive note, research efforts have increasingly focused on optimizing irrigation and water management, critical areas for Venezuelan agriculture, which is grappling with the impacts of climate change.

The conclusions presented in the previous paragraphs highlight that climate change undeniably poses a significant challenge to agriculture in Venezuela, affecting agricultural productivity, food security, and the livelihoods of farmers. To mitigate these effects and adapt to evolving climatic conditions, it is essential to adopt a forward-thinking approach to agriculture. Below are some key strategies and actions to achieve this goal:

7.7.1 Strengthening Agricultural Research and Innovation

(a) Enhancing agricultural research institutions such as universities and the National Institute of Agricultural Research (INIA). This includes strengthening postgraduate programs in relevant agricultural disciplines directly linked to climate change adaptation.
(b) Research in crop genetic improvement to develop varieties resistant or tolerant to changes in temperature, water availability, and pests and diseases.
(c) Research on climate variability and its effects on agricultural production to analyze how climate variability (temperature, precipitation, extreme events) impacts agriculture at the regional level, and to develop adaptive practices, such as adjusting planting dates.
(d) Mathematical models of crop production: These models serve as tools to assess crops and agricultural practices aimed at adapting to anticipated climate changes. Develop production models for permanent tropical crops, such as coffee and cocoa, as well as various fruit crops that offer comparative and competitive advantages for export.
(e) Precision agriculture: This field focuses on developing knowledge and technologies designed to optimize agricultural production by addressing the spatial and temporal variability of soils and crops within farms.
(f) Sustainable soil management: This involves soil management practices that enhance carbon capture and storage, including conservation tillage, the use of cover crops, and the improvement of soil organic matter.

7.7.2 Agroforestry and Agrosilvopastoral Systems

Promote agroforestry and agrosilvopastoral systems that integrate trees and crops on one hand, and livestock on the other. These systems not only contribute to improving biodiversity and soil health, but also function as carbon sinks and help reduce CH_4 emissions from enteric fermentation in ruminants, thereby aiding both climate change mitigation and adaptation.

7.7.3 Sustainable Water Management

Implement efficient irrigation techniques to optimize water use in agriculture. This chapter includes reports on adapting to water deficits and improving irrigation water efficiency across various crops. These findings can be incorporated into technology transfer programs, especially in regions where water resource shortages are already beginning to emerge.

7.7.4 Promotion of Organic and Regenerative Agriculture

Promoting organic agriculture is essential to reduce dependence on chemical inputs while enhancing soil health and biodiversity. Organic farming practices contribute to the production of healthier food and foster environmental conservation. In parallel, implementing regenerative agriculture practices is critical to restore degraded agroecosystems. These approaches not only improve soil health but also enhance carbon sequestration and increase biodiversity, creating more resilient and sustainable agricultural systems in the face of climate change.

7.7.5 Training and Empowerment for Farmers

Developing comprehensive training programs for farmers across different regions of the country is a priority. These programs should focus on climate change and sustainable adaptation practices, equipping farmers with the knowledge and skills necessary to implement effective strategies in their production systems. Additionally, it is crucial to encourage active community involvement by promoting the participation of agricultural producer associations in planning and decision-making processes related to climate change adaptation. Such engagement strengthens collective capacity and ensures that adaptation measures are locally relevant and effectively adopted.

7.7.6 Policies and Regulatory Framework

Designing and implementing policies that promote sustainable agricultural practices and adaptation to climate change are fundamental. These policies should be accompanied by efforts to strengthen Venezuela's existing regulatory framework to protect natural resources and ensure the effective implementation of adaptive agricultural practices. A robust policy and regulatory environment will provide the necessary support for farmers and stakeholders to transition toward more resilient and sustainable agricultural systems.

7.7.7 Access to Agricultural Financing and Insurance

Facilitating access to funding for agricultural projects related to climate change through available international financial mechanisms is essential. Improving financing opportunities for Venezuelan farmers, particularly small- and medium-sized producers, will enable them to invest in technologies and adaptation measures that increase their resilience to climate impacts. In addition, promoting agricultural insurance programs is vital to protect farmers from economic losses resulting from extreme weather events, thus reducing vulnerability and enhancing their adaptive capacity.

7.7.8 National, Regional, and International Cooperation

Encouraging the active participation of the private sector, including agricultural producer associations, cooperatives, clusters, agro-food companies, and non-governmental organizations, in climate change mitigation and adaptation efforts is key to achieving transformative results in agriculture. Furthermore, fostering strategic alliances with international organizations, academic institutions, and other countries will facilitate the exchange of knowledge, technologies, and best practices. Such cooperation will strengthen Venezuela's capacity to build a sustainable and climate-resilient agricultural sector.

References

Almeida J, Ríos AR, Prager S, Schiek B, González Rodríguez CE (2020) Cinco medidas de adaptación para reducir la vulnerabilidad al cambio climático del sector agrícola en América Latina y el Caribe. Banco Interamericano de Desarrollo (BID). https://n9.cl/nx64ra

Ávila-Lovera E, Flores H, Móvil O, Santiago L, Wilmer T (2021) Shade tree species affect gas exchange and hydraulic conductivity of cacao cultivars in an agroforestry system. Tree Physiol 41:240–253. https://doi.org/10.1093/treephys/tpaa119

Bello V, Machado Castillo J P, Tornes Olivera N (2019) Evaluación de tres momentos de riego en elcultivo de la sábila en el Municipio Colina Estado Falcón Venezuela. Koinonía vol. IV, núm. 7: 280–292. Fundación Koinonía. https://www.redalyc.org/journal/5768/576869546015/html/

Blanco EL, Rada F, Paolini J, Guerrero JA (2021) Effects of induced water deficit and biofertilization on growth dynamics and bulb yield of onion (*Allium cepa* L.) in a Neotropical semi-arid environment. Can J Soil Sci 101(3):494–506. https://doi.org/10.1139/cjss-2021-0011

Brouwer C, Goffeau A, Heibloem M (1985) Irrigation water management training manual no. 1 chapter 6 drainage. Food and Agriculture Organization of the United Nations (FAO), Rome. http://www.fao.org/docrep/r4082e/r4082e00.htm#Contents

Carrero DL, Villalobos MR, Ferrer G, Colmenares C (2020) Efecto del riego deficitario controlado sobre el rendimiento del cilantro y su asociación con el cebollín. Bioagro 32(1):23–30. https://revistas.uclave.org/index.php/bioagro/article/view/2683

Comerma J, Machado D, Izquierdo A, Araujo J (1999) Evaluación de tierras para la producción de cacao a nivel nacional Venezuela. In: Memorias 14° Congreso Latinoamericano de Suelos. Sociedad Latinoamericana de la Ciencia del Suelo. 8-12 noviembre 1999 Pucon, Chile

Corrales W, Baritto F, García P, Iskandar L, Rosales A (2023) Estrategias para la Agroexportación en el Marco de Escenarios del Desarrollo Futuro de Venezuela. In: Grupo Orinoco para el Desarrollo Sostenible. Estudio realizado con el patrocinio del Instituto Interamericano de Cooperación para la Agricultura (IICA), Caracas. https://grupoorinoco.org/2024/03/14/

Cristiani M (2023) El sector agropecuario extensivo en América Latina: oportunidades y desafíos del cambio climático y la protección de la biodiversidad. Corporación Andina de Fomento (CAF) Banco de Desarrollo de América Latina y el Caribe. Paper No. 20. https://n9.cl/37pixs

Datta S, Taghvaeian S (2023) Soil water sensors for irrigation scheduling in the United States: a systematic review of literature. Agric Water Manag 278:108148. https://doi.org/10.1016/j.agwat.2023.108148

Escalante EE, Messa HF (2023) Silvopastoral systems with native tree species in Venezuela. In: Silvopastoral systems of Meso America and northern South America. Springer International Publishing, Cham, pp 239–261. https://doi.org/10.1007/978-3-031-43063-3

FAO (2010) Climate smart agriculture: policies, practices and financing for food security, adaptation and mitigation. Food and Agriculture Organization of the United Nations (FAO), Rome, Italy. https://www.fao.org/4/i1881e/i1881e00.pdf

FAO (2013) Captación y almacenamiento de agua de lluvia. Opciones técnicas para la agricultura familiar en América Latina y el Caribe. 1ed. Food and Agriculture Organization of the United Nations, Santiago, Chile, p 272. http://www.fao.org/docrep/019/i3247s/i3247s.pdf

FAO (2014) Feasibility study of rainwater harvesting for agriculture in the Caribbean subregion. Food and Agriculture Organization of the United Nations, Santiago, p 74. http://www.fao.org/3/a-bq747e.pdf

FAO (2021) Climate-smart agriculture case studies 2021. Projects from around the world. Food and Agriculture Organization of the United Nations, Rome. https://doi.org/10.4060/cb5359en

FAO (2025) Sustainable soil/and Management for Climate-Smart Agriculture. In: Climate-Smart Agriculture Sourcebook, Production and resources, Rome. https://n9.cl/1n8vzg

Fedeagro (2019) La Catástrofe Agrícola del 2019. Confederación de Asociaciones de Productores Agropecuarios de Venezuela. Fedeagro. Documentos. https://fedeagro.org/resultados-de-la-agricultura-vegetal-del-2018/. Accessed 08 Mar 2022

Gil Marín JA, Montaño-Mata NJ, Pérez Córcega J (2020) Efectos de regímenes de riego sobre el rendimiento y el uso del agua en Berenjena (*Solanum melongena* L.), en condiciones de campo. Apthapi 6(3) https://n9.cl/2v2xt

Gil-Marín JA, Rodriguez MXC, Mata NJM (2021) Efectos de los regímenes de riego sobre el rendimiento y el uso del agua del calabacín (*Cucurbita pepo* L.) en condiciones de campo. In *Anales Científicos*, *82*(2), 237–250. https://doi.org/10.21704/ac.v82i2.1786

Gómez D, Pulido LE (2019) Hongos micorrícicos arburculares y Trichoderma harzianum R: Alternativas ecológicas para la producción de posturas de café (Coffea arabica L.) en el estado Táchira, Venezuela. Univ Cienc 8:12–28. https://n9.cl/o4u69z

González GH, González PAF, Atencio J, Soto A (2021) Evaluación de calidad de suelos plataneros a través de la actividad microbiana en el sur del lago de Maracaibo, estado Zulia, Venezuela. Rev Fac Agron 38:216–240. https://n9.cl/6vwar0

Goyo Y, Zambrano C (2017). Uso de biocontroladores y la bioestimulación del cultivo de la soya en Venezuela. Edición especial No 73 de la Revista de la Facultad de Agronomía de la Universidad Central de Venezuela y la Fundación DANAC. https://n9.cl/jlvlw

Gu Z, Qi Z, Burghate R, Yuan S, Jiao X, Xu J (2020) Irrigation scheduling approaches and applications: a review. J Irrig Drain Eng 146(6):04020007.10.11648/j.bio.20241206.11

Gurovich L, Oyarce P (2015) New approaches to agricultural land drainage: a review. Irrig Drain Syst Eng 4. https://doi.org/10.4172/2168-9768.1000135

Lobo D, Lozano Z, Toro M, Valencia IH, Rey JC (2022) Soil biodiversity under different uses and managements in Venezuela. In: Reyes-Sánchez LB, Horn R, Costantini EAC (eds) Sustainable soil management as a key to preserving soil biodiversity and stopping its degradation. International Union of Soil Sciences (IUSS). Vienna, sustainable soil management as a key to preserve soil biodiversity and stop its degradation. Austria, p 240. https://acortar.link/ddoEd9

Leroy D (2019) Farmers' perceptions of and adaptations to water scarcity in Colombian and Venezuelan Páramos in the context of climate change. Mt Res Dev 39(2):R21–R34. https://doi.org/10.1659/MRD-JOURNAL-D-18-00062.1

Lozano Z, Bravo C, Rivero C, Hernández-Hernández RM, Mora R, Aular L (2018) Efecto de los cultivos de cobertura sobre la materia orgánica y la estabilidad estructural de un suelo de sabana bajo sistemas de manejo conservacionista. Rev Fac Agron 44(2):62–73. Universidad Central de Venezuela (Maracay). http://saber.ucv.ve/ojs/index.php/rev_agro/article/view/22125

Lozano-Povis A, Álvarez-Montalván CE, Moggiano N (2021) El cambio climático en los Andes y su impacto en la agricultura: una revisión sistemática. Sci Agropecu 12(1):101–108. https://doi.org/10.17268/sci.agropecu.2021.012

Llambí L D, Garcés A (2021) Adaptación al cambio climático en Los Andes: Vacíos y prioridades para la gestión del conocimiento. *Consorcio para el Desarrollo Sostenible de la Ecorregión Andina* CONDESAN. Quito-Ecuador. https://n9.cl/oz8g2

Márquez M, Cáceres A, Hernández-Valencia I, Romero L (2019) Parámetros micorrízicos y calidad del suelo en cultivos de papas nativas e introducidas (Mérida, Venezuela). Acta Biol Venez 39(1):91–105. http://saber.ucv.ve/ojs/index.php/revista_abv/article/view/19383

Mercado L, Muschler R, Rosales A, Villareyna R, Vargas A, Chavez EI (2018) Training Manual on Climate Smart Agriculture Practices for Belize. Costa Rica, CATIE (Tropical Agriculture Research and Higher Education Center)-PNUD. https://www.researchgate.net/publication/346973127

Mercado A, Vessuri H, Córdova K, Sánchez Rose I, López M S (2023) La publicación científica en Venezuela: crisis y transformaciones en pos de la sobrevivencia. Dosier: Debates contemporáneos en torno a las revistas científicas: miradas latinoamericanas a problemáticas globales. Palabra Clave Vol. 12 Núm. 2. doi:https://doi.org/10.24215/18539912e184

Mogollón JP, Rivas W (2021) Determinación de la fertilidad del suelo mediante un sistema de información geográfica en el municipio Falcón de la Península de Paraguaná, Venezuela. Observador del Conocimiento 6(2):87–106. https://n9.cl/5exz5

Mora E, Lopez-Hernández D, Toro M (2019) Arbuscular Mycorrhizae and PGPR applications in tropical savannas. In: Zúñiga-Dávila D, González-Andrés F, Ormeño-Orrillo E (eds) Microbial probiotics for agricultural systems. Sustainability in plant and crop protection. Springer, Cham. https://doi.org/10.1007/978-3-030-17597-9_11

Mora Guerrero A F (2022) Prácticas de gestión de los productores agrícolas de Venezuela. Rev Cient UCSA Vol. 9 Iss: 2 p 66–76. Universidad del Cono Sur de las Américas. Paraguay. http://scielo.iics.una.py/pdf/ucsa/v9n2/2409-8752-ucsa-9-02-66.pdf

Morales-Márquez JA, Jiménez-Ballesta R, Hernández-Hernández RM, Sánchez GK, Lozano Z, Castro I (2022) Effect of maize conservation crops associated with two vegetal covers on the edaphic macrofauna in a well-drained savanna of Venezuela. Land 11:464. https://doi.org/10.3390/land11040464

Nicholls C I, Altieri M A, Henao A, Montalba R, Talavera E (2015) Agroecología y el diseño de sistemas agrícolas resilientes al cambio climático. Red Iberoamericana para el Desarrollo de Sistemas AgrícolasResilientes al Cambio Climático (REDAGRES) y la Sociedad Científica Latinoamericana de Agroecología (SOCLA). Lima, Perú. http://celia.agroeco.org/wp-content/uploads/2019/02/Diseño-de-sistemas-agricolas-resilientes-1.pdf

Olivares BO, Araya-Alman M, Acevedo-Opazo C, Rey JC, Cañete-Salinas P, Kurina FG, Gómez JA (2020) Relationship between soil properties and banana productivity in the two main cultivation areas in Venezuela. J Soil Sci Plant Nutr 20:2512–2524. https://doi.org/10.1007/s42729-020-00317-8

Olivares BO, Rey JC, Perichi G, Lobo D (2022) Relationship of microbial activity with soil properties in banana plantations in Venezuela. Sustainability 14(20):1313531. https://doi.org/10.3390/su142013531

Olivares Campos BO (2023) Potential use of soil morphological properties to differentiate levels of banana productivity. In: Banana production in Venezuela: novel solutions to productivity and plant health. Springer Nature Switzerland, Cham, pp 141–164. https://doi.org/10.1007/978-3-031-34475-6

Pardo-Plaza YJ, Paolini GJE, Cantero-Guevara ME (2019) Biomasa microbiana y respiración basal del suelo bajo sistemas agroforestales con cultivos de café. Rev U D C A Actual Divulg Cient 22(1):e1144. https://doi.org/10.31910/rudca.v22.n1.2019.1144

Peña H, Diánez F, Ramírez B, Sulbarán J, Arias K, Huertas V, Santos M (2025) Compost and Vermicompost as substrates enriched with Trichoderma asperellum for the production of basic potato seed in the Venezuelan Andes. Horticulture 11(2):124. https://www.mdpi.com/2311-7524/11/2/124

Pereyra G (2012) Potential yield of Venezuelan maize varieties under variable water supply. Thesis of master of science in agricultural sciences. University of Hohenheim. Faculty of Agricultural Sciences Institute for Plant Production and Agroecology in the Tropics and Subtropics, Hohenheim, Stuttgart, Alemania, p 71. https://n9.cl/6ond8

Peters RT, Neibling H, Stroh R (2016) Low energy precision application (LEPA) and low elevation spray application (LESA) trials in the Pacific northwest. In: Proceedings California alfalfa and forage symposium, Reno, NV, Nov 29-Dec 1, 2016. UC Cooperative Extension, Plant Sciences Department, University of California, Davis, CA 95616. https://n9.cl/9o6uk

Pezo D, Muschler R, Tobar D, Pulido A (2019) Intervenciones y tecnologías ambientalmente racionales (TAR) para la adaptación al cambio climático del sector agropecuario de América Latina y el Caribe. Banco Interamericano de Desarrollo. https://doi.org/10.18235/0002086

Rodríguez-Yzquierdo G A, Basso-de-Figuera C A, Díaz-Reyes G, León-Pacheco R I (2020) Riego deficitario controlado, su efecto sobre la nutrición, productividad y calidad de la fruta maracuyá. Agron Mesoam vol. 31 núm. 2, p 405–418. https://www.redalyc.org/journal/437/43762994009/html/

Rodríguez Padrón AR (2020) Manejo de riego en el cultivo de pimentón (*Capsicumannuum* L.). En: Luis G. Ramírez (ed) Agrobiología Una visión general y sus aplicaciones. 2020. Mérida Publishers CC-BY 4.0. https://www.Meridapublishers.com/libroagrobio.php?pag=capitulo1

Romero A, Díaz E, Colmenares C (2014) Medidas de adaptación agrícola al cambio climático en la cuenca del río Pao, Carabobo Venezuela. UNELLEZ Cienc Tecnol 32:65–73. https://n9.cl/yjrhn

Romero A, Flores Y (2020) La producción de yuca y la resiliencia al cambio climático en el estado Cojedes, Venezuela. Ambientis Occidentales 01 Enero-Junio/2020:39–46. https://n9.cl/b6bqs

Rondon T, Hernández RM, Guzmán M (2021) Soil organic carbon, physical fractions of the macro-organic matter, and soil stability relationship in lacustrine soils under banana crop. PLoS One 16(7):e0254121. https://doi.org/10.1371/journal.pone.0254121

Roobroeck D, Van Asten P, Jama B, Harawa R, Vanlauwe B (2016) Integrated soil fertility management: contributions of framework and practice to climate-smart agriculture. GACSA Alliance for Climate-Smart Agriculture. https://n9.cl/wtlf9y

Senninger Agriculture Irrigation (2025) LEPA. In: Hunter. https://agriculture.hunterirrigation.com/guide-senningerr-wobblerr-technology

Toro M, Andrade G (2021) Arbuscular mycorrhizae: beneficial microorganisms for sustainable agriculture. In: Leal Filho W, Azul AM, Brandli L, Lange Salvia A, Wall T (eds) Life on land. Encyclopedia of the UN sustainable development goals. Springer, Cham. https://doi.org/10.1007/978-3-319-95981-8_122

Villafañe R, Pacheco X, Gorrín R, Basso C (2018) Riego deficitario combinado con riego no deficitario para uso eficiente del agua en el cultivo de Stevia. Bioagro 30(3):201–208. https://revistas.uclave.org/index.php/bioagro/article/view/2718/1700

World Resources Institute (WRI) (2024) Aqueduct food: water risk and agricultural production platform–Venezuela. https://n9.cl/7et9nz

Chapter 8
From Evidence to Action: Conclusions and Future Outlook

8.1 Introduction

This research carries an undeniable social relevance, as it addresses Venezuelan agriculture as a fundamental pillar for food security, rural employment generation, and community cohesion, especially within contexts of prolonged crises and territorial inequalities. The analysis provides tangible impacts by revealing the climatic challenges affecting regions and social actors differently, offering benefits that transcend academia and directly contribute to the well-being of rural populations. Its social projection is broad, as it equips stakeholders with tools to strengthen community resilience, local development, and climate justice, reaffirming that agriculture cannot be separated from processes of human and environmental development.

From a contemporary relevance perspective, this book examines first of all a current situation characterized by the convergence of economic, political, and climatic crises, which have weakened agricultural systems across the country. This contextual focus enables an understanding that climate variability and change are not future threats but realities already transforming production, water availability, and ecosystem health. By analyzing contemporary dynamics such as droughts, floods, emerging pests, and institutional deterioration, this research provides a rigorous and updated diagnosis, essential for formulating public and community strategies relevant to the Venezuelan and broader Latin American reality.

Regarding its methodological utility and convenience, the book offers new approaches to study the interaction between agriculture and climate change, integrating assessments of biophysical impacts, sociopolitical dimensions, institutional management, and future scenario building. This systemic and transdisciplinary approach becomes a valuable methodological tool for researchers, planners, and decision-makers, as it facilitates a comprehensive understanding of complex phenomena and their translation into practical adaptation and mitigation measures. The convenience of this research lies in the applicability of its results to rural

A. Rosales Hernández et al., *Agriculture Under Climate Change*, The Latin American Studies Book Series, https://doi.org/10.1007/978-3-032-06368-7_8

development programs, agro-environmental policies, and climate action plans, both in Venezuela and in countries with similar contexts.

The practical justifications and theoretical value of this work are evident in its contributions to designing innovative solutions such as climate-smart agriculture, soil restoration using local inputs, and regionally adapted productive diversification. These contributions not only expand existing knowledge on vulnerabilities and response capacities in agriculture but also strengthen theoretical foundations related to socioecological resilience, adaptive governance, and climate agroecology. Furthermore, its projection is wide-reaching, as its findings and approaches can be replicated and adapted by educational institutions, rural communities, and international organizations working toward agricultural sustainability and food security in the face of global change.

Ultimately, this book is not merely a compilation of diagnoses and projections, but an urgent and hopeful call to recognize that Venezuelan agriculture, with its wealth of ancestral knowledge, biological diversity, land resources, and innovative potential, can become a driving force for social and environmental transformation. Overcoming the challenges of climate change requires a collective and courageous vision, capable of uniting science, policy, and community action under a common purpose: ensuring a dignified, productive, and sustainable future for present and future generations. The task is complex, but the opportunity to reimagine agriculture as a source of life, resilience, and climate justice lies in our hands, and its realization will depend on the determination and commitment we cultivate today.

8.2 Conclusions: Evidence, Gaps, and Opportunities

Venezuelan agriculture finds itself at a pivotal moment, confronted by the converging forces of economic and political instability, institutional decline, and the escalating impacts of global climate change. These contests have not only undermined the efficiency and the productive capacities of the sector but have also exposed its structural weaknesses, including technological obsolescence, insufficient rural infrastructure, and limited access to agricultural inputs and financing. Over the past two decades, agriculture in Venezuela has undergone a marked deterioration, evolving from a strategic pillar of national development to a marginalized and under-resourced sector. This decline has had profound repercussions, not only on food security and rural employment but also on the resilience of communities and ecosystems in the face of climate variability.

Climate change, far from being a distant or abstract threat, is already revealing measurable and disruptive ways across the Venezuelan territory. From the drought-prone Llanos to the fragile mountain ecosystems of the Andes and the vulnerable coastal and marine zones, evidence of climatic disruption is mounting. Rising average temperatures, altered precipitation regimes, increasing frequency of droughts and floods, and the resurgence or emergence of crop pests and diseases are reshaping the country's agricultural landscapes. The impacts are not homogeneous; rather, they are deeply influenced by regional climatic conditions, agroecological contexts, and levels of institutional capacity to respond.

This book compiles and analyzes a wide range of scientific studies conducted mainly between 2018 and 2024, illuminating both the severity of the threat and the efforts underway to understand and address it. Although the volume of scientific production remains modest, it reflects a growing awareness of the need to reduce greenhouse gas (GHG) emissions, enhance carbon sequestration, and develop resilient agricultural practices. Particularly noteworthy are the emerging strategies related to sustainable land management, energy efficiency, soil restoration, land irrigation, and the promotion of agroforestry and agrosilvopastoral systems, all of which aim to integrate productivity with environmental stewardship.

However, the analysis also reveals some limitations. Research is unevenly distributed across the country, with significant gaps in underrepresented regions such as the eastern plains, the Venezuelan Guayana Shield, and some coastal territories. Thematic gaps are also evident, especially in areas such as crop genetic improvement, the development of climate-resilient seed varieties, impacts and modeling of pest and disease behavior under changing climates, impacts of climate change on forest lands, and assessments of the socioeconomic implications of climate adaptation strategies. Despite the potential of these fields to enhance long-term resilience, they remain critically underexplored.

Compounding these scientific gaps is a broader institutional crisis that has severely weakened the country's agricultural research setup/infrastructure. The exodus of qualified researchers, reduction of public investment in science and technology, and lack of operational continuity in research programs have hindered Venezuela's ability to generate, systematize, and apply knowledge. Furthermore, the absence of cohesive national climate change policy and strategies, a fragmented regulatory framework, and limited integration between public institutions and local communities have all constrained the effective implementation of climate-smart practices.

Nonetheless, the findings compiled in this book confirm that there is a strong foundation upon which to build. The existence of localized success stories, such as the application of controlled deficit irrigation in semi-arid zones, the use of native mycorrhizae to enhance soil health, and agroecological innovations in mountainous coffee systems, demonstrates the creativity and adaptability of Venezuelan producers and researchers. These practices, although not yet widespread, provide valuable insights into context-sensitive solutions that align with both environmental needs and local capacities.

The conclusions from each chapter also converge on a common understanding: adaptation and mitigation cannot be treated in isolation. Instead, they must be integrated into a broader vision of sustainable rural development. This vision requires a multisectoral, inter-institutional, and participatory approach that bridges science, policy, and practice. It also demands a renewed recognition of the central role of small and medium-scale producers, indigenous communities, and local organizations as agents of change.

Outstandingly, Venezuelan agriculture should not be viewed solely through the lens of vulnerability. The country's vast biodiversity, varied agroecological zones, and accumulated traditional knowledge offer unique opportunities for developing innovative, low-carbon agricultural models adapted to local realities. If properly supported, these models can contribute not only to national adaptation and mitigation goals but also to broader objectives of food sovereignty, ecosystem conservation, and climate justice.

In sum, this book confirms that while the challenges are significant, they are not insurmountable. What is needed is a strategic shift, one that recognizes agriculture as a cornerstone of climate resilience and as a vehicle for inclusive and sustainable development. Robust institutions, adequate funding, regionalized knowledge, and the political will to prioritize the long-term well-being of both people and the environment must support this shift.

8.3 Future Outlook: Strategic Pathways

Building a climate-resilient agricultural sector in Venezuela is not a task that can be postponed. It requires deliberate and sustained action across multiple dimensions, political, scientific, economic, social, technical, and institutional. Drawing on the collective insights of the chapters in this book, the following strategic directions are proposed as a roadmap for the future:

1. Strengthening research and innovation capacities: Scientific research must be reinvigorated and geographically diversified. It is imperative to rebuild the country's public research institutions and enhance academic university programs focused on climate-smart agriculture, sustainability, adaptation, mitigation practices, and technological innovation. Priorities should include crop genetic improvement for climate resilience, land irrigation to cope with the foreseeable water deficit and extreme climate-associated events, agroforestry systems tailored to Venezuelan biomes, climate modeling, and socioenvironmental impact assessment tools. In parallel, the expansion of precision agriculture, remote sensing technologies, climate-crop modeling, and big data analytics can provide farmers with actionable information to manage climate risks more effectively.
2. Mainstreaming climate-smart agricultural practices: National agricultural policies should prioritize the large-scale adoption of proven practices such as conservation tillage, organic fertilization, integrated pest management, and crop-livestock-forest integration. These strategies not only reduce emissions but also restore degraded soils, improve water retention, and foster biodiversity. Extension services and farmer training programs, both public and private, must be strengthened to support the dissemination and contextualization of these practices.
3. Promoting water security and sustainable irrigation: As water scarcity intensifies, investment in efficient irrigation systems becomes critical. Technologies such as drip irrigation, moisture sensors, and climate-based scheduling should

be promoted, especially in the most water-stressed areas, such as the country's eastern region. Watershed management plans that integrate upstream and downstream users can help reduce conflict and ensure equitable distribution, such as soil and water conservation and reducing deforestation in the watersheds. Water governance must also be improved through legal reforms, institutional restructuring, investment in public services, and stakeholder participation.

4. Reforming legal and policy frameworks: It is essential to align national legislation with international climate agreements, incorporating adaptation and mitigation objectives into land-use planning, environmental regulation, and agricultural development. Legal recognition of ecosystem services, incentives for sustainable practices, sanctions against deforestation, and land degradation are necessary components of an updated framework, and incentives for initiatives to protect forests and reforest degraded forests, as well as for soil conservation.
5. Conserving critical ecosystems and biodiversity: Natural ecosystems such as mangroves, páramos, floodplains, and cloud forests are vital for climate regulation, biodiversity conservation, and agricultural sustainability. These ecosystems must be protected through zoning, conservation easements, ecological restoration and regeneration programs, and community stewardship initiatives. Their preservation is not only a moral and ecological imperative but also a practical necessity for food production and water security.
6. Encouragement local empowerment and inclusive governance: Effective climate responses require the active participation of farmers, indigenous peoples, women, and youth in decision-making processes. Decentralized governance structures that include community councils, cooperatives, watershed agencies and producer associations can ensure that local knowledge and priorities shape adaptation strategies. Inclusive governance enhances transparency, equity, and sustainability.
7. Enhancing monitoring, evaluation, and learning systems: Appropriate mechanisms are needed to track climate trends, assess the effectiveness of interventions, and adapt strategies in real time. This includes expanding and modernizing/upgrading the hydrometeorological network, developing performance indicators for adaptation and mitigation, and facilitating knowledge exchange platforms. Documenting and scaling successful experiences can accelerate innovation and avoid duplication of efforts.
8. Integrating artificial intelligence and predictive technologies: Harnessing artificial intelligence (AI), big data, machine learning (ML), and deep learning (DL) represents a transformative opportunity to build climate-resilient agricultural systems in Venezuela. The development of predictive models capable of integrating climatic, edaphic, and socioeconomic variables can support early warning systems, optimize resource allocation, and guide strategic planning for crop production under uncertainty. Implementing big data frameworks and federated data spaces that consolidate diverse datasets, from meteorological networks to soil health indicators and remote sensing outputs, will enable researchers and policymakers to perform advanced analytics with higher spatial and temporal

resolution. Strengthening national capacities in data science, fostering partnerships with international AI research centers, and integrating these technologies within climate-smart agricultural policies are essential strategic pathways to accelerate innovation, close knowledge gaps, and promote inclusive, sustainable, and technologically empowered agricultural development in the face of escalating climate risks.

9. Facilitating access to finance and risk management instruments: Climate adaptation requires resources. It is crucial to increase access to climate finance from international mechanisms such as the Green Climate Fund, as well as from regional development banks and bilateral cooperation. National financial institutions should also be restructured to provide credit, insurance, and risk-sharing instruments tailored to the needs of smallholders and agroecological projects.
10. Strengthening international cooperation and technology transfer: Strategic alliances with international research centers, UN agencies, and climate funds can provide technical expertise, funding, and policy support. Participation in regional climate platforms and networks enhances South-South collaboration, positioning Venezuela to benefit from global experiences in climate-resilient agriculture.
11. Promoting education, communication, and cultural change: Adaptation is as much a cultural process as it is a technical one. Embedding sustainability and climate awareness into national curricula, rural education programs, and public media campaigns is key to fostering a new agricultural culture. Intergenerational dialogue, respect for ancestral knowledge, and youth engagement should be at the heart of this transformation.

Index

A. Rosales Hernández et al., *Agriculture Under Climate Change*, The Latin American Studies Book Series, https://doi.org/10.1007/978-3-032-06368-7

MIX
Papier aus verantwortungsvollen Quellen
Paper from responsible sources
FSC® C105338

If you have any concerns about our products,
you can contact us on
ProductSafety@springernature.com

In case Publisher is established outside the EU,
the EU authorized representative is:
Springer Nature Customer Service Center GmbH
Europaplatz 3, 69115 Heidelberg, Germany

Printed by Libri Plureos GmbH
in Hamburg, Germany